About Island Press

Island Press, a nonprofit organization, publishes, markets, and distributes the most advanced thinking on the conservation of our natural resources—books about soil, land, water, forests, wildlife, and hazardous and toxic wastes. These books are practical tools used by public officials, business and industry leaders, natural resource managers, and concerned citizens working to solve both local and global resource problems.

Founded in 1978, Island Press reorganized in 1984 to meet the increasing demand for substantive books on all resource-related issues. Island Press publishes and distributes under its own imprint and offers these services to other nonprofit organizations.

Funding to support Island Press is provided by Apple Computers, Inc., The Mary Reynolds Babcock Foundation, The Education Foundation of America, The Charles Engelhard Foundation, The Ford Foundation, The George Gund Foundation, The William and Flora Hewlett Foundation, The Joyce Foundation, The J. M. Kaplan Fund, The John D. and Catherine T. MacArthur Foundation, The Andrew W. Mellon Foundation, Northwest Area Foundation, The Jessie Smith Noyes Foundation, The J. N. Pew, Jr., Charitable Trust, The Rockefeller Brothers Fund, The Florence and John Schumann Foundation, and The Tides Foundation.

About the Sierra Club Legal Defense Fund

The Sierra Club Legal Defense Fund is a nonprofit environmental law firm founded in 1971. Its purpose is to provide conservation groups with expert legal representation and advice and to make maximum use of the law to protect natural resources and combat pollution. Toward those ends its 25 lawyers devote their full time and attention, pursuing litigation to protect the environment from the Virgin Islands to Alaska, from Maine to Guam. The services of Legal Defense Fund lawyers are available without cost to all national, regional, and local citizen conservation groups, regardless of size.

The Sierra Club Legal Defense Fund is best known for its protection of particular places, especially public lands and wildlife habitat. In recent years, however, Legal Defense Fund attorneys have taken the lead in forcing reductions in air pollution and the emissions that are the source of acid rain. They have brought suits to penalize polluters of our country's rivers and coasts, to curb the use of pesticides and herbicides that endanger human health, and to control the handling of radioactive wastes. Uniting all of these diverse activities is the desire to empower citizen groups by providing the legal talent without which citizens would be barred from effective access to the courts and administrative agencies that enforce environmental laws.

The Sierra Club Legal Defense Fund receives no government support. It is funded almost entirely by private individuals and foundations, whose contributions are deductible under section 501(c)(3) of the Internal Revenue Code.

THE
POISONED
WELL

This book is dedicated to the memory of Alfred S. Forsyth, lawyer, conservationist, and pioneer in the field of environmental law. Al served as a trustee of the Sierra Club Legal Defense Fund from 1975 until his death in 1985 and was one of the first to realize the threat of groundwater pollution. His devotion to conservation, his wisdom, and his sense of humor are missed, but they are not forgotten.

THE POISONED WELL

NEW STRATEGIES FOR GROUNDWATER PROTECTION

Sierra Club Legal Defense Fund
Eric P. Jorgensen, Editor

ISLAND PRESS
Washington, D.C. □ Covelo, California

"This publication is designed to provide accurate and authoritative information in regard to the subject matter covered. It is sold with the understanding that the publisher and the author are not engaged in rendering legal, accounting, or other professional service. If legal advice or other expert assistance is required, the services of a competent professional person should be sought."

From a Declaration of Principles jointly adopted by a Committee of the American Bar Association and a Committee of Publishers and Associations.

Text design by Irving Perkins Associates
Cover design by Ben Santora

Library of Congress Cataloging-in-Publication Data

The Poisoned well : new strategies for groundwater protection / Sierra
    Club Legal Defense Fund ; edited by Eric P. Jorgensen.
        p.   cm.
    Includes index.
    ISBN 0-933280-56-4 : $31.95.—ISBN 0-933280-55-6 (pbk.) : $19.95
    1. Water, Underground—Pollution—Handbooks, manuals, etc.
2. Water, Underground—Pollution—United States—Citizen
participation—Handbooks, manuals, etc.   I. Jorgensen, Eric P.,
1957– . II. Sierra Club. Legal Defense Fund.
TD426.P64   1989
363.7'394—dc19   89-1940
                                                                                    CIP

Printed on recycled, acid-free paper

Manufactured in the United States of America

10   9   8   7   6   5   4   3   2

Acknowledgments

THIS BOOK WAS a team effort. It involved the collaborative efforts of Sierra Club Legal Defense Fund staff, special project staff, and numerous law clerks.

Staff Attorneys

Eric Jorgensen supervised the project from the early development stages through the completion of the manuscript, served as general editor of the book, with particular responsibility for Parts I–III, and wrote or substantively edited several chapters in Parts II and III (Chapters 6–8, 10–14, 18, and 21). Howard Fox wrote Chapters 16 and 20, edited several other chapters, and supervised the preparation of the final manuscript and graphic materials. Durwood Zaelke, who conceived the original idea for this project, wrote Chapters 9 and 15, assisted with the editing of Chapters 16 and 19, and helped supervise the preparation of Part IV. Robert Dreher wrote Chapter 17. Vawter Parker assisted in the editing of Chapter 19 and also in the coordination of the final manuscript preparation.

Law Associates

Elizabeth Ulmer wrote and edited several chapters of Part IV, coordinated the peer review process for the book and, in consultation with Howard Fox, managed the preparation of the final manuscript and graphic material. Wendy Dinner and Sandy Goldberg assisted with the editing of several chapters and with the response to reviewer suggestions.

Special Project Staff

Pam Stone helped with the early development stages of the project, served as researcher and writer for Part I and Chapter 9, and did general research for the entire project. Laura Polacheck, an attorney, also played an important role in the early development stages, researched federal and state groundwater programs, wrote drafts of most of the chapters of Parts II, III, IV, compiled graphic material, and contributed in a variety of ways to the project. Barry Lampke helped research state and local programs and wrote some preliminary drafts for Part IV. Gerald Fox assisted with the details of references, graphics, anecdotes, and general production.

Law Clerks

Several generations of Legal Defense Fund law clerks made important contributions ranging from cite checking to substantial research, editing, and rewriting, particularly for Part IV.

Bill Carpenter, Julie Cheng, David Cozad, Lory Gordon, Tom Grasso, Jennifer Kohout, Gwen Logan, Rebecca Lord, Elizabeth Napier, Turner Odell, Wendy Pulling, Pam Reiman, Deborah Smith, Sherryl Statland, and Doug Wertheimer were all helpful.

Other Staff

Dallas Motlagh, Rebecca Via-Hall, and Janice Whittington typed countless versions of the book and assisted in the preparation of the final manuscript.

Many other people were helpful to us in the researching and writing of this book, including Will Collette, Morgan Gopnik, Don Hickman, Suzi Ruhl, Velma Smith, and Amy Svoboda, who provided particularly useful advice on a variety of topics.

Our manuscript reviewers made many constructive suggestions for changes and improvements. Our thanks go to Steven Baer, Dick Bean, Jane Bloom, Will Collette, Jay Feldman, John Gale, Patti Goldman, Morgan Gopnik, Jason Gray, Francis Hallahan, Laura Kosloff, Lane Krahl, Dave Lennett, Jim Lyon, Jan McAlpine, Ed Mazzullo, Fred Millar, Meg Nagle, Erik Olson, Suzi Ruhl, Lisa St. Amand, Kathy Stein, Andrew Stevenson, and Jackie Warren.

Finally, the Sierra Club Legal Defense Fund and all users of this book owe a special debt of gratitude to the Jessie Smith Noyes Foundation, whose commitment to future generations and generous support made the book possible.

CONTENTS

List of Figures   xv
Foreword   xvii
Introduction   xix

PART I
GROUNDWATER AND CONTAMINATION

CHAPTER 1.   Groundwater Basics   3
Hydrologic Cycle   3
Aquifers   3
Groundwater and Contaminant Movement   6

CHAPTER 2.   Health and Groundwater Contamination   12
Testing the Health Effects of a Contaminant   14
Health Effects of Common Groundwater
    Contaminants   15
Factors Complicating the Link Between Health Effects
    and Contaminants   18
Do Legal Limits on Contaminants Protect
    Your Health?   20

CHAPTER 3.   Sources of Groundwater Contamination   24
Landfills   24
Surface Impoundments   25
Underground Storage Tanks   27
Waste Disposal Wells   29
Agricultural Practices   32
Land Application   34
Household Hazardous Wastes and Products   35
Military Toxics   36
Mining   40
Septic Systems   41
Transportation   45

CHAPTER 4. Testing Groundwater Quality   51
            Monitoring Underground Sources of Water   51
            Monitoring Water at the Tap   56
            Scrutinizing Water Test Results   66

CHAPTER 5. Mapping Aquifers and Contamination   71
            Getting Started   72
            Compiling Your Own Water Supply Maps   72
            Seeking the Help of Experts   76

PART II
CITIZEN ACTION

CHAPTER 6. Freedom of Information Acts   83
            How to Make a Request   83
            Appeal Procedures   85

CHAPTER 7. Action in the Administrative Process   87
            General Advice   87
            Participation in the Rulemaking Process   88
            Using the Permit Process   90
            Making the Agency Enforce the Law   95
            Review of Agency Action by the Courts   98

CHAPTER 8. Taking the Polluter to Court   101
            Citizen Suits   102
            Common-Law Actions   106

CHAPTER 9. Grass-Roots Action   111
            Organizing for Community Action   112
            Taking Political Action   113
            Encouraging Waste Reduction and Recycling   114
            Successful Citizen Action   115

PART III
GROUNDWATER AND FEDERAL PROGRAMS

CHAPTER 10. An Introduction to Federal Groundwater Protection 123
             Overview of the Federal Statutory Framework   123
             *Citizen Oversight of State Agency Enforcement of
                Federal Laws   125*
             Finding the Law   128
             EPA's Office of Groundwater Protection   129
             New Federal Legislation   132

CHAPTER 11. The Safe Drinking Water Act Water Quality
             Programs   134
             *Checking Compliance with Monitoring Requirements
                 and Contamination Limits 135*
             Policing Variances and Exemptions 146
             Participating in Enforcement 148
             Imminent Hazard Situations 150

CHAPTER 12. Aquifer Protection Programs   152
             Wellhead Protection Program 152
             Sole Source Aquifer Program 159

CHAPTER 13. Action Under Superfund   162
             *How to Identify Spills and Contaminated Sites Subject
                 to Superfund Action 162*
             *How to Get Action to Clean Up Spills and Other
                 Emergencies 164*
             Participating in Superfund Cleanup Actions 165
             Getting Potential Superfund Sites Evaluated by EPA 166
             *Participating in Actions Against Immediate Hazards:
                 Removal Actions 170*
             Remedial Actions: Participating in Permanent Cleanups 173
             Participating in Cleanups Involving the Polluter 188
             Challenging Agency Removal and Remedial Decisions 189
             Taking Your Own Action 191

CHAPTER 14. Imminent Hazard Actions   193
             When Is an Imminent Hazard Action Appropriate? 193
             How to Produce Action 196
             Bringing a Citizen Suit 197

CHAPTER 15. The National Environmental Policy Act   199
             Federal Action 199
             Major Actions with Significant Environmental Effects 200
             Alternatives to the Action 200
             Mitigation Measures 200
             Exemptions for EPA Actions 201
             How to Use NEPA 201

CHAPTER 16. Waste Disposal Facilities   204
             Gathering Information on Facilities 206
             Remedying Releases into Groundwater 209
             *Investigating and Correcting Violations of Regulatory
                 Requirements 210*
             Setting Regulatory Requirements 220

Nonhazardous Solid Waste Facilities   226

CHAPTER 17.   Industrial and Commercial Sources   230
              Identifying Possible Industrial Sources   231
              Responding to Spills of Chemicals That May
                 Threaten Groundwater   236
              The Resource Conservation and Recovery Act (RCRA)   240
              The Clean Water Act (CWA)   242
              The Toxic Substances Control Act (TSCA)   245

CHAPTER 18.   Waste Disposal Wells   248
              How to Find Disposal Wells   248
              Taking Action Against Releases and Contamination   250
              Ensuring Compliance with Operating Regulations   252
              How to Use Agency Enforcement Procedures   257
              Citizen Suits   259
              Preventing Future Problems by Using the Permit Process   259
              Preventing Aquifer Exemptions   262

CHAPTER 19.   Underground Storage Tanks   264
              Discovering Locations   265
              Taking Action Against Immediate Threats   267
              Enforcing LUST Regulations   270
              Developing an Effective State Program   274

CHAPTER 20.   Pesticides and Other Agricultural Problems   278
              Identifying the Problem   278
              Remedying Current Contamination   281
              Preventing Future Contamination   282
              Fertilizers and Feedlots   289

CHAPTER 21.   Mining   291
              Using the Clean Water Act   291
              Mining on Federal Lands   292
              Coal Mining   293

CHAPTER 22.   Transportation of Hazardous Materials   305
              Obtaining Information About Transportation and
                 Potential Problems   305
              Responding to Transportation Spills and Other Releases   306
              Preventing Transportation-Related Contamination   307

PART IV
STATE AND LOCAL PROGRAMS FOR GROUNDWATER PROTECTION

CHAPTER 23. Comprehensive State Groundwater Protection
Programs 315
General State Groundwater Policies and Strategy 315
Groundwater Classification Systems 317
Groundwater Quality Standards 318
Protecting Groundwater Through Control of Nonpoint
Pollution Sources 320

CHAPTER 24. State and Local Land Use Controls 322
Local Land Use Controls 322
State Development Legislation 326
Sensitive Areas Protection 327

CHAPTER 25. State Superfund Programs 330
Identifying and Ranking Sites 330
Participating in Cleanup Decisions 331
Going After Responsible Parties 332
Special Liability Rules 332

CHAPTER 26. State Environmental Policy and Protection Acts 334
State Environmental Policy Acts 334
Environmental Protection Acts 335

CHAPTER 27. State Regulation of Solid and Hazardous Waste
Disposal 337
Solid Waste Disposal 337
Hazardous Waste Disposal 338
Local and State Attempts to Restrict Waste Disposal
by Origin 341

CHAPTER 28. State Regulation of Commercial and Industrial
Facilities 344
State Pollution Discharge Permit Programs 344
Hazardous Materials Controls 345
Emergency Preparedness and Response Programs 346
Reducing Waste at the Source 346
Property Transfer and Cleanup Statutes 347

CHAPTER 29. State Regulation of Waste Disposal Wells 351
Siting 351
Waste Characterization 352
Monitoring 353

Inspection   353
Class V Injection Wells   354
Class II Injection Wells   355
Additional Regulations   355

CHAPTER 30.   State Regulation of Underground Storage Tanks   357
*Locating Underground Storage Tanks in Your
    Community   357*
Regulations Addressing UST Leaks and Cleanup   358
*Regulations Governing Underground Storage
    Tank Operation   359*

CHAPTER 31.   State Regulation of Agricultural Sources   364
Finding the Information   364
*Private Agricultural Sources of Groundwater
    Contamination   365*
Government Application of Pesticides   368

CHAPTER 32.   State Regulation of Mining   370
Groundwater Permits   370
State Mining Permits   372
Land Use Controls   373
State Environmental Policy and Protection Acts   373

CHAPTER 33.   State Regulation of Septic Systems   375
Siting   375
Inspections and Maintenance   376
Septage Disposal   377
Septic Disposal Wells   378
Large Residential Septic Systems   378
Commercial and Industrial Septic Systems   378

CHAPTER 34.   State Regulation of Transportation of Hazardous
              Materials and Highway Runoff   381
Hazardous Materials Transportation   381
Highway Runoff   383

Appendixes
     A.   *EPA Regional Offices   385*
     B.   *Government Sources for Reference Materials   386*
     C.   *Organizations Providing Information and Services   388*
     D.   *Hotlines, Emergency Assistance, and Other Important Phone
            Numbers   397*
     E.   *Agencies Designated to Receive Notifications of Underground
            Storage Tanks   399*

     Index   405

LIST OF FIGURES

1.1 Diagram of the Hydrologic Cycle 4

1.2 How Groundwater Occurs in Rocks 5

1.3 Diagram of Geologic Strata and Various Types of Wells 6

1.4 Idealized Pattern of Groundwater Flow from Recharge to
    Discharge Areas 7

1.5 Plume Formation of Groundwater Contaminants 8

1.6 Cone of Depression 9

1.7 Area of Influence 10

2.1 Potential Health Effects of Superfund Contaminants 16

3.1 Hazardous Waste Landfill 26

3.2 Double Wall Tank and Leak Detection System 28

3.3 Deep-Well Injection of Liquid Waste 30

3.4 Aquifer Contamination Through Improperly Constructed or
    Abandoned Wells 31

3.5 Groundwater Contamination at Selected DOE Facilities 39

3.6 A Typical Septic System 42

4.1 Well Placement and Effective Monitoring 52

4.2 Effect of Hydrology on Monitoring Well Effectiveness 52

4.3 Effect of Geological Features on Groundwater Monitoring 53

4.4 Effect of Pollutant Density on Groundwater Monitoring 54

4.5 Regional Aquifer-System Analysis 55

4.6 Office of Technology Assessment Summary of Federal Groundwater
    Monitoring Provisions and Objectives 58

5.1 Watershed Boundaries 74

5.2 Parameters for D.R.A.S.T.I.C. System of Aquifer Assessment 78

7.1 EPA Permit and Appeal Process 92

8.1 Citizen Suit Provisions Under Various Statutes Protecting
    Groundwater 103

9.1   Household Hazardous Waste Chart   116

10.1   Major Federal Laws Relevant to Groundwater Protection   129

10.2   Major Groundwater Legislation Introduced in the 100th Congress   130

11.1   National Primary Drinking Water Regulations   136

11.2   Contaminants Regulated Since June 1989   141

11.3   Contaminants to Be Regulated by 1997   142

11.4   Monitoring for Unregulated Contaminants   144

11.5   Notification Requirements for Public Water Systems   147

11.6   National Secondary Drinking Water Regulations   150

12.1   Zones of Influence and Contribution   155

12.2   Designated Sole Source Aquifers—Nationally (Status as
       of 12/31/88)   158

13.1   The ATSDR/EPA List of the 100 Most Hazardous Contaminants   169

13.2   CERCLIS-National G.M.1 National Priorities List (NPL) Sites   175

13.3   Hazardous Waste Ranking System Factors   176

13.4   Superfund Scorecard   178

13.5   Examples of Slurry Walls   185

14.1   EPA Imminent Hazard Authorities   194

16.1   Design and Operating Requirements   212

18.1   Regulatory Programs for States with Active Hazardous Waste
       Injection Wells   255

19.1   Leak Detection Alternatives   272

20.1   EPA's List of Pesticides Most Likely to Leach into Groundwater   280

20.2   Potential Sources of Pesticide Contamination of Groundwater   283

20.3   Pesticides for Which Health Advisories Have Been Prepared
       by the EPA Office of Drinking Water   286

21.1   Coal Fields of the United States   295

21.2   The SMCRA Enforcement Process   299

28.1   Industry Waste Reduction: Some Success Stories   348

FOREWORD

GROUNDWATER IS A precious resource for the United States. Nearly half of the population of this country, and over 90 percent of the population in rural areas, depend on groundwater as their primary source of water. We all expect that groundwater will be clean and pure, and, in particular, free from the contaminants that threaten our health. The frightening reality is, however, that groundwater is not safe from contamination.

Groundwater contamination is a national problem and the sources of potential groundwater contamination are everywhere. In this country, there are over 100,000 landfills, many of which contain hazardous materials, and many of which leak directly into groundwater. As many as 10 million underground storage tanks for petroleum and industrial chemicals threaten groundwater quality. Already, 41 states report groundwater contamination from such tanks. Wells are used to inject hundreds of millions of tons of toxic, hazardous, and other liquid wastes into the ground every year. Twenty-five states report groundwater contamination from pesticides that are applied in this country at a rate of 700 million pounds a year. Even the average American household uses about 25 gallons of hazardous chemical products each year to clean, paint, repair, and weed the garden. Many of these products end up down the drain or in the garbage and, eventually, in the groundwater.

The challenge for all of us is to act to protect the groundwater we all need from the contamination sources in all our communities. In Congress, we are responding with new legislation. In the 100th Congress, I introduced several significant bills that would strengthen federal laws protecting groundwater. The legislation I proposed would create a comprehensive structure to provide the same kind of protection to groundwater that we provide today to our nation's lakes and rivers under the Clean Water Act, I also introduced bills to expand research into the problem of groundwater contamination and to protect groundwater from pesticide contamination. Many of my colleagues in the Senate and in the House of Representatives support these efforts and some have introduced similar legislation. I hope and expect that groundwater protection legislation will be a high priority for Congress in the coming year.

Groundwater contamination is not, however, a problem that can be solved on the federal level alone. It will require strong and effective action in state legislatures and city councils, too. And, perhaps most of all, protection of our vulnerable groundwater resource will require that alert and concerned people around the country take action to protect the groundwater they and their neighbors use every day. Governments will not be able to solve this problem without the help and prodding of local citizen groups.

I strongly support the efforts of these groups to organize and educate their neighbors about groundwater and I encourage citizens to learn about and get involved in federal, state, and local groundwater programs. This kind of citizen action is what *The Poisoned Well* is all about. The book discusses in a practical way how to use the laws and regulations that are in place to stop ongoing contamination and get effective cleanup response from polluters and government agencies and to prevent contamination in the future. *The Poisoned Well* will help concerned people become active and effective advocates for groundwater protection.

Protecting our groundwater for our children is a challenge we all must accept and struggle with together if we are to succeed. The time for action is now.

David F. Durenberger
United States Senator from Minnesota

INTRODUCTION

RESIDENTS OF COLUMBIA, Mississippi, learned that barrels of benzene left at an abandoned chemical plant were leaking and had contaminated the groundwater supplying their town with drinking water. Faced with this serious health threat, the citizens of the town pressured the Environmental Protection Agency (EPA) into initiating a cleanup action.

In Lake Charles, Louisiana, people discovered that hazardous wastes were being injected into deep wells outside their town. Toxic contaminants were found in the regional drinking water aquifer. Citizens organized and forced some action at the site, but their fight to complete the cleanup continues.

In Massachusetts, an abandoned chemical pesticide plant caused the contamination of groundwater 60 feet below the surface, leading the town to close several drinking water wells. Local residents formed a citizens' group, spread information about the problem, and with the assistance of a national group, worked with EPA to develop a long-term cleanup plan.

These are just a few examples of the countless stories from around the United States about groundwater contamination. Groundwater contamination is no longer an isolated problem limited to a few heavy-industry states. Groundwater contamination is everywhere—cases range from pesticides polluting groundwater in farm states like Iowa to solvents from storage tanks leaking into aquifers in California's Silicon Valley. And because groundwater is such an important source of drinking and other water supplies—over 90 percent of the water used in rural areas, for example, comes from groundwater—contamination of groundwater can be a serious threat to human health. Pollutants commonly found in groundwater have been linked to illnesses ranging from bacterial infections to cancer. Individuals and citizen groups around the country are waking up to news stories about groundwater contamination in their own community and are beginning to worry about their health and their children's health. The question is, what can people do to protect their families from groundwater contamination?

This book is designed to help citizens to help themselves. This book is based on the idea that the best way for citizens to protect their health is to go out and wage their own fight against groundwater contamination. People can make a difference if they know how to use the legal and other tools that are available to them. The primary focus of this book is advice about how to use the various federal laws and programs that can affect groundwater quality. There is no comprehensive federal groundwater protection scheme, but there are many laws and programs that can be used to protect groundwater quality. The discussion of these programs is directed toward the citizen who is neither a lawyer nor a scientist and it communicates the information about these programs that anyone who wants to use them needs to know.

We also recognize that often citizens can be most effective in dealing with local groundwater problems by working at the local level. As a result, we have included a substantial discussion of programs that are being tried by state and local governments and used by citizen groups to combat groundwater contamination. Because of the number and variety of such programs around the country we cannot cover them all, nor can we provide the sort of step-by-step advice that we offer in our presentation of federal programs. Instead, the state and local programs section of the book highlights some of the approaches in use and provides general suggestions for citizen action. It is up to the reader to take the ideas we present and investigate similar options wherever he or she may live.

All of the programs discussed in this book are similar in one respect. They provide only a framework for action. The laws themselves do not protect groundwater. It takes aggressive and effective enforcement action to ensure that the law is translated into clean groundwater.

Many people assume that government agencies alone are responsible for enforcing the law and providing the protection the law envisions; but all too often agencies do not fulfill their role. This inaction may happen for a variety of reasons ranging from budget shortfalls and meager staffs to agency policies or personnel opposed to aggressive enforcement.

Whatever the reason, when agencies do not do their job, citizens have to step in and do the work themselves. In the end, the laws and programs discussed in this book work only as well as you make them work.

HOW TO USE THIS BOOK

This book is split into four parts, each with a different purpose. To make the best use of this book, you need to know how it is organized. Part I contains basic information about groundwater, how it is contaminated, how you might be affected, and how to find out whether your groundwater is polluted. Read Part I to lay the foundation for the suggestions for action made in later chapters.

Part II describes in a general way the nuts and bolts of using the laws. It explains how to obtain information from the government, how to work with administrative agencies, and how to use the courts. It also provides some basic advice about organizing and using nonlegal political tools as part of an overall strategy.

Part III gives step-by-step advice about using specific federal programs. It is organized primarily by the potential sources of groundwater contamination, such as underground storage tanks or injection wells, but it also describes the programs generally applicable to a variety of potential sources. This part is the heart of the book and is designed to provide specific suggestions for action.

Part IV summarizes state and local programs. It is organized to parallel the structure of Part III, but it provides more general information about the types of state and local programs in existence and how citizens can use them.

To use the book, then, most readers will want to study Part I to gain an understanding of the basic problem and read Part II to learn about the basic processes they will become involved in and use. Parts III and IV can be read more selectively depending on the contamination problem in a particular locality. A reader concerned primarily about underground storage tanks, for example, should read the chapters devoted particularly to regulation of underground storage tanks under federal and state law (Chapters 19 and 30). Material describing general federal and state programs may also be helpful and should be reviewed (see Chapters 10 to 15 and Chapters 23 to 26). The other source-specific chapters, however, need not be studied, unless of course the reader discovers that the problem is broader or deeper than storage tanks.

PART I

GROUNDWATER AND CONTAMINATION

This part of *The Poisoned Well* provides the background information a citizen needs to implement the specific suggestions for action discussed in later chapters of the manual. This part begins with a brief introduction to groundwater—what it is and how it moves. Chapter 2 then explains how contaminants found in groundwater can affect your health, how those effects are measured, and how legal limits on contamination are established. Chapter 3 summarizes basic information about all of the major sources of groundwater contamination—what they are, what contaminants they produce, and how big a problem they create.

This part concludes with two chapters that will help citizens gather more basic information about the problem in their area. Chapter 4 explains how groundwater is monitored and tested, including how citizens can test their own water. Chapter 5 discusses techniques that can identify problem areas through the use of maps of groundwater and contamination sources.

Becoming an activist for groundwater protection means starting a continuing process of learning. The information contained in Part I will give the reader a solid foundation from which to begin taking concrete steps to preserve groundwater quality.

CHAPTER 1

Groundwater Basics

HYDROLOGIC CYCLE

WATER IS ALWAYS on the move. The sun evaporates it from oceans, lakes, ponds, streams, and the leaves of plants. It falls to earth as rain, snow, sleet, and hail. Gravity pulls it down to rivers and into the ground. Hydrologists call the total system the hydrologic cycle. (See Figure 1.1.) Groundwater is one of the less visible components of the cycle, but the global volume of groundwater is second only to the oceans and polar ice caps, and of all available fresh water in the United States, 96 percent is groundwater.

Groundwater is basically precipitation that has percolated down into soil and filled the spaces in the rock below in the same way that water fills a sponge. The first water entering soil from rainfall or snowmelt replaces water previously evaporated or used by plants during dryer periods. Some of this new water quickly repeats the hydrologic cycle: it evaporates, is taken up and transpired by plants, or runs off into streams. (See Figure 1.2.) Any remaining precipitation, or water that leaches from surface water bodies into the soil, travels through an upper portion of soil and rock that hydrologists call the unsaturated zone. While the degree of saturation varies with the amount of precipitation, the unsaturated zone is generally characterized as containing water and air in the smaller pores or spaces of rocks and soil. Any water in this area that is not left clinging to soil due to molecular attraction will drain from the unsaturated zone down to the water table. The water table is a seasonally fluctuating boundary between the unsaturated zone and the saturated zone. In the saturated zone the pores and cracks in rocks and soils are filled only with water.

AQUIFERS

Underground saturated rock formations that yield usable water are called aquifers. The minimum water content necessary to qualify a rock formation as an aquifer is a relative concept depending on the availability of

3

other water sources in the region; what is one person's rock may be
another person's aquifer. A geologic formation's ability to yield water to
wells is dependent on its porosity combined with its permeability. Porosity refers to the pores (spaces or cracks) in rocks, or the percentage of
the rock's volume that is not occupied by the rock itself. The quantity of
water that any type of rock can contain depends on the rock's porosity.
Permeability refers to the degree to which underground pores are interconnected with each other, that is, the degree to which water can flow
freely from one pore to another. The importance of permeability is illustrated by the substance clay, which, though it can have the same porosity

FIGURE 1.1
DIAGRAM OF THE HYDROLOGIC CYCLE

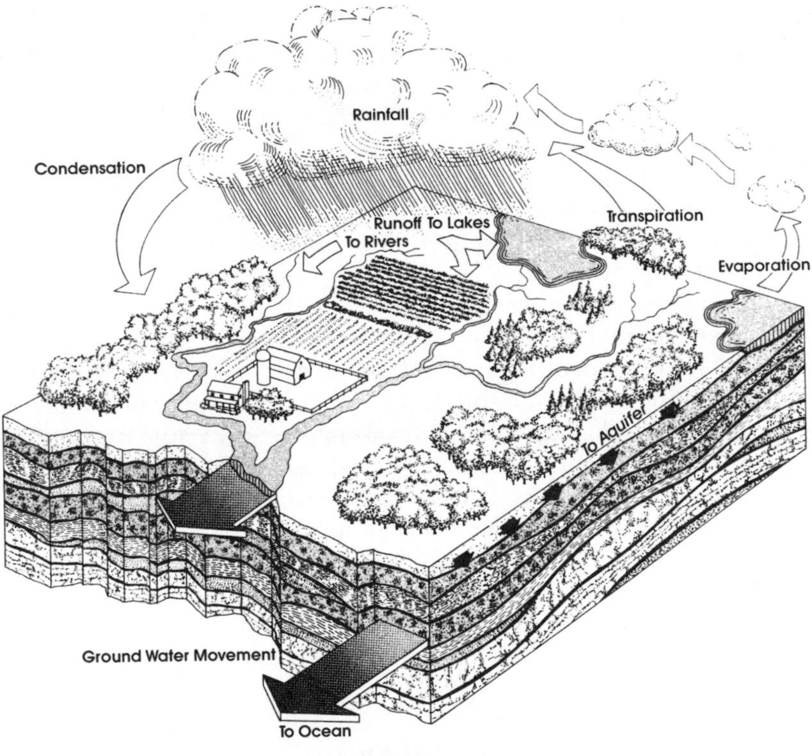

*Groundwater is part of the hydrologic cycle, groundwater and surface water
are interconnected. Reprinted with permission of the Chemical Manufacturers Association, from Ground Water, 1983, p. 1.*

FIGURE 1.2
HOW GROUNDWATER OCCURS IN ROCKS

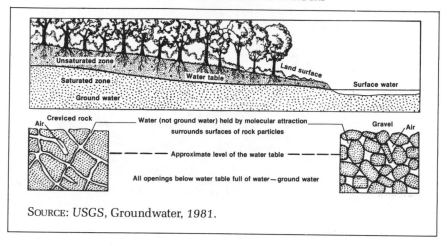

SOURCE: USGS, Groundwater, *1981.*

as coarse gravel, will not be as good an aquifer as the gravel due to the clay's lack of permeability.

Varying layers of permeable and impermeable materials in the earth create different types of aquifers. The most familiar type is the unconsolidated aquifer, in which water is contained in the spongelike pore spaces of sand and gravel (this is known as "primary porosity"). All unconsolidated aquifers are underlain by a layer of impermeable material—called an aquitard—that prevents the water from flowing further down into the earth. One subcategory of unconsolidated aquifer, called an unconfined or surficial aquifer, has an aquitard below but none above. Thus, water is free to percolate into the aquifer from the earth's surface and the unsaturated zone.

The other subcategory of unconsolidated aquifer, called a confined or artesian aquifer, has aquitards below and above. The upper aquitard severely limits water from entering the aquifer from directly above; instead, water enters laterally by sideways motion through the aquifer. Because it is sandwiched between two layers of impermeable material, this kind of aquifer may be under great pressure, and may spurt substantially above the earth's surface when tapped by a well.

In some areas unconsolidated aquifers occur stacked in layers, with an unconfined aquifer on top and one or more confined aquifers beneath it, as illustrated in Figure 1.3.

The second major category of aquifer is the bedrock or consolidated aquifer, which occurs in areas of nonporous rock that lacks the capacity to absorb water. In such an aquifer, water is not found in pore spaces, but

FIGURE 1.3
DIAGRAM OF GEOLOGIC STRATA AND VARIOUS TYPES OF WELLS

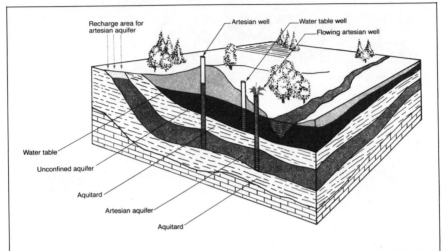

Reprinted with permission from "Groundwater Information Pamphlet."
Copyright 1983, the American Chemical Society.

rather in fractures or holes in the rock (this is known as "secondary porosity").

One example of a consolidated aquifer is a hard crystalline bedrock where the water resides in fractures or cracks. The well yield in such an aquifer will depend on the size and frequency of the water-bearing fractures intersected by the well.

Another example of a consolidated aquifer is karst limestone, which occurs in areas of soft limestone rock. As a result of millions of years of erosion caused by underground water flow, karst limestone formations have been cut through with a Swiss cheese network of fissures and holes, which, in some cases, are large enough to form underground caverns and caves.

GROUNDWATER AND CONTAMINANT MOVEMENT

Recharge and Discharge Areas

Any area of land allowing water to pass through it and into an aquifer is called a recharge area. Water moves from the recharge area through the aquifer and out to the discharge area. Discharge areas can be wells, lakes,

springs, geysers, rivers, or oceans. The uniting of groundwater and surface water in recharge and discharge areas is extremely important. (See
Figure 1.4.) Recharge areas are the conduits between surface contamination and groundwater supplies. (The reverse is also true. The discharge of
contaminated groundwater may affect the more than 30 percent of our
nation's streamflow that comes from groundwater.)

In an unconfined or surficial aquifer, the recharge area is generally
located immediately above and adjacent to the point at which drinking
water wells have been drilled, so that pollution occurring near the wellhead can have a devastating effect on groundwater quality. By contrast, a
confined or artesian aquifer is protected by an overlying aquitard and
thus may be less vulnerable to pollution entering the ground near the
wellhead. The recharge areas for such an aquifer can be located at substantial distances from the wellhead, making water quality vulnerable to

FIGURE 1.4

**IDEALIZED PATTERN OF GROUNDWATER FLOW FROM RECHARGE TO DISCHARGE
AREAS**

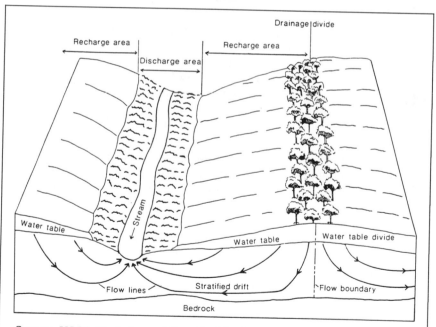

SOURCE: *USGS, Hydrology of Stratified-Drift Aquifers and Water Quality in
the Nashua Regional Planning Commission Area, South-Central New
Hampshire, 1987.*

FIGURE 1.5
PLUME FORMATION OF GROUNDWATER CONTAMINANTS

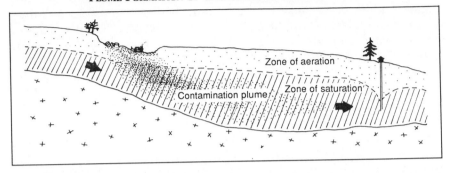

the effects of faraway land uses. Recharge areas for bedrock and karst limestone aquifers can be located either near to or far from the wellhead, or both.

Direction of Flow

The direction of flow from areas of recharge to areas of discharge is dependent on gravity, pressure, and friction. Generally, groundwater moves in response to a hydraulic gradient from points of high elevation and pressure to points of lower elevation and pressure. The high and low elevations must be taken into account over large areas of land, because groundwater flow does not correspond precisely with surface topography. The points of higher elevation usually serve as watershed boundaries, called drainage divides. Watersheds, also called drainage basins, are those areas of land which drain runoff water to surface water bodies. Aquifers are often found beneath the surface of drainage basins; the high elevations serving as watershed boundaries may also be aquifer boundaries.

   If a pollution source contaminates groundwater, it most often affects only that portion of the aquifer downgradient of the site, that is the lower elevations and lower pressure areas, rather than that portion of the aquifer upgradient of the site. Tracking contaminant movement is not, however, always as simple as determining an upgradient-downgradient direction. Even in unconsolidated aquifers, where movement is most predictable, water can be diverted from its normal downgradient course by a deposit of impermeable material that obstructs flow. In bedrock or karst limestone aquifers, unpredictable flow patterns are the rule rather than the exception, and water will go wherever the often irregular underground cracks and holes lead it.

Speed of Flow

In unconsolidated aquifers, groundwater generally travels very slowly. Where stream velocity is measured by feet per second, groundwater velocity can be measured in feet or inches per day or year. Groundwater has a laminar flow pattern, meaning that it is subject to little mixing and follows distinctive paths; many contaminants entering groundwater will behave the same way. Such contaminants remain in concentrated masses called plumes, which resemble clouds or fingers, as illustrated in Figure 1.5. Unlike contaminants in surface water, contaminants in groundwater are subject to very little dispersion by mixing, sun exposure, temperature differentials, and variations in bacterial life-forms. Thus there is very little physical, chemical, or biological breakdown of contaminants on a short-term basis. The shape and concentration of the plume is dependent only on local geology, elevation profiles, physical and chemical properties of the contaminant, rate of pollution by the contaminating source, and modifications in flow from wells or pumping.

In karst limestone aquifers, flow is not laminar, but rather corresponds more closely with the flow pattern one might expect in a surface stream. This means that water can travel much faster than in an unconsolidated aquifer, reaching speeds as great as several miles per day.

Human Influence on Flow

The natural path of groundwater and its flow rate can change dramatically through groundwater well pumping. Wells will draw in groundwater and

FIGURE 1.6
CONE OF DEPRESSION

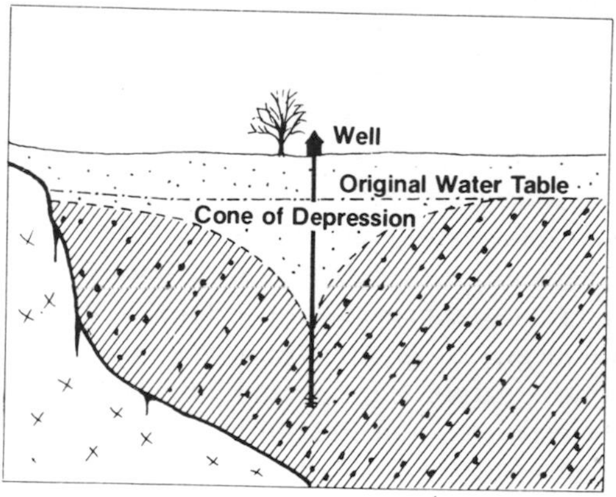

Pumping a public-supply well lowers the water table around it, creating a cone of depression. Reprinted with permission from the Massachusetts Audubon Society.

FIGURE 1.7
AREA OF INFLUENCE

The land area above the cone of depression is called the area of influence. Reprinted with permission from the Massachusetts Audubon Society.

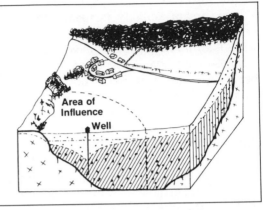

contaminants from all directions, and can substantially increase the flow rate. Wells actually create a false discharge area for contaminants and water.

The drawing-in action of wells creates a cone of depression around well sites. It is called a cone because, when the well withdraws groundwater, the water table surrounding the well lowers, creating slopes that become increasingly steep closer to the well. (See Figure 1.6.) The geologic characteristics of the aquifer and the rate and duration of pumping will affect the size and shape of the cone. For example, the cone will be much greater around large public wells than small private wells.

The land area above the cone of depression is called the area of influence. (See Figure 1.7.) Pollutants discharged within this area can have a devastating affect on the quality of water withdrawn by the well. The area of influence is an important recharge area for individual wells drawing from the surficial aquifer. Any pollutants discharged within the area will be pulled directly to the well. The closer the source of contamination to the well, the faster the contaminants will be drawn into the well. It is important to monitor possible sources of pollution within these areas as well as all other land uses. The area of influence will shift if, for example, a parking lot creates a relatively impermeable surface where there was once direct recharge; in such circumstances, land previously outside the boundaries of the area of influence will become important recharge areas for the well. (In karst limestone aquifers, the cone of depression and area of influence may be difficult to define; and in confined aquifers, pollutants released in the immediate vicinity of the well will not necessarily enter the aquifer).

REFERENCES

GROUNDWATER BASICS

American Chemical Society. *Groundwater Information Pamphlet.* Washington, D.C. 1983.

Concern, Inc. *Groundwater: A Community Action Guide.* Washington, D.C. 1984.

Conservation Foundation. *A Guide to Groundwater Pollution: Problems, Causes, and Government Responses.* Washington, D.C. 1987.

DiNovo, Frank, and Martin Jaffe. *Local Groundwater Protection: Midwest Region.* American Planning Association. Chicago, Ill. 1984.

DiNovo, Frank, and Martin Jaffe. *Local Groundwater Protection.* American Planning Association. Chicago, Ill. 1987.

Environmental Defense Fund. *Dumpsite Cleanups: A Citizen's Guide to the Superfund Program.* Washington, D.C. 1983.

EPA, Office of Groundwater Protection. *Overview of State Groundwater Program Summaries.* Washington, D.C. 1985.

Feliciano, Donald V. *Groundwater: What It Is and How It Is Being Protected.* Environment and Natural Resources Policy Division, U.S. Congressional Research Service. Washington, D.C. 1985.

Gordon, Wendy. *A Citizen's Handbook on Groundwater Protection.* Natural Resources Defense Council. New York, N.Y. 1984.

Patrick, Ruth. *Groundwater Overview.* Address to Third National Water Conference. Philadelphia, Pa. 1987.

Scalf, Marion R., James F. McNabb, William Dunlap, and Roger L. Cosby. *Manual of Groundwater Quality Sampling Procedures.* Environmental Protection Agency. Washington, D.C. 1982.

Senate Committee on Environment and Public Works; Subcommittee on Toxic Substances and Environmental Oversight. *Hearings: Groundwater Contamination and Protection.* Part 1: June 17, June 20, July 16, 1985. Part 2: October 24, November 14, and December 12, 1985.

U.S. Geological Survey. *Groundwater.* Washington, D.C. 1981.

U.S. Geological Survey. *Water Fact Sheet: Regional Aquifer Systems of the U.S.* Washington, D.C. 1984.

White, Lyn. *An Introduction to Groundwater and Aquifers.* Massachusetts Audubon Society. Lincoln, Mass. 1985.

White, Lyn. *Groundwater and Contamination: From the Watershed Into the Well.* Massachusetts Audubon Society. Lincoln, Mass. 1985.

CHAPTER 2

Health and Groundwater Contamination

WE ARE ALL concerned about what contaminated groundwater can do to our health and our children's health. As a society we have grown dependent on a myriad of potentially dangerous technologies and chemical substances, but it is only recently that we have begun to understand how our past and present handling of these substances could affect us for generations to come. It is alert citizens, not government agencies, who are usually first to realize that illnesses are occurring at unusual levels in their community, and who are first to begin the search for the cause. All the puzzle pieces start to fit together in a process that usually goes something like this:

First I heard about Barbara, and of course I felt real bad about it.

Not too long after that, I heard about Danny. I saw his wife, Donna, at the drugstore picking up a prescription. She told me Danny had brain surgery and lost part of his hearing, just like Barbara.

Then I remembered Beth Saner who died so young of cancer. And other things started happening little by little. It wasn't long after that, that Sylvia Valdez was really sick with lupus.

I started inquiring. I heard about a lot of cases of lupus. It seemed that so many in our age group were coming down with cancer or other problems.

Being so close to Hughes and the Air National Guard, I thought maybe we were exposed to radiation, something being dropped in the desert that we didn't know about. We talked about compiling a list and trying to find out what had caused it.

Then a couple of years later the story broke on the water, and I realized, "Aha, this could be it."

I started remembering back about Linda Moore's mother who died so young of breast cancer after I graduated from high school. Then a cousin on Calle Bocina had to have a hysterectomy at a very young age because of cancer.

12

She lived right across the street from Joe Burchell's brother, who had died of leukemia. Laura Castro Urias, who had lupus, lived on Elvira, one street south of Bocina, and that was two blocks south of where Beth Saner had lived.

I think that's odd. There's just too many of them.[1]

Melinda, the narrator here, was talking about the suspected effects of a common industrial solvent, trichloroethylene (TCE), in the public water wells of Tucson, Arizona. Five of the women who were pom-pom girls and cheerleaders with Melinda in high school 20 years ago had serious illnesses: brain tumors, lupus, multiple sclerosis, rare tumors, and arthritis. Melinda was not alone in her concerns. Other residents were turning the bits and pieces of information over in their minds, wondering what was going on.

Social worker Carol Roos almost went to the county health department years before she heard of TCE contamination, when she thought she had a third teenager with testicular cancer in her program. "Testicular cancer is almost unheard of in young people. It's very rare. When I looked it up, the most common ages were between 29 and 35. These kids were 16," she said. When the third case turned out to be a prostate problem and not testicular cancer, Carol decided not to seek help from the county health department. However, her concern over what could be causing "an awful lot of childhood leukemia" and other cancers did not end.

In 1981 Tucson was shaken to learn that its only water supply was threatened by toxic chemicals. Slowly the story unfolded to residents that for over 25 years Hughes Aircraft Company, in its work for the U.S. Air Force, had dumped toxic industrial wastes into the surrounding desert. The city's aquifer was very vulnerable to contamination due to the highly permeable desert sands overlying it. More than four years after discovery of contamination in area water supplies, however, city, county, state, and federal officials continued assuring area residents that the water they drank was safe and that, by the time the water reached area homes, the TCE level was below the state's guidelines. It took a six-month investigation by the *Arizona Daily Star* to learn that these officials were wrong.

In some ways Tucsonians are lucky. At least they know their aquifer is contaminated and can take steps to protect themselves. Many people in similar situations are unaware that contamination exists at all, much less by which sources and which chemicals. However, Tucsonians are still faced with a problem common to all communities suffering toxic exposure: how has that contamination affected their health?

---

[1] (Kay, Jane, " 'Too Many' Are Ill in Area Touched by TCE," *Arizona Daily Star*, Tucson, Arizona, May 1985.)

This chapter will introduce you to some basic information about how health effects are measured, typical health problems associated with types of contaminants, and how the uncertainties about health problems affect the regulatory process. The basic message of the discussion is that although the problem of how health is affected is complex, it is one that you cannot afford to leave entirely in the hands of the scientists and regulatory agencies.

TESTING THE HEALTH EFFECTS OF A CONTAMINANT

Today in the U.S., industry uses over 60,000 chemicals and produces between 500 and 1000 new chemicals per year. Yet only a tiny fraction of these chemicals have been adequately tested for their effects on our health and environment. What little we do know about the possible effects of chemical contaminants on the body is the result of animal, epidemiological, and microbiological laboratory studies.

Animal tests are premised on the idea that there are basic similarities in the way human and animal cells, tissues, and organs will respond to a chemical contaminant. Scientists commonly select rats, mice, rabbits, dogs, monkeys, or baboons based on the needs of the study. Adequately designed animal studies are widely accepted by the scientific community, and most authorities agree that, for example, any compound which has caused cancer in an animal should be assumed to be capable of causing cancer in humans. However, some debate still surrounds the extrapolation of potential human health effects from animal effects, especially the use of data from the high chemical doses that are usually given to lab animals to draw conclusions about the health effects from human exposure to low doses.

Epidemiological studies involve the observation of human populations where the observed groups either have been exposed to the same contaminant or have the same disease. If the community was exposed to the same contaminant, researchers will observe those people to monitor the types of disease that develop. If the group has the same disease, researchers will collect data on their past and present activities to find common factors that may have caused the disease. Both groups are compared with a "control" population, in which there is either no known exposure to the same compound, or no similar disease. Epidemiological studies can often make statistical links between an exposure and a disease, but they fall short of proving that the particular exposure actually *caused* the disease, or that the same phenomena would occur again in another population. This is due partly to the nature of modern human communities, and partly to the inherent limits of epidemiological inference: we are individually exposed to thousands of different possible causes of disease and it is unlikely that the same environmental conditions and levels of exposure experienced by one community will be experienced again by another.

Contaminants can also be tested in individual cells or microorganisms. One such test exposes bacteria to chemicals in order to test the chemicals' effects on DNA, one of the basic building blocks of cells—including human cells. If a contaminant adversely affects DNA, then human exposure to that contaminant might lead to birth defects and cancer.

HEALTH EFFECTS OF COMMON GROUNDWATER CONTAMINANTS

In 1984 the Office of Technology Assessment (OTA) compiled a list of over 200 contaminants known to occur in groundwater. (See Figure 2.1.) Due simply to increased monitoring efforts, we know the list of contaminants found today is over three times larger. Though our ability to produce contaminants and release them into the environment far outstrips our ability to adequately measure—through animal, epidemiological, or laboratory tests—exactly what these substances will do to us, in this section we will discuss some of the health effects commonly associated with a few of the more prominent and well-studied groundwater contaminants.[2]

Biological contaminants in groundwater include bacteria, viruses, algae, and other microscopic creatures. While groundwater usually has fewer microorganisms than surface water, substantial numbers of them do occur, due primarily to the natural decomposition of plants, animals, and animal wastes. Once in the aquifer, some contaminants may flourish due to the lack of light and air and to the presence of moisture. Dangerously high levels of bacteria and viruses in well water can be caused by leaching septic tanks and animal feedlots. This leaching of fecal material near well water can result in outbreaks of gastrointestinal illnesses, typhoid, infectious hepatitis, cholera, and tuberculosis. Statistics show biological contamination to be the most common form of groundwater contamination affecting human health, but this conclusion could change as we become more aware of the effects of chemical contamination.

Inorganic contaminants, such as metals and salts, reach groundwater from natural and human sources. Adverse health effects from heavy

---

[2] Federal law requires businesses to have on hand information concerning health effects, and to report this information to government authorities. See the discussion of Material Safety Data Sheets in Chapter 17.

Figure 2.1

POTENTIAL HEALTH EFFECTS OF SUPERFUND CONTAMINANTS

Toxic Effects

Most commonly found groundwater contaminants at hazardous waste sites

Contaminant	Carcinogenicity	Mutagenicity	Teratogenicity	CNS effects	Cardiovascular effects	Kidney damage	Liver damage	Peripheral nervous system effects	Immunological effects	Gastrointestinal effects	Reproductive effects	Embryotoxicity	Lung/respiratory effects	Endocrine effects	Blood cell disorders	Upper respiratory tract irritation	Skin damage	Visual damage	Allergic sensitization	Eye irritation	Skin irritation
1. Trichloroethylene	●			●		●	●													●	●
2. Benzene	●	●		●			●								●					●	●
3. Toluene				●									●		●						●
4. Lead		●		●		●									●						
5. Chloroform	●	●		●		●	●			●										●	●
6. 1,1,1-trichloroethane				●			●													●	●
7. Tetrachloroethylene	●			●			●				●									●	●
8. Arsenic	●	●	●				●							●							
9. Phenol						●	●														
10. Trans-1,2-dichloroethylene	●	●		●	●		●				●	●								●	●
11. PCBs	●	●	●		●		●				●	●				●					●
12. Ethylbenzene	Data not yet available—may pose a cancer risk to humans																				
13. Methylene Chloride	●			●			●				●									●	
14. Chromium	●	●		●	●					●									●	●	
15. Xylene				●						●											
16. Cadmium	●	●		●	●					●	●									●	
17. Zinc and compounds																					
18. Vinyl Chloride	●	●		●			●				●		●							●	●
19. 1,1-dichloroethane		●		●			●								●						
20. Mercury		●		●														●			

SOURCE: *Adapted from ENVIRON Corporation, Approaches to the Assessment of Health Impacts of Groundwater Contaminants (Washington, D.C.: Office of Technology Assessment, 1983). Reprinted courtesy of The Conservation Foundation.*

metals are the best documented results of such contaminants. Common heavy metal contaminants include arsenic, cadmium, chromium, copper, iron, lead, mercury, nickel, and zinc. Because these metals do not break down, are excreted very slowly, and build up in the body over time, they can cause a range of severe conditions including central nervous system disorders, lung, kidney, and liver damage, gastrointestinal disturbances, birth defects, and cancer.

Nitrates from septic tanks, animal wastes, fertilizers, landfills, decomposing vegetation, and geologic deposits may be the single most common cause of groundwater contamination. Nitrates in drinking water supplies have been linked to nervous system impairments, cancer, birth defects, and methemoglobinemia, or blue-baby syndrome.

Organic compounds (compounds that contain carbon), particularly those that are man-made, are an increasingly common type of contaminant found in groundwater: of the OTA's 1984 list of approximately 200 groundwater contaminants, 175—fully seven-eighths—were organic chemicals. In general, many of the synthetic organic compounds, especially volatile organic compounds, are more mobile and less susceptible to biological and chemical degradation than microbiological and inorganic contaminants.

Common organic contaminants found in groundwater include pesticides and solvents. In a recent survey, EPA found at least 17 different pesticides contaminating the groundwater supplies of 23 states simply from normal agricultural use. Solvents are leaking into groundwater at alarming rates from improper storage and disposal methods used by plating and electronic industries, military contractors, airports, chemical waste handlers, gas stations, dry cleaners, and a myriad of other businesses. An EPA survey found that 80 percent of the reported incidents of synthetic organic contamination in groundwater included the solvent TCE, or trichloroethylene, leaking primarily from underground storage tanks.

Although health information on organic chemicals is just beginning to emerge, exposure to common solvents such as TCE, benzene, toluene, methylene chloride, and acetone has been linked to impairments of the central nervous and circulatory systems as well as to skin, nose, throat, and lung damage. Long-term exposure to certain organic pesticides has been linked to liver damage, birth defects, sterility, genetic mutations, spontaneous abortions, and cancer.

An ironic twist to the contamination of water by organic compounds is that chlorination, often used to disinfect public water supplies, may cause reactions with organic contaminants and lead to the formation of new dangerous compounds. The most frequently detected by-products of chlorination, the trihalomethanes, are associated with a wide array of subtle health effects, such as fatigue, irritability, nausea, and irritation of the eyes, lungs, and skin that could easily be attributed by medical personnel to many other causes.

Radionuclides of various types occur naturally in almost all aquifers. The levels of these substances and the types of radiation they emit vary with the nature of the geology surrounding the aquifer. Radionuclides do occur in groundwater as a result of man-made sources. Activities associated with the production of nuclear power, testing of nuclear weapons, medical and scientific research, phosphate mining, and the production of phosphoric acid, can leach radionuclides into groundwater.

The health effects associated with the approximately 20 radionuclides detected in groundwater vary, depending partially on the type of radiation (alpha, beta, or gamma) the radionuclide emits. In general, human exposure to radiation can cause a wide range of carcinogenic, mutagenic, teratogenic, and reproductive effects.

FACTORS COMPLICATING THE LINK BETWEEN HEALTH EFFECTS AND CONTAMINANTS

Though it is possible to discuss in general some of the health effects linked to certain types of contaminants, it is very difficult to predict the health effects that might be caused by a particular contaminant in a particular case of groundwater contamination. There are many factors that influence a chemical's effects on humans, and these factors are often difficult to quantify with any formula. Scientists thus may only be able to guess at the potential effects of groundwater contamination on the health of the general public, even if there are testing data available for particular contaminants. The relevant factors include:

- How much of the chemical a person was exposed to
- How often the person was exposed
- How long each exposure lasted
- The route by which the contaminant entered the body
- The biological susceptibility of the exposed individual

Humans may experience a multitude of varying health effects depending on the quantity, frequency, and duration of exposure to contaminants. Groundwater contamination usually exposes individuals to low levels of contaminants over long periods of time, as was the case with TCE in Tucson. This type of exposure is called chronic exposure. However, most of the known health effects associated with contaminants result from exposure at a very high level over a brief period of time, called acute exposure. The known effects from acute exposure may not be a good indicator of the effects from chronic exposure to the same chemical; acute exposure effects are usually immediately apparent and are sometimes of a violent nature, while chronic exposure effects often may be dormant for many years before making their presence known.

Chemicals that may not cause an acute reaction can, when taken into our bodies day after day in small amounts, cause a chronic effect. For example, some chemicals can prove to be carcinogens that cause uncontrolled cell growth, producing tumors, leukemias, and other forms of cancer. Beginning the chain of events that leads to a cancer may only require the slightest alteration in the genetic molecule; therefore, many knowledgeable experts believe there is no threshold or safe level for any carcinogen. The biggest problem is that twenty to forty years may pass between the time of exposure and the time of cancer diagnosis. Therefore, linking the cancer back to a small exposure to one particular carcinogen may prove impossible.

Some chemicals are teratogens, which cause nonhereditary birth defects in the fetus after exposure by the mother. Still others are mutagens that cause heritable malformations by permanently affecting the exposed individual's genetic material (DNA). Growing evidence suggests most mutagens are carcinogenic, and most carcinogens are mutagenic.

The route of entry for contaminants in groundwater can be through ingestion, as when drinking the water; through inhalation of water vapor, as when taking a shower; and also through the skin, as when bathing and swimming. The route of a contaminant through the body can influence which organs and tissues are adversely affected. Thus, the symptoms arising from exposure to the contaminant in air may be completely different from the symptoms that arise when the contaminant is found in drinking water. In addition, not many of the currently available test results indicate what will happen when the victim's exposure occurs through the multiple pathways (ingestion, inhalation, and absorption) that contaminants in water can take.

The biological susceptibility of the individual makes it even more difficult to estimate accurately the health effects of a contaminant. When exposed to chemicals, individuals experience differing reactions based on their anatomical and physical differences, the state of their health, age, sex, and diet. Those typically most susceptible to harmful effects from chemical exposure include those who are very young, old, ill, or pregnant. If exposed to the same level of contaminant, chemically sensitive people may become ill while those who are less susceptible may experience no effects at all.

In addition, it is rare that groundwater is polluted by only one kind of contaminant. How does the body react when exposed to multiple chemicals present in groundwater? What types of health effects do these contaminants cause if they react with each other? If one chemical can cause multiple health effects, what effects are caused by several contaminants?

Finally, the link between contaminants and disease may often go undetected because family physicians are led astray by the close similarity between the symptoms of chemical contamination and those of more common diseases: for example, early pesticide poisoning symptoms are often misdiagnosed as flu. And the physician is a victim of the same lack of information as everyone else. Health effects information on acute exposure, which is all that is available for many chemicals, is not likely to be of much help in discovering chronic exposure. Doctors may rely on information provided by manufacturers or the local health department, which may be deficient or even misleading. The physician's lack of experience with toxicology can lead him to ask the wrong questions or order the wrong lab tests, and thereby miss an important clue.

DO LEGAL LIMITS ON CONTAMINANTS PROTECT YOUR HEALTH?

In furtherance of the public health, many laws require agencies to set numerical limits or standards governing the amount of certain contaminants that will be allowed in groundwater. But how do agencies develop these standards in the face of the many problems associated with determining the health effects of chemicals? What should it mean to you when a public official reassuringly tells you that your health effects are not linked to groundwater contamination because no standards have been violated?

Legal standards are designed to protect the public health. However, standard setting is a tedious regulatory process that does not rely solely on health data. Under many of our environmental laws, the standard-setting process must balance health and economic needs through use of analytical methods called risk assessment and risk management. Risk assessments calculate the potential of an adverse health effect occurring after exposure to the contaminant. The process involves quantitatively integrating all available and relevant data on the contaminant to determine the increased chances an individual has of experiencing a health effect if exposed to the contaminant at a specified level. At low exposure levels, some harmful chemicals may not cause measurable health effects. The highest level of exposure that does not cause a health effect is the chemical's threshold level. Above this level, the chemical will adversely affect health. There is growing evidence that many chemicals, especially carcinogens and mutagens, have no threshold level: any exposure has the potential to cause irreversible harm.

Risk management is the process of developing actual regulatory standards to control the risks described in a risk assessment. Risk management often involves a cost-benefit analysis that balances health risks along with the cost of controlling these risks, the effectiveness of the controls, the benefits of controlling the exposure and the benefit of the polluting activity.

The use of risk assessment and risk management means that the standard-setting process involves compromises that could adversely affect your health. The compromises are based on the idea that our society should protect not only the health, but also the economic well-being of its citizens. If the compromises are based on accurate and openly debated statistics and assumptions, you may understand and support the compromise. However, risk assessment and risk management are flawed tools frequently touted as "pure science." You should be aware of their shortcomings.

Risk assessments are based on health data, but nevertheless the process requires scientists to make value-laden judgment calls. In fact, the National Academy of Sciences estimates that as many as 60 assumptions must be made by the scientist trying to develop a risk assessment for a carcinogen. By manipulating the risk assessment formulas, plugging in the same contaminant exposure and health effects data, but changing the assumptions made during the process, radically different projections of individual risk can result.

Risk management is much more obviously subject to value judgments. It is essentially a process of determining how to reduce health risks to an "acceptable level" given the limits of technology and cost. Deciding how much is too much to spend for an additional level of protection is not susceptible to rigid mathematical analysis. It is the kind of decision you can and should be involved in making.

CONCLUSION

This chapter describes many of the uncertainties surrounding the relationship between health problems and contaminants in groundwater. When your community is affected by a contamination threat, you need to encourage public discussion and to ask the questions that expose these uncertainties, questions such as:

- How can you measure the effects of exposure to several chemicals that may mix and recombine in groundwater and be taken into the body through many different routes?

- What happens when this exposure is combined with the health effects from daily exposures to the multitudes of chemicals currently associated with our lifestyles?
- How can we trace the vague symptoms associated with chronic toxicity exposures back to groundwater contamination?
- How can we account for different individuals' varying susceptibilities to the same quantity of contaminants?
- If we exhibit one of the many subtle symptoms of toxic exposure, will our health professionals think to make the connections between health and groundwater?
- Will regulatory decisions be stringent enough to protect our health?

By familiarizing yourself with the controversial aspects of the health effects problem, you can gain more control over the debate. While it is not your job to prove what science cannot, you can be more effective if you have the basic knowledge to ask the right questions. The key issues raised in this chapter will play an important part in many of the opportunities for citizen action discussed in the following sections of this manual, from commenting on the conditions of a permit to influencing how thoroughly EPA cleans up a dumpsite. You can work to see that these problems are addressed openly and that, as much as possible, your health is protected.

REFERENCES

HEALTH AND GROUNDWATER CONTAMINATION

Bacon, J. Maichle, and William A. Oleckno. "Groundwater Contamination: A National Problem with Implication for State and Local Environmental Health Personnel." *Journal of Environmental Health*, v. 48, n.3. 1985.

Citizen's Clearinghouse for Hazardous Waste. *Environmental Testing.* Arlington, Va. 1985.

Citizen's Clearinghouse for Hazardous Waste. *Will a Community Health Survey Work for You?* Arlington, Va. 1984.

Conservation Foundation. *A Guide to Groundwater Pollution: Problems, Causes, and Government Responses.* Washington, D.C. 1987.

The Delicate Balance. Quarterly publication. For information, contact: Center for Environmental Health Strategies, 1100 Rural Avenue, Voorhees, N.J. 08043. (609) 429-5358.

Ecological Illness Law Report. Quarterly publication. For information, contact: Earon Davis, editor, P.O. Box 6099, Wilmette, Ill. 60091. (312) 256-3760.

Environ. *Elements of Toxicology and Chemical Risk Assessment: A Handbook for Nonscientists, Attorneys, and Decision Makers.* Washington, D.C. 1986.

Environmental Defense Fund. *Dumpsite Cleanups: A Citizen's Guide to the Superfund Program.* Washington, D.C. 1983.

Gordon, Wendy. *A Citizen's Handbook on Groundwater Protection.* Natural Resources Defense Council. New York, N.Y. 1984.

Hattis, Dale, and David Kennedy. "Assessing Risks From Health Hazards: An Imperfect Science." *Technology Review,* May/June 1986.

King, Jonathan. *Troubled Water.* Rodale Press. Emmaus, Pa. 1985.

Legator, Marvin S., Barbara L. Harper, and Michael Scott, eds. *The Health Detectives Handbook: A Guide to the Investigation of Environmental Health Hazards by Nonprofessionals.* Johns Hopkins University Press. Baltimore, Md. 1985.

Moore, Andrew Owens. *Making Polluters Pay: A Citizens' Guide to Legal Action and Organizing.* Environmental Action Foundation. Washington, D.C. 1987.

National Institute of Environmental Health Sciences. *Environmental Health Perspectives.* Scientific journal focusing on the health hazards of environmental agents. Order by contacting Superintendent of Documents, U.S. Government Printing Office, Washington, D.C. 20402.

The Reactor. Bimonthly publication. For information, contact: Susan Molloy, editor, P.O. Box 575, Corte Madera, Calif. 94925.

Rural Community Assistance Program. *From Watershed to Well.* Vol. 1, no. 2. Winchendon, Mass. 1987.

Sherry, Susan, et al. *High Tech and Toxics: A Guide for Local Communities.* Golden Empire Health Planning Center. Sacramento, Calif. 1985.

Sittig, Marshal. *Handbook of Toxic and Hazardous Chemicals and Carcinogens.* Noyes Publications. Park Ridge, N.J. 1985.

Wilson, Reid. *Testing For Toxics: A Guide to Investigating Drinking Water Quality.* U.S. Public Interest Research Group. Washington, D.C. 1986.

World Health Organization and United Nations Environment Programme. *Environmental Health Criteria.* Volumes available for approximately 60 contaminants. Order by contacting World Health Organization, Distribution and Sales, 1211 Geneva 27, Switzerland.

CHAPTER 3

Sources of Groundwater Contamination

THIS CHAPTER DESCRIBES most of the significant threats to groundwater quality. Each section discusses how a particular source of contamination may degrade groundwater quality and provides the latest information about the scope of the problem. Together with the suggestions in the next two chapters about groundwater monitoring and mapping, the information here will help lay the foundation for taking action against particular contamination sources.

LANDFILLS

Landfills today may be built with elaborate leak prevention systems, but most, particularly the older ones, are simply large holes in the ground filled with wastes and covered with dirt. Originally designed to reduce the air pollution and unsightly trash that accompanied open dumping and burning, landfills became the disposal method for every conceivable type of waste. Unfortunately, landfills were not a problem-free solution. It is now widely accepted that almost all landfills, even those applying the best available design technology, will eventually leak wastes into surrounding soils and groundwater.

The United States is now forced to deal with the threats posed by an estimated 75,000 on-site industrial landfills, 18,500 municipal landfills, and perhaps 24,000 to 36,000 closed or abandoned municipal landfills. The EPA estimates that 12,000 to 18,000 municipal (also called "sanitary") landfills, though not officially designated as hazardous waste disposal sites, may contain hazardous wastes. Whatever the official numbers, the distinction between municipal landfills and hazardous waste landfills is a legal one and can be misleading. Municipal landfills legally accepted large quantities of hazardous wastes in the past and currently accept the three to four-and-one-half million tons of hazardous waste that so-called small-quantity generators (see Chapter 17 for definition) produce, as well as substantial quantities of household hazardous waste. Plainly, "sanitary" landfills can pose a serious risk of groundwater contamination.

Landfills contaminate groundwater because most cannot offer a perpetually impermeable barrier to the toxic substances they hold. Chemicals in the landfill escape their containers and mix with other, often incompatible, chemicals. Rain or runoff water will also enter the landfill and pick up the contaminants along the way. This percolation of liquids through the landfill creates leachate, which eventually seeps downward to the water table.

Unfortunately, during the period when most landfills began operating, their sites were based on convenience and price rather than the hydrogeological vulnerability of the area. In fact, landfills were commonly placed in some of the worst possible hydrogeologic locations—such as marshlands, abandoned strip mines and gravel pits, or limestone sinkholes—because many believed landfills would help reclaim "waste" lands. The "disposal technology" simply involved filling the hole with liquid, solid, hazardous, and nonhazardous wastes, compacting the wastes with a bulldozer, then covering the day's work with a layer of soil. It is estimated that at least 75 percent of these older landfills have already developed leaks.

Today new techniques have been designed to prevent groundwater contamination. Figure 3.1 details some of these techniques.

Any landfill, whether the old hole in the ground or the "state-of-the-art," is extremely vulnerable to problems that can cause irreversible groundwater contamination. For example, burrowing animals can dig through any liner, plastic or clay; plastic liners can be torn by bulldozers or dissolved by wastes; collection pipes can become clogged; pipes can be crushed by the weight of waste above; mounded cover may crack or slump inward as bulk solids settle and barrels disintegrate; and when mounds crack, allowing rainwater in landfill, the collection system can overload and the landfill can overflow.

SURFACE IMPOUNDMENTS

Surface impoundments are often called pits, ponds, or lagoons. Ranging in size from a few square feet to several thousand acres, surface impoundments serve as disposal or temporary storage sites for hazardous and nonhazardous wastes. They are designed to accept purely liquid wastes, or mixed solids and liquids that separate in the impoundment. Liquids in the impoundment are either treated and discharged, allowed to seep into

FIGURE 3.1
HAZARDOUS WASTE LANDFILL

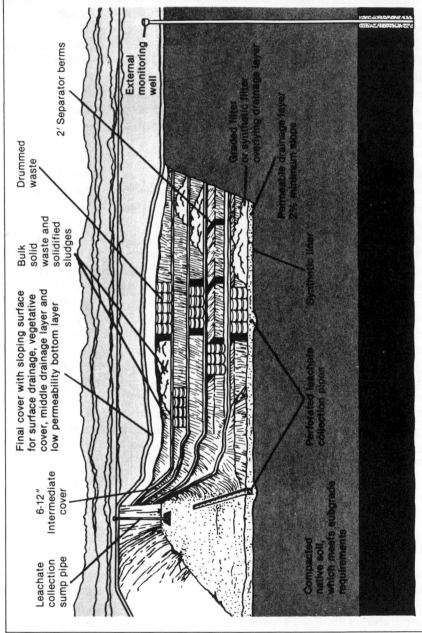

SOURCE: *United States Congress, Office of Technology Assessment, Technologies and Strategies for Hazardous Waste Control (March 1983).*

the soil, or evaporate. Surface impoundments are commonly used by municipal wastewater and sewage treatment operations, animal feedlots and farms, and by many industries including oil and gas, mining, paper, and chemical operations.

In 1982 EPA identified over 180,000 waste impoundments at 80,000 sites. Of the industrial sites evaluated, 95 percent were within one mile of drinking water wells, 70 percent were unlined, and 50 percent were on top of aquifers. Only 7 percent were located in areas posing little or no threat to groundwater. There are also many municipal impoundments that accept both industrial and domestic wastes.

These frightening statistics help add up to a total of 36 states with groundwater contamination caused by surface impoundments. Though many impoundments are unlined and therefore allow waste seepage, few impoundment locations were chosen with area hydrogeology in mind. A 1984 OTA study, based on very conservative leakage rates for such impoundments, reached the startling conclusion that approximately 1,800 billion gallons of liquid wastes were leaching into U.S. soil every year. In addition to chronic slow leakage into area groundwater, surface impoundments have caused hazardous emergencies and contaminated water when heavy rains or broken dikes allowed sudden overflows of wastes.

UNDERGROUND STORAGE TANKS

Underground tanks are ubiquitous. While most often associated with gasoline service stations, these tanks are also used by America's manufacturers, farmers, government agencies, and homes as storage facilities for gasoline, oil, hazardous chemicals, and chemical waste products. Estimates of the number of tanks—those abandoned and those currently in use—reach as high as ten million. Not surprisingly, a recent EPA survey found that 41 states suffered groundwater contamination from faulty underground tanks.

The most frequent cause of contamination of groundwater by underground tanks is the leaking of stored materials due to internal and external corrosion. Since the 1950s, when the use of underground tanks began to increase dramatically, the most popular type of storage tank has been bare steel. We now know that the life expectancy for such tanks is only 15 to 20 years. Many bare steel tanks installed in the 1950s and 1960s are still used today or have been abandoned and forgotten. In a recent survey of motor fuel storage tanks, the EPA found that 35 percent of the estimated 796,000 such tanks leaked. Abandoned tanks were found at 14 percent of the surveyed establishments, but EPA did not conduct tests for leakage on these abandoned tanks.

FIGURE 3.2
DOUBLE WALL TANK AND LEAK DETECTION SYSTEM

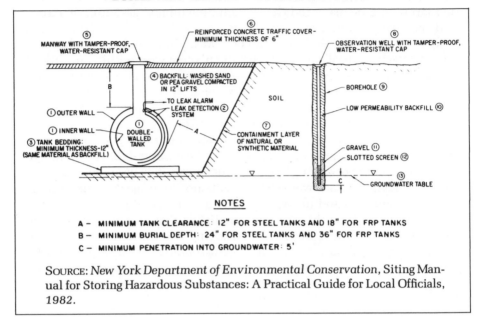

NOTES

A — MINIMUM TANK CLEARANCE: 12" FOR STEEL TANKS AND 18" FOR FRP TANKS
B — MINIMUM BURIAL DEPTH: 24" FOR STEEL TANKS AND 36" FOR FRP TANKS
C — MINIMUM PENETRATION INTO GROUNDWATER: 5'

SOURCE: *New York Department of Environmental Conservation*, Siting Manual for Storing Hazardous Substances: A Practical Guide for Local Officials, 1982.

External tank corrosion occurs most often when the tank is situated in acidic soil or near saline groundwater. No one is certain exactly how many underground tanks are actually located below the water table. No matter what the surrounding medium, a small nick or flaw on the outside of the tank induces rapid corrosion and eventually the tank becomes pitted or rusted out. Even a pinpoint hole can cause major problems. Internal corrosion can be caused by reactions from stored chemicals such as solvents. Other causes of groundwater contamination by underground tanks include improper installation, loose fittings, and corrosion of connecting distribution lines or vent and fill pipes, as well as such common practices as pumping fuel into service station tanks at high velocities, or repeatedly dropping the dipstick directly below the tank's opening.

Recent developments to prevent leakage include the use of fiberglass tanks, coated steel tanks, synthetic underground containment liners, double-walled tanks, and placement of tanks in subgrade vaults. (See Figure 3.2.) The cost increase associated with these preventive measures varies dramatically.

The susceptibility of underground storage tanks to leakage, combined with their widespread use, means that every community dependent on groundwater needs to be concerned about this potential contamination source. The problem is compounded in rural areas, where underground tanks are extensively used by farmers, small businesses, and schools, and where groundwater is often the primary drinking water source.

WASTE DISPOSAL WELLS

Every year in the United States hundreds of millions of tons of toxic, hazardous, radioactive, and other liquid wastes are dumped directly into the earth through hundreds of thousands of waste disposal wells. This practice, most commonly utilized by the chemical, petroleum, metals, minerals, aerospace, and wood-preserving industries, has contaminated groundwater in at least 18 states.

Underground injection began in the oil fields of the 1930s. Oil drillers had already been injecting fluids into the ground in order to push petroleum toward their wells. They also knew from experience that the water they brought up from deep underground was much more saline than the water they were drinking from area wells and that oil had remained underground for millions of years. Thus, oil drillers came upon the idea that they could reinject the byproducts of oil extraction without harming their own water. When the oil industry diversified into petrochemical products during the 1950s, oil field waste practices were continued with new toxic wastes. Industrial use of injection disposal grew and is rapidly becoming the most popular form of industrial disposal.

Waste disposal wells come in all shapes and sizes—from a shallow dug pit just a few feet deep collecting drainage from a field to deep injection wells designed to force liquids into rock formations thousands of feet below the surface. By far the most common wells, and those with the least protection against groundwater contamination, are the simple shallow wells that are either dug or drilled to contain agricultural, household, and industrial liquid wastes. There may be as many as one million of these wells around the country. These shallow wells are often located without regard to the protection of groundwater, sometimes directly over vulnerable aquifers, and should be viewed as a major threat to groundwater quality. Of the many categories of such wells, the following pose the greatest danger of contamination:

- Agricultural drainage wells
- Raw-sewage disposal wells
- Septic system wells
- Automotive service station waste wells

The most disturbing recent development is probably the rapid growth in the use of deep injection wells—perhaps due to their low relative cost,

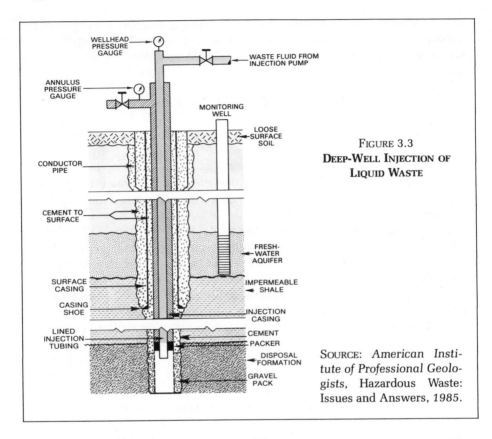

FIGURE 3.3
DEEP-WELL INJECTION OF
LIQUID WASTE

SOURCE: *American Institute of Professional Geologists*, Hazardous Waste: Issues and Answers, *1985.*

particularly for hazardous wastes. A state-of-the-art deep injection well (illustrated in Figure 3.3) is constructed by first drilling past the depth of any area drinking water sources. A steel casing or pipe is placed in the resulting borehole and, to seal the pipe in the hole, cement is poured outside the pipe's entire length. At this point, drinking water sources should be protected by a layer of steel and cement.

Next, drilling continues below the first casing until an appropriate geologic formation for holding the wastes, an "injection zone," is reached. Injection zones are characteristically permeable, brackish sandstone or limestone/dolomite deposits confined by layers of relatively impermeable geologic deposits. Water contained within the injection zone deposits should be unpotable and untreatable.

After the well reaches the injection zone, a steel casing is installed from the surface down to the bottom of the zone. Another layer of concrete is poured along the outside of the second casing. Drinking water sources are now protected by a second layer of steel and cement. Finally, a small pipe, called the injection tubing, is placed within the inner casing and secured in place at the surface. The space, called the "annulus," between the injection tubing and steel casing is filled with a noncorrosive pressurized fluid, contained within the space by a "packer" or "plug" at the base of the annulus.

Pressure levels of the annulus fluid and injected wastes are monitored for changes indicative of leakage or system failures. Wastes are pumped through the injection tubing at pressure levels high enough to displace saline fluids occupying geologic pores where wastes will be stored. At the same time, the pressure cannot be so high as to create geologic fractures, which might allow wastes to escape from the injection zone.

The surface of injection well areas is usually occupied not only by wellheads and monitoring equipment, but also by pipelines carrying wastes to the wells, and surface tanks or ponds where waste is stored before pumping.

As the citizens whose water has been contaminated by injection can attest, even this state-of-the-art technology can fail to protect ground-water. (See Figure 3.4.) Wastes injected into the ground can reach area water supplies through small cracks in well casings or corrosion caused by injected chemicals' adverse reactions with the casings. Injected

FIGURE 3.4
AQUIFER CONTAMINATION THROUGH IMPROPERLY CONSTRUCTED OR
ABANDONED WELLS

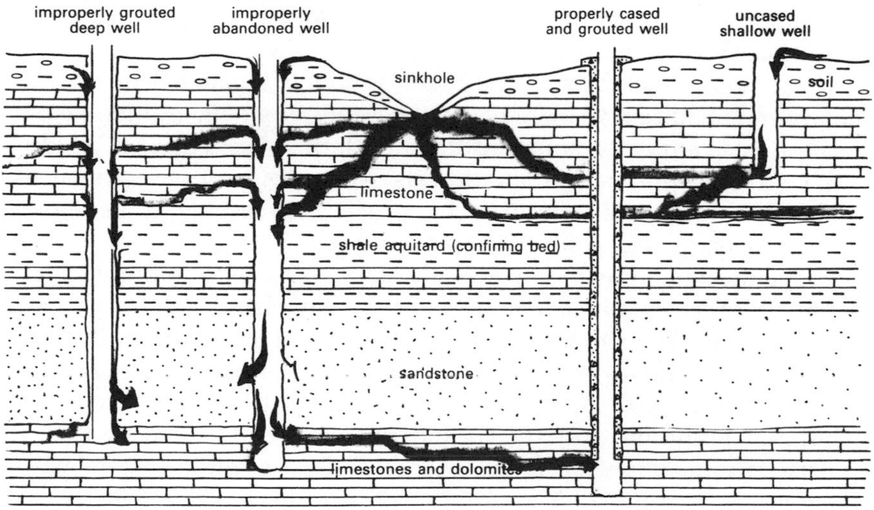

SOURCE: Minnesota Extension Service, University of Minnesota.

chemicals have also dissolved the rock formations supposedly containing them, thereby escaping upward to area aquifers. Gas by-products from chemical reactions among injected fluids or between fluids and rock formations have caused "blowouts," or explosions, which spew the underground wastes into the air, causing widespread area contamination. Previously undetected faults and fractures in confining rock formations may allow upward migration of wastes. Some underground injection operations have caused earthquakes, which can, in turn, cause new fractures in confining rocks. Wastes may also migrate up to aquifers along the outside of the injection well casing or through one of the hundreds of thousands of unplugged "dry holes" left over from unsuccessful drilling attempts for oil and gas. Finally, aquifers can become contaminated by the saline water previously located in pores now filled with waste, as well as through leaks in above-ground storage areas associated with underground injection operations.

AGRICULTURAL PRACTICES

Pesticides

Pesticides were first identified in groundwater less than ten years ago, but now over 25 states report groundwater contaminated by pesticides. Not all of these states are considered agricultural. Pesticides have been widely used for many agricultural and nonagricultural purposes and for each use there are a wide variety of products. Recent limited groundwater monitoring efforts are only beginning to tell the story of decades of sometimes indiscriminate pesticide use.

There are 50,000 different pesticide products registered in the U.S., composed of 600 active ingredients. We apply about 700 million pounds of pesticides annually, not all of them on farms. Pesticides are used on golf courses, lawns and gardens, forests, roadsides, parks, electric utility and railroad rights of way, oil and gas drill well-casings, home foundations, finished wood products, and so on.

Pesticide contamination in groundwater can come from point or nonpoint sources. A point source, where the contaminant flows in a fairly distinct plume from an identifiable source, could be caused by the spilling or general mismanagement of pesticides in areas where bulk quantities are manufactured, stored, disposed, or distributed in commerce. Nonpoint pollution, widespread and relatively difficult to trace to its precise origin, might be found over an entire farming region, where regular low-level pesticide applications continuously leach into groundwater and build chemical concentrations up to dangerous levels. Runoff from pesticide-treated land areas can also contaminate groundwater through uncapped or improperly constructed wells or sinkholes, and it can pollute nearby surface water.

Whether a particular pesticide will contaminate groundwater depends on many chemical and environmental variables. The EPA now requires pesticide manufacturers to supply information on their product's leaching potential during registration of new, and reregistration of old, pesticide products. The basic factors that influence leaching potential are the degree to which the pesticide dissolves in water, disperses in air, and adheres to soil particles. Of the pesticides currently used, EPA testing has identified over 100 potential leachers based on these types of tests.

Environmental factors that might affect whether a chemical reaches groundwater supplies include the texture, moisture content, and porosity of area soil, the depth from topsoil to groundwater, and precipitation. The method of pesticide application, the amount and time of application in relation to environmental conditions, and local irrigation and cultivation practices, are other important factors determining the impact of pesticide use on underlying aquifers.

Fertilizers

Unsound agricultural practices commonly include an overuse of fertilizers. Nitrogen, potash, and phosphate are the three basic fertilizers, but nitrogen represents over half of the total used. Unfortunately, of the three fertilizers, nitrogen is the most likely to leach. Between 1975 and 1981 there was a 38 percent increase in the use of nitrogen on U.S. agricultural lands, increasing the total to over 10 million metric tons. Such an overuse of nitrogen causes an accumulation ready to leach into groundwater. In a recent USGS survey of 124,000 wells, over 24,000, or 20 percent, had a nitrate/nitrogen concentration of over 3 mg/1, the level indicative of human activity; 8,200, or 6 percent, had a nitrate/nitrogen concentration exceeding EPA's 10mg/1 limit for drinking water.

Livestock

The United States contains an estimated 170 million cattle, sheep, and hogs (1985 figure) and over 8 billion chickens and turkeys (1984 figure). Their production of 160 million dry-weight tons of manure annually contaminates underlying aquifers with nitrogen, bacteria, viruses, hormones, and salts. Although groundwater can be contaminated by relatively small livestock operations if they are located above porous soils, the most obvious threat stems from animal feedlots, where dense livestock populations are confined to small areas. A typical feedlot yields about 23 metric tons of nitrate per acre per year. Facilities that treat or dispose of animal wastes likewise pose a threat to groundwater.

Irrigation and Chemigation

Modern irrigation practices can lead to salt contamination of underlying aquifers. Irrigation water contains small quantities of salt which, because they are not transpired by crops or evaporated from soil, build up within the soil and eventually leach into groundwater. In some areas of the country excess irrigation water is applied to rid the root zone of potentially crop-devastating salt buildup. Though it may maintain crop productivity, this practice degrades underlying groundwater supplies.

Chemigation, the direct mixing of pesticides and fertilizers with irrigation water, poses a substantial threat to groundwater. When irrigation pumps are turned off or are malfunctioning, the chemical feeder may allow chemicals to be siphoned directly into groundwater wells. While backflow valves designed to prevent this are available, they may not be required and are expensive.

LAND APPLICATION

Land application is a treatment and disposal method also called land treatment and land farming. The practice involves spreading over farm and forest lands the waste sludges and wastewater generated by public treatment works, industrial operations such as paper, pulp and textile mills, tanneries and canneries, livestock farms, and oil and gas exploration and extraction operations. If properly designed and operated, land application recycles nutrients and waters to the soil and aquifer. However, at least seven states report groundwater contamination due to land-application practices.

Land application contaminates groundwater when heavy metals, toxic chemicals, nitrogen, and pathogens found in sludges and wastewater leach to underlying aquifers rather than remain in upper soil layers. This occurs if the sludge or wastewater has not received adequate pretreatment, and if the site's soil types and the depth to groundwater are not appropriately considered. The most obvious cause of contamination from land application occurs when hazardous materials that do not degrade are applied to the land. Significant groundwater problems are resulting from this practice, yet in California, for example, 40 percent of the state's hazardous waste handled onsite is "treated" by land farming.

New land-application methods that utilize underground monitoring and other protection methods are rapidly growing in popularity for the disposal of wastewater and sludge from public sewage-treatment plants. In this case, land application is used in lieu of advanced treatment at the plant to remove greater levels of nitrogen than would be economically feasible using other technology and chemical treatment. Crops that will receive the wastewater are selected based on their ability to take up nitrogen.

The newer methods are viewed as innovative or alternative because they are helping to slow the decline of large bay and lake ecosystems greatly suffering from excess nitrogen levels added through public treatment effluent. However, effects on local aquifers must be determined before the practice can be embraced as an environmentally safe alternative.

HOUSEHOLD HAZARDOUS WASTES AND PRODUCTS

The industrial and commercial sources of potential groundwater contamination exist because our society has grown dependent on the products and services they provide. Yet, it is often easy to distance our personal role in these potential contamination sources. When, for example, do you demonstrate more concern about the adverse effects of pesticides—when you select only blemish-free fruit at the grocery or when you purchase a pesticide at the hardware store to cut back on weeds in the garden? Although you have a role in potential contamination with each purchase, your awareness is most heightened when you actually have the chemical directly in hand. Though no one can offer exact figures, Americans use about 25 gallons per person per year of hazardous chemical products in the home. Thus we, like industrial interests, are potential sources of toxic threats to our communities. It's easy to forget what happens to that old bottle of paint thinner once it leaves the house, yet taken cumulatively, such discarded household products can cause tragedies within our communities. To take one example, New York's Nassau County Department of Health estimated that each year 83,000 gallons of organic chemicals from household hazardous products end up in the county's groundwater supply.

What Are Hazardous Household Products?

The basic categories of household hazardous products and waste include:

Drain openers
Wood and metal cleaners and
    polishes
Grease and rust solvents
Air conditioning refrigerants
Paint thinners
Adhesives
Pesticides
Paint

Oven cleaners
Automotive oil and fuel
    additives
Carburetor and fuel injection
    cleaners
Starter fluids
Paint strippers and removers
Herbicides
Fungicides/wood preservatives

Some products falling into the above categories might not be hazardous. Many, however, contain constituents that, if used in an industrial setting, would be subject to strict disposal regulations. Poor labeling of household products is common; it is rare that the name of every chemical in the product appears on the label, and even less likely that the label lists the percentages of each chemical constituent contained in the product or provides chronic health-effect data and disposal information. Therefore, all chemical products that fall into the above categories should be handled and disposed of with caution.

How Do Household Hazardous Wastes Contaminate Groundwater?

Household hazardous waste is most frequently poured down the drain or put out on the curb with the rest of the trash. While usually sold in more diluted formulations than used in industry, household hazardous products contain chemical constituents that persist in the environment. The accumulation of small quantities of hazardous wastes from thousands or even millions of households eventually reaches a dangerous level and becomes a vital public concern. Household chemicals placed in municipal landfills can make the town dump almost as great a threat to groundwater as an industrial landfill, because household chemicals, just like industrial chemicals, leach when placed in landfills. In 1987, municipal landfills, often called "sanitary landfills," made up 20 percent of the Superfund National Priority List (EPA's list of priority cleanup sites).

Household chemicals poured down the drain can pose an immediate health threat, especially if the end of the drain is a septic system and the homeowner relies on local groundwater for drinking supplies. Even for homes serviced by a public treatment works, the chemicals can still contaminate drinking water and the environment because the treatment that the wastewater receives is often not designed to remove hazardous chemicals.

Reducing the Household-Waste Contamination of Groundwater

If you can't pour it down the drain or put it out with the trash, how can you safely dispose of household hazardous waste? One answer is household hazardous waste collection programs, offered by communities in most states, that arrange for proper disposal of waste. Another is to reduce the use of toxic products in the home and to recycle those that must be used. (For more information on reducing and recycling, see Chapter 9.)

MILITARY TOXICS

Are you surprised to find the military included in our description of groundwater-contamination sources? You are not alone. Many find it hard to believe that in its efforts to protect our country from outside threats, the military has itself become a serious threat to citizens across the country. According to the Citizen's Clearinghouse for Hazardous Waste, the U.S. military branches are the largest generator of hazardous waste in the country, producing over 1 billion pounds per year, more than the top five civilian chemical companies combined. CCHW believes that any military installation, no matter how large or small or the branch of the service, is a likely candidate for a contamination problem.

The Department of Defense produces and utilizes a myriad of toxic and hazardous substances, not only in its munitions operations, but also in such day-to-day activities as painting ships and dry-cleaning uniforms. DOD daily operations involving toxics include large amounts of maintenance and cleaning for machines and vehicles; these activities use or produce such substances as waste fuel, solvents, oils, paint thinners and strippers, heavy metals, and acids. DOD maintains a large stockpile of highly toxic chemical weapons, and the Department of Energy produces nuclear material for weapons and other defense purposes—one of the most dangerous industrial operations in the world. DOE facilities produce radioactive waste and are also the cause of nonradioactive, but highly toxic, chemical contamination.

While the military was developing the latest in high technology for defense, its storage and disposal practices remained as antiquated as those in the rest of the country: deadly chemicals were simply burned, poured over the land, or placed in leaking pits, ponds, lagoons, and landfills. During times of fast-paced military development and war, records of disposal areas were either not kept at all, or were carelessly maintained. For many years afterwards, the veil of national security hid not only the locations where wastes were disposed, but also the kinds of wastes that were placed there.

To some extent, the shortcomings of the military's waste-management practices simply reflect the tension inherent in "the fox watching the chicken coop": until very recently, the production, use, handling, storage, disposal, and cleanup of military toxics were entirely self-policed by military officers.

We remain uncertain about the extent of contamination by military facilities. Many older investigations of installations have been found to be completely inaccurate, and many new investigations are just getting underway. However, we do know that:

- 7,200 hazardous waste sites exist on the Department of Defense's 759 installations.
- 470 Defense Department installations produced hazardous wastes in 1985.

- Approximately 25,360 underground storage tanks exist on Defense Department land.
- On Superfund's National Priority List, there are now 32 Federal facilities, and seven more are proposed.
- The Department of Energy produces nuclear material at 18 major sites across the U.S.; hundreds of waste sites exist at these and other DOE facilities across the country.
- Until 1986, one naval ship would typically utilize 150,000 hazardous chemicals on board.
- Toxic gases and hazardous materials have killed at least 31 sailors and civilian employees on navy vessels between 1981 and 1986, and 1,200 sailors were involved in 2,100 incidents involving hazardous wastes during the same time period.
- Within the borders of the Department of Defense's Rocky Mountain Arsenal lies what the *Sacramento Bee* termed "the most toxic square mile on earth."

The military's toxic threat does not end at the facility boundary. Consider the following:

- The military has contracted with thousands of independent businesses to manufacture products for its use. Many of these manufacturing procedures produce hazardous waste, and many such facilities have contaminated groundwater.
- The military has paid private landfills across the country to accept its toxic wastes, but records of most contracts no longer exist—some of those that do have already been traced to landfills leaking toxic substances into community water supplies.
- The military owns facilities all over the country, which it pays private corporations to operate (called GOCOs for government owned-contractor operated). Whether the military or the contractor is ultimately responsible for waste cleanup, many of these GOCOs have contaminated communities with toxics.
- Once the military discovers toxic waste on its property, it often pays private corporations to haul the waste away. Some of these corporations have contaminated host communities.
- Finally, the military has sold to unsuspecting buyers "excess" federal land that turned out to be contaminated with hazardous waste. Such inexpensive excess land has, in some cases, been purchased by local communities to build schools.

Of major concern are the activities of the Department of Energy, which utilizes products and generates wastes laced with radioactivity. For over 40 years one of the principal jobs of the DOE (originally called the Atomic Energy Commission, then the Energy Research and Development Administration) has been developing and manufacturing nuclear materials for weapons and other defense purposes. It currently performs these opera-

FIGURE 3.5
GROUNDWATER CONTAMINATION AT SELECTED DOE FACILITIES[a]

Facility	Major type(s) of contamination	Level of contamination
Feed Materials Production Center, Ohio	nitrates and chloride	Nitrates and chloride have been reported above drinking water standards.
Fuel fabrication plant, South Carolina	solvents[b] and nitrates	Solvents have been reported at levels over 30,000 times the proposed drinking water standards. Nitrates have been reported at levels over 10 times the drinking water standards.
Los Alamos National Laboratory, New Mexico	none[c]	No contaminants resulting from the laboratory's operations have been reported that exceed drinking water standards.
Mound Laboratory, Ohio	tritium[d]	Although in 1976 tritium concentrations were above the drinking water standards, continual remedial actions keep the levels below the standards.
N-Reactor, Washington State	strontium-90,[e] tritium, and nitrates	Strontium-90 has been detected at levels over 400 times higher than the drinking water standards. Tritium and nitrates are slightly above drinking water standards.
Rocky Flats plant, Colorado	solvents, cadmium, and selenium	Solvents have been reported as high as 1,000 times the proposed drinking water standards. Cadmium and selenium have been detected at or slightly above the drinking water standards.
Reprocessing plant, South Carolina	tritium, nitrates, and mercury	Tritium has been reported over 2,500 times the drinking water standards. Nitrates and mercury have been detected at levels slightly above drinking water standards.
Reprocessing plant, Washington State	tritium, iodine-129,[f] and nitrates	Tritium concentrations have been reported over 25 times higher than the drinking water standards. Iodine-129 and nitrates have both been reported above the drinking water standards.
Y-12 plant, Tennessee	Solvents, nitrates, mercury, arsenic, and chromium	Solvents have been detected over 1,000 times greater than proposed drinking water standards. Nitrate concentrations have been reported at levels 1,000 times the drinking water standards. Mercury has been detected at levels 500 times the drinking water standards. Arsenic has been detected at levels 60 times the drinking water standards. Chromium has been detected at levels over 30 times the drinking water standards.

[a]The table shows only major contaminants (e.g., those that exceed existing and/or proposed drinking water standards). At many facilities, other contaminants are in the groundwater above background levels.

[b]Solvents are cleaning agents, such as trichloroethylene, 1,1,1, trichloroethane, and/or tetrachloroethylene. These are classified as hazardous waste and toxic pollutants.

[c]According to DOE, at this facility arsenic and fluoride occur naturally in the groundwater.

[d]Tritium is a radioactive isotope of hydrogen. It has a half-life of 12.5 years.

[e]Strontium-90 is a radioactive isotope of strontium with a half-life of 30 years.

[f]Iodine-129 is a radioactive isotope of iodine with a half-life of over 15 million years.

SOURCE: U.S. General Accounting Office.

tions at 18 major sites. In a 1986 GAO study of nine diverse DOE facilities, groundwater contamination was discovered at eight of the nine sites, in levels sometimes thousands of times greater than national drinking-water standards. (See Figure 3.5, on preceding page.)

Many facilities have multiple problems. DOE's Idaho National Engineering Laboratory, for example, is an immense tract of land containing 20 operable nuclear reactors and 379 waste-disposal units, 27 of which are considered hazardous. Although the facility has contaminated groundwater, DOE nevertheless recently elected to haul 150 tons of highly radioactive waste from Three Mile Island's 1979 disaster to the facility for research purposes. Though the DOE states that there is little threat to man or to the environment because most contamination remains on federal property, some off-site migration of substances such as uranium and tritium was revealed in the GAO study.

MINING

The construction techniques, products, and by-products of mining operations are serious threats to the quality and quantity of nearby aquifers. Millions of acres of U.S. land have been mined for coal, copper, uranium, and other minerals. It is not only active mines that cause contamination. Inactive and abandoned mines, of whose existence many communities may be unaware, can also be steady and serious polluters; there are an estimated 67,000 inactive or abandoned mines in the United States.

Over 4 million acres in the United States have been disturbed by active or abandoned coal mines. Both the techniques for mining the coal and the chemical by-products of coal mining can greatly affect the quality of groundwater.

The two basic approaches to coal mining are surface (strip) and underground (deep) mining. Surface mining is performed when coal is close enough to the surface for it to be feasible and economical to remove any overlying soil and bedrock in order to reach the coal. Underground mining involves tunneling into coal seams located deep in the earth. Both of these techniques can affect the natural flow of an aquifer and create new channels for water movement. Aquifers may actually be intercepted by mining operations (and then sometimes pumped out at high rates). Blasting operations can fracture rock formations, and underground mining sometimes causes subsidence (the collapse of rock and soil above mines); both types of procedure can alter water movement. These new flow routes may permit water contaminated by mining operations to flow directly into or leach into the aquifer. Coal mining can contaminate water with coal particles, minerals, bacteria from disrupted septic tanks and drainage fields, high sulfate levels, and other materials.

The new channels of flow created by mining also allow water and air to contact previously unexposed chemical elements, causing reactions that release new chemical compounds, some of which may make their way into the aquifer. Some new by-products may only be a nuisance, discoloring water and staining clothes. Others are much more harmful. Acid mine drainage is a common term for one of these more harmful products. It is caused by a chemical reaction occurring when iron- and sulfur-containing minerals, specifically pyrite and marcasite (found as part of the coal and in bedrock surrounding coal) come into contact with oxygen and water, producing sulfuric acid. The runoff water's resulting high acidity helps release trace metals and minerals in surrounding soil and bedrock, many of which are toxic, such as aluminum, arsenic, barium, cadmium, nickel, lead, and mercury.

Groundwater contamination results not only from coal mines, but also from the nation's many copper, silver, gold, asbestos, uranium, zinc, and lead mines. Jonathan King's *Troubled Water* describes as "one of the worst sites of mining waste contamination . . . a tri-state area where the Oklahoma, Missouri, and Kansas borders join. There, at the so-called Tar Creek site, wastes running from abandoned zinc and lead mines have dumped cadmium and lead into 40 square miles of shallow groundwater running from northeastern Oklahoma into southwestern Missouri. The pollution also threatens the deeper Roubidoux aquifer, which supplies drinking water to more than 40,000 people in four states."

Uranium mining and milling has contaminated groundwater in many southwestern states, especially New Mexico and Colorado. There are actually more leftover wastes from uranium mining than there is final product. These wastes contain high levels of radioactivity and heavy metals and are commonly dumped into waste ponds or piled up in tailings heaps. In this way, contaminants become concentrated and commonly leach into groundwater aquifers.

SEPTIC SYSTEMS

Approximately 22 million septic systems are operating in the United States today, and about one-half million new systems are installed every year. These systems serve nearly one-third of the nation's population. Combined, domestic and commercial septic systems discharge more than one trillion gallons of effluent per year into soil and groundwater.

If properly sited, installed, and maintained, septic systems provide a safe and effective means for domestic waste disposal. Under improper environmental conditions, however, or with poor system maintenance, septic systems represent a serious threat to groundwater quality and to public health. Groundwater contamination from septic systems is a geographically widespread problem; 36 states have reported such contamination. The Centers for Disease Control estimate that between 1946 and 1980 the majority of illnesses attributable to groundwater were caused by microorganisms of the sort found in septic systems. Moreover, according to EPA, septic systems represent the largest reported cause of groundwater-induced disease outbreaks in the United States. And septic system contaminants are not limited to infectious microbes, but can also include toxic chemicals that may threaten the public health.

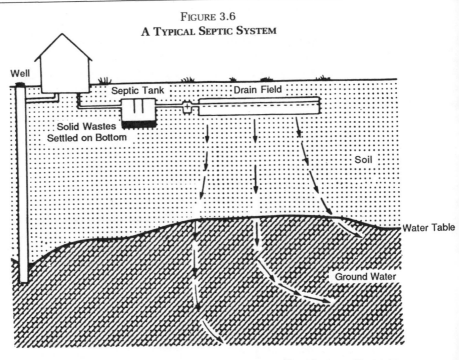

FIGURE 3.6
A TYPICAL SEPTIC SYSTEM

There are a few rules of thumb that tell us generally when septic systems are most likely to function properly and minimize groundwater contamination.

• Good soil makes a good system. If the soil is appropriate, it should facilitate treatment and disposal of septic system wastewater; for exam-

Typical Septic Systems

Septic systems generally are composed of a septic tank and a drain field into which effluent flows from the tank. Within the tank, physical processes separate the inflow into sludge (which accumulates on the bottom of the tank), wastewater, and scum (which forms on top of the wastewater). The tank eventually becomes filled with the layers of sludge and scum, preventing wastewater from passing through the tank and into the drain field. Once a tank reaches a certain percentage of its capacity, the sludge and scum, called septage, must be pumped out of the tank, so that the tank will continue to function properly. If the tank is not pumped when it reaches these designated capacity levels, sludge might escape into the drain field. (See Figure 3.6.)

ple, soils made of a combination of sand, silt, and clay work well. On the other hand, if too much clay is in the soil, the waste may not percolate through; if the soil contains too much sand and large particles, wastewater may pass through to the groundwater untreated.

- Proper design and use mean everything. Each septic system is designed to treat and dispose of a specific volume and type of wastewater in the conditions found at the site. The system must not be overloaded by disposing of a greater volume of different type of wastewater than the septic system was designed to handle. Hazardous chemicals or grease should not be disposed of in septic systems. Water conservation will help prevent overloading and extend the life of the system.

- Waste treatment in the soil occurs more readily above the water table where the soil is relatively dry and contains plenty of oxygen. The greater the depth to water, the longer the wastewater remains in the unsaturated soil, where it can be treated most effectively.

- Septic systems need enough space to do their job well. Not all microorganisms and chemicals are removed from wastewater as it travels through the soil. Even a properly operating system will discharge nutrients (phosphates and nitrates) and some bacteria or viruses to the groundwater. To avoid heavy loading of groundwater with septic system effluent, systems should be installed on lots that provide enough space. The proper amount of space varies for individual lots, and many factors need to be considered.

- Routine maintenance is critical to preventing system failures. Septic tanks must be pumped regularly. Over time, sludge and scum accumulate in the tank and, if allowed to remain for too long, will eventually cause the tank to overflow and clog the soil absorption system.

SOURCE: EPA.

If a system is operating properly, wastewater flows out of the tank through pipes and into the drain field, where wastes are treated by soil filtration and biological degradation. It is imperative that the septic system be sited properly in soils suitable for onsite wastewater disposal. If the soil is permeable, allowing the wastewater to flow immediately downward without having enough time to become absorbed by the soil, the wastewater might flow into and thereby contaminate a groundwater aquifer.

Contamination by Home Septic Systems

Serious system failures are usually quite evident because wastes will surface and flood the drainage field (not only causing an odor, but also exposing people to pathogenic bacteria and viruses). Unfortunately, we cannot see or smell contaminants from underground systems that leach into aquifers. Years may pass before contamination emanating from poorly designed systems is detected.

Human wastes leaching into groundwater commonly contain bacteria, viruses, and other microorganisms from human skin and digestive and respiratory tracts that have caused outbreaks of diseases such as acute gastrointestinal illness, hepatitis A, and typhoid. High levels of nitrate in groundwater have been positively linked to methemoglobinemia, or "blue-baby" syndrome.

Microorganisms are not the only threat to groundwater from home use of septic systems. Many household products are toxic or hazardous and should not be disposed of in the system (see Household Hazardous Wastes, this chapter). Contamination by organic chemicals such as chloroform and trichloroethylene also results from home use of septic system "cleaners," which purport to save purchasers the cost of pumping by dissolving scum layers that clog outlets, pipes, and drain fields. However, many experts believe the solvents are actually counterproductive: toxic organic chemicals commonly found in these products kill the beneficial bacteria that biologically degrade wastes.

The septage that is pumped from septic tanks presents its own serious health threats. Septage generally is high in bacteria, ammonia, and organic nitrogen. It also might contain pathogenic bacteria or viruses, as well as various waste chemicals, such as paint thinner or cleaning solvents that have been dumped into household drains. Septage should be disposed of only in approved sites, such as special drying beds or sewage-treatment plants designed to handle highly concentrated wastes. Illegal dumping of septage creates a serious health hazard by exposing the public to untreated waste.

Contamination by Commercial and Industrial Septic Systems

Commercial and industrial septic systems present unique and potentially more severe problems to groundwater contamination than domestic systems because of the nature of the wastes disposed of in these systems. For instance, EPA has identified several commercially used septic systems as sources of chemical contamination at sites around the nation designated for cleanup under the federal Superfund law. Chemicals that might be dumped in these systems and that might contaminate groundwater include nitrates, heavy metals (such as lead, copper, and zinc), and certain synthetic organic chemicals (such as toluene, trichlorethylene, chloroform, and tetrachloroethylene).

Septic systems are designed to treat human wastes, not toxic chemicals. Such chemicals leach into soil and groundwater as if there were no treatment at all. Though we are often tempted to associate groundwater problems solely with large industrial enterprises, many small businesses contaminate groundwater by using septic systems improperly. EPA lists the following types of business and the likely pollutants:

- Laundries and laundromats—soil and stain removers
- Paint dealers and hardware stores—solvents and cleaning products
- Restaurants—large quantities of grease and cleaning products
- Gasoline and service stations—waste oils, degreasers, and automotive fluids
- Laboratories—a variety of waste chemicals
- Beauty shops—products such as dyes

TRANSPORTATION

Every year billions of tons of hazardous wastes and toxic chemicals are transported across the country, thousands of accidents occur, and hundreds of people die or are injured as a result. In a report on the transportation of hazardous materials, the Congressional Office of Technology Assessment found that 62 percent of accidents are caused by human error. But whether caused by human error or design flaws, the spills, fires, and explosions associated with the "toxics merry-go-round," a seemingly never-ending cycle between the generation, use, storage, treatment, and disposal of hazardous substances, place every community's water supply at risk.

A community does not have to be home to a company that generates or handles toxic materials to be a thoroughfare for the trucks, trains, or ships that transport these substances, and a major accident does not have to occur in order for a spill to cause serious groundwater contamination. Uncontrolled leaks and spills can occur any time during the storage, transportation, and transfer of hazardous chemicals. For example, faulty couplings between transfer lines and storage tanks can allow leakage, and loading and unloading procedures can cause overflows. The greater the distance a material travels, the more it is transferred from one type of container to another, the greater the likelihood an accident will happen. Unfortunately, such long-distance transportation is an integral part of the waste business: hazardous wastes arriving at a landfill in South Carolina come from every state and from Puerto Rico.

The transportation of chemical products presents an even greater threat than hazardous waste transportation. U.S. Department of Transportation (DOT) data reveal an average of 12 "incidents" involving hazardous materials transportation each day. An incident is an accident causing death or hospitalization, or involving $50,000 in damages to carrier or property. The DOT, however, relies on voluntary reporting of accidents by carriers. EPA estimates that the voluntary system leaves 80 percent of hazardous materials transportation incidents unreported.

As people become more aware of this threat, new measures are being adopted to respond to the problem of toxic and hazardous substances on the move. Small communities have successfully reduced the threat presented by speeding chemical trucks on their bumpy country roads, major cities are working to control the movement of hazardous substances through the busy streets, and "midnight dumpers" are getting caught by alert citizens.

REFERENCES

SOURCES OF GROUNDWATER CONTAMINATION

Landfills:

American Institute of Professional Geologists. *Hazardous Waste: Issues and Answers.* Arvado, Colo. 1986.

Gibbs, Lois Marie, Will Collette et al. *Solid Waste Action Project Guidebook.* Citizen's Clearinghouse for Hazardous Waste. Arlington, Va. 1987.

Colburn, Betsey A. *Landfills and Groundwater Protection.* Massachusetts Audubon Society. Lincoln, Mass. 1987.

EPA, Office of Groundwater Protection. *Census of State and Territorial Subtitle D Non-Hazardous Waste Programs.* Washington, D.C. 1986.

EPA, Office of Groundwater Protection. *State Groundwater Program Summaries.* Washington, D.C. 1985.

Feliciano, Donald V. *Groundwater: What It Is, and How It Is Being Protected.* Environment and Natural Resource Policy Division, U.S. Congressional Research Service. Washington, D.C. 1985.

Gordon, Wendy. *A Citizen's Handbook on Groundwater Protection.* Natural Resources Defense Council. New York, N.Y. 1984.

King, Jonathan. *Troubled Water.* Rodale Press. Emmaus, Pa. 1985.

Stone, P.J., et al. *Protecting the Nation's Groundwater From Contamination.* U.S. Congress, Office of Technology Assessment. Washington, D.C. 1984.

Surface Impoundments:

Conservation Foundation. *A Guide to Groundwater Pollution: Problems, Causes, and Government Responses.* Washington, D.C. 1987.

EPA, Office of Groundwater Protection. *Overview of State Groundwater Program Summaries.* Washington, D.C. 1985.

Feliciano, Donald V. *Groundwater: What It Is, and How It Is Being Protected.* Environment and Natural Resources Policy Division, U.S. Congressional Research Service. Washington, D.C. 1985.

Sherry, Susan, et al. *High Tech and Toxics: A Guide for Local Communities.* Golden Empire Health Planning Center. Sacramento, Calif. 1985.

Underground Storage Tanks:

DiNovo, Frank, and Martin Jaffe. *Local Groundwater Protection: Midwest Region.* American Planning Association. Chicago, Ill. 1984.

DiNovo, Frank, and Martin Jaffe. *Local Groundwater Protection.* American Planning Association. Chicago, Ill. 1987.

EPA, Office of Groundwater Protection. *State Groundwater Program Summaries.* Washington, D.C. 1985.

EPA, Office of Underground Storage Tanks and Office of Toxic Substances. *Underground Motor Fuel Storage Tanks: National Survey.* Washington, D.C. 1986.

Feliciano, Donald V. *Groundwater: What It Is, and How It Is Being Protected.* Environment and Natural Resource Policy Division, U.S. Congressional Research Service. Washington, D.C. 1985.

Gordon, Wendy. *A Citizen's Handbook on Groundwater Protection.* Natural Resources Defense Council. New York, N.Y. 1984.

King, Jonathan. *Troubled Water.* Rodale Press. Emmaus, Pa. 1985.

Millman, Joel. "Tank Trouble." *Technology Review,* February/March 1986.

Waste Disposal Wells:

DiNovo, Frank, and Martin Jaffe. *Local Groundwater Protection.* American Planning Association. Chicago, Ill. 1987.

EPA, Office of Drinking Water. *Report to Congress on the Injection of Hazardous Waste.* Washington, D.C. 1985.

EPA, Office of Groundwater Protection. *State Groundwater Program Summaries.* Washington, D.C. 1985.

Feliciano, Donald V. *Groundwater: What It Is, and How It Is Being Protected.* Environment and Natural Resource Policy Division, U.S. Congressional Research Service. Washington, D.C. 1985.

Gordon, Wendy, and Jane Bloom. *Deeper Problems: Limits to Underground Injection As A Hazardous Waste Disposal Method.* Natural Resources Defense Council. New York, N.Y. 1986.

Agriculture:

Bolduc, John. *Pesticides and Groundwater Protection.* Massachusetts Audubon Society. Lincoln, Mass. 1986.

Conservation Foundation. *A Guide to Groundwater Pollution: Problems, Causes, and Government Responses.* Washington, D.C. 1987.

EPA, Office of Groundwater Protection. *Pesticides in Groundwater: Background Document.* Washington, D.C. 1986.

Gottlieb, Bob, and Peter Wiley. "Pesticides Hit Home." *Sierra,* January/February 1986.

Northwest Coalition for Alternatives to Pesticides. *Journal of Pesticide Reform.* Vol. 7, nos. 1 and 2.

U.S. General Accounting Office. *Nonagricultural Pesticides: Risks and Regulation.* Washington, D.C. 1986.

Household Hazardous Waste:

Bureau of National Affairs. *Environment Reporter.* Washington, D.C. May 15, 1987.

Clean Water Action Project. *Consumer Guide to Toxic and Non-Toxic Household Products.* Amherst, Mass. 1986.

EPA, Office of Solid Waste and Emergency Response. *A Survey of Household Hazardous Wastes and Related Collection Programs.* Washington, D.C. 1986.

EPA, Office of Solid Waste and Emergency Response. *Household Hazardous Waste: Bibliography of Useful References and List of State Experts.* Washington, D.C. 1988.

Tufts University, Center for Environmental Management. *Household Hazardous Waste Collection Programs: Resource Packet from the First National Conference on Household Hazardous Waste Collection Programs.* Medford, Mass. 1987.

Land Application:

Conservation Foundation. *A Guide to Groundwater Pollution: Problems, Causes, and Government Responses.* Washington, D.C. 1987.

EPA, Office of Groundwater Protection. *Census of State and Territorial Subtitle D Non-Hazardous Waste Programs.* Washington, D.C. 1986.

Feliciano, Donald V. *Groundwater: What It Is, And How It Is Being Protected.* Environment and Natural Resource Policy Division, U.S. Congressional Research Service. Washington, D.C. 1985.

Gordon, Wendy. *A Citizen's Handbook on Groundwater Protection.* National Resources Defense Council. New York, N.Y. 1984.

Toxics Coordinating Project. *Toxics Watchdog.* Sacramento, Calif. June 1987.

Military:

Citizen's Clearinghouse for Hazardous Wastes. *Dealing With Military Toxics.* Arlington, Va. 1987.

Donnelly, Michael, and James G. Van Ness. "The Warrior and the Druid—The DOD and Environmental Law." *Federal Bar News and Journal*, Vol. 33, no. 1. 1986.

Environmental Law Reporter. *Superfund Deskbook*. Environmental Law Institute. Washington, D.C. 1986.

National Research Council. *Disposal of Chemical Munitions and Agents*. National Academy Press. Washington, D.C. 1984.

Office of Deputy Assistant Secretary of Defense—Environment. *Status of the Department of Defense Installation Restoration Program For Fiscal Year 1986*. Washington, D.C. 1987.

Pollad, Stephanie, and Seth Shulman. "Pollution and the Pentagon." *Science for the People*, Vol. 19, no. 3. 1987.

Stever, Donald W. *Perspectives on the Problem of Federal Facility Liability for Environmental Contamination*. 17 Environmental Law Reporter. (Envtl. L. Inst.) 10114. April 1987.

U.S. General Accounting Office. *DOD Efforts to Preclude Disposal of Contaminated Property Need Improvement*. Washington, D.C. 1986.

U.S. General Accounting Office. *Environmental Issues at DOE's Nuclear Defense Facilities*. Washington, D.C. 1986.

U.S. General Accounting Office. *Federal Agency Hazardous Waste Disposal at Kettleman Hills*. Washington, D.C. 1985.

U.S. General Accounting Office. *Impact of Savannah River Plant's Radioactive Waste Management Practices*. Washington, D.C. 1986.

Mining:

Conservation Foundation. *A Guide to Groundwater Pollution: Problems, Causes, and Government Responses*. Washington, D.C. 1987.

King, Jonathan. *Troubled Water*. Rodale Press. Emmaus, Pa. 1985.

Student Environmental Health Project, Center for Health Services, Vanderbilt University. *Water Problems and Coal Mining: A Citizen's Handbook*. Nashville, Tenn. 1984.

Septic Systems:

EPA, Office of Groundwater Protection. *Septic Systems and Groundwater Protection I: A Program Manager's Guide and Reference Book*. Washington, D.C. 1986.

EPA, Office of Groundwater Protection. *Septic Systems and Groundwater Protection II: An Executive's Guide*. Washington, D.C. 1986.

EPA, Office of Groundwater Protection. *State Groundwater Program Summaries*. Washington, D.C. 1985.

Transportation:

Citizen's Clearinghouse for Hazardous Wastes. *Transportation: Danger On The Road*. Arlington, Va. 1986.

Council on Economic Priorities. *Hazardous Waste Management: Reducing the Risk*. Island Press. Washington, D.C. 1986.

Department of Transportation. *Emergency Response Guidebook for Initial Response to Hazardous Materials Incidents*. Washington, D.C. 1987.

ICF, Incorporated. *Lessons Learned: A Report on the Lessons Learned From State and Local Experiences in Accident Prevention and Response Planning For Hazardous Materials Transportation*. Report for U.S. Department of Transportation and U.S. Environmental Protection Agency. Washington, D.C. 1985.

Johnston, Lucy. "Accidents Will Happen." *Environmental Action*. Washington, D.C. January/February 1987.

National Response Team. *Hazardous Materials Emergency Planning Guide*. Washington, D.C. 1987.

CHAPTER 4

Testing Groundwater Quality

VARIOUS FEDERAL, STATE, and local officials collect water samples for laboratory testing in order to determine whether groundwater is contaminated with chemicals that could adversely affect your health. Many private citizens also test their own water for the same purpose. Water samples commonly are collected at three places: when the water is still in its underground environment, when it is receiving above-ground treatment at the water utility, or after it comes out of home water taps.

MONITORING UNDERGROUND SOURCES OF WATER

Collecting samples while the water is still in the ground is done primarily by drilling monitoring wells. The placement of these wells can be designed to collect either ambient or point-of-contamination information. Ambient monitoring defines the preexisting state of the aquifer, providing information on such basic aquifer characteristics as quality, flow, and vulnerability to contamination from overlying land uses. Point-of-contamination monitoring locates and defines the extent of contamination from specific sources. The two types of monitoring can be used to determine, for example, the suitability of a site proposed for a new hazardous waste facility, whether an existing hazardous waste facility has leached chemicals into the aquifer, the movement of a known contaminant plume, or the adequacy of existing or proposed groundwater-protection strategies.

However, collecting accurate information from water that is still underground is a difficult task. The unique characteristics of groundwater flow and contaminant plume movement (see Chapter 1) make it very difficult to place monitoring wells where they will intersect a contaminant plume and/or to trace its movement. Extensive testing and information-gathering is essential before actually selecting the well location; these activities include studying area geology, gathering soil maps, drilling soil borings and rock corings, and conducting geophysical tests. Geophysical tests use

51

equipment that transmits radio waves or electrical impulses through the soil to measure variations indicative of soil types, or of the presence of a contaminant plume or unseen buried wastes. Even with such expensive pretesting of the area before siting a monitoring well, however, many problems can develop and hinder the well's collection of accurate information.

Figure 4.1 shows a landfill with three wells placed downgradient of the landfill and one well placed upgradient. The upgradient well is installed

FIGURE 4.1
WELL PLACEMENT AND EFFECTIVE MONITORING

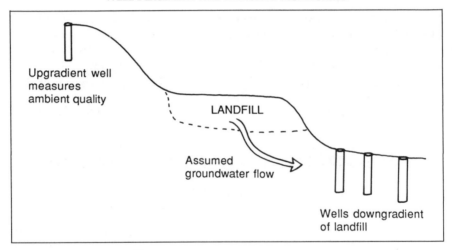

FIGURE 4.2
EFFECT OF HYDROLOGY ON MONITORING WELL EFFECTIVENESS

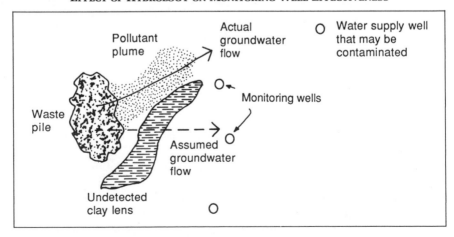

FIGURE 4.3
EFFECT OF GEOLOGICAL FEATURES ON GROUNDWATER MONITORING

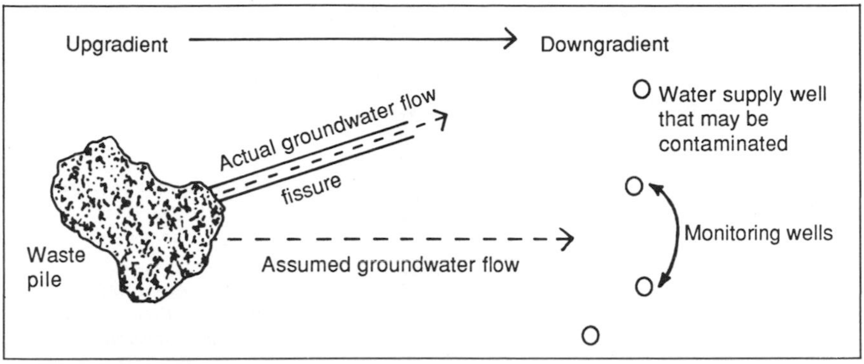

to measure the ambient quality of the aquifer before it comes into contact with the landfill. Comparing the differences between the upgradient and downgradient sample results should indicate the contamination directly attributable to the facility.

Even with the system described in Figure 4.1, it is easy to collect information not indicative of what is really happening underground. In Figure 4.2, the three downgradient wells are totally missing the contaminant plume. Unanticipated barriers to flow or other hydrogeological conditions may allow the plume to move at a different depth or in a different direction than expected. Undocumented well-pumping activity in the area can also lead to unexpected changes in plume movement.

A similar problem can occur in bedrock aquifers, where cracks in underground rock formations can divert water away from its normal downgradient flow path. Figure 4.3 illustrates this problem. Generally, a larger number of wells are necessary in order to track contamination in such aquifers. (In karst limestone aquifers also, water can flow in unexpected directions, but karst aquifers are so unpredictable that monitoring wells may be of very little assistance in detecting contaminant movement). See Chapter 1 for an explanation of these aquifer types.

Another common problem leading to inaccurate information is caused by the density of varying contaminants. Test results from the well samples may show there is contamination by one type of contaminant less dense than water, called a "floater," which spreads across the surface of the water table, while not indicating presence of a contaminant more dense than water, called a "sinker," which migrates down to the top of the next impermeable layer. Figure 4.4 illustrates the effect of pollutant density on groundwater monitoring.

To increase the likelihood that monitoring wells will detect contamina-

FIGURE 4.4
EFFECT OF POLLUTANT DENSITY ON GROUNDWATER MONITORING

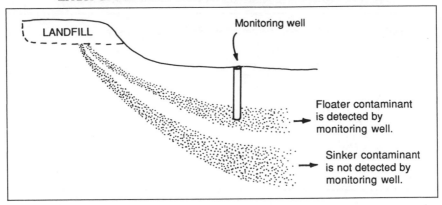

tion, the wells are commonly installed in clusters. Instead of placing three wells downgradient, for example, you could install three clusters of five wells each. Generally, each cluster would contain wells at varying depths to decrease the probability that contaminants will escape detection. However, each additional well will greatly add to project expenses and can still yield inaccurate results.

Ironically, monitoring wells may actually contribute to groundwater contamination—a problem that worsens as the number of wells increases. Wells can create routes for rapid contaminant migration in several ways. If, for example, the well is installed with construction materials that cannot withstand the potentially corrosive chemicals in a contaminant plume, the well will allow the chemicals to leak through to other layers of soil or aquifer. An improperly sealed well creates routes for contaminant migration alongside the well bore. This idea can also work in reverse; an improperly installed or sealed well can allow surface contaminants to enter a previously uncontaminated aquifer. In some cases, even the well construction materials can contaminate a once-clean aquifer. (See Figure 3.4, on page 31.)

The problems and expense of cluster well drilling have led researchers to experiment with new well designs. These alternatives include:

- Installing one well with screened openings at varying depths, thereby eliminating the need to install a different well for each sampling depth. With this method, extra precautions must be taken to ensure that the samples collected from varying depths do not mix together within the well.
- Installing one well with a long length of screened entry collecting water from its entire length. The problem with this method is that the water from different depths will mix together to yield only an *average* of underground water quality.

• Installing a single well that collects samples from varying depths by halting the installation drilling at appropriate depths to collect samples. With this method, researchers must refrain from collecting the water sample until they have allowed enough time for the evaporation and collection of drilling wastes that could contaminate the sample.

Who Is Monitoring Groundwater Quality in Your Area?

Many different federal, state, and local agencies, as well as private entities, are monitoring groundwater. Who should you turn to for information on the quality of your water? Large-scale ambient monitoring of groundwater is performed primarily by the United States Geological Survey (USGS) and by states entering into cooperative agreements with the USGS. All fifty states have some degree of participation in this program. The USGS may be monitoring your aquifer as part of its Regional Aquifer System Analysis (RASA) Program, which studies the nation's ambient groundwater quality on a regional scale. (See Figure 4.5.) EPA and some

FIGURE 4.5
REGIONAL AQUIFER-SYSTEM ANALYSIS

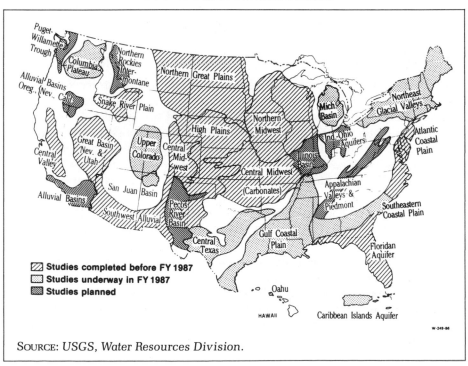

SOURCE: *USGS, Water Resources Division.*

other federal agencies also occasionally initiate ambient monitoring projects. Local colleges or universities may also be studying groundwater through monitoring projects.

To find out if any ambient monitoring activities are taking place in your area, we suggest you contact a local or regional USGS office and a regional EPA office. Because so many agencies are gathering information, it often remains fragmented. Both USGS and EPA are trying to consolidate their groundwater monitoring data into a better format that is more easily accessible. Currently, however, no one agency is likely to be able to tell you exactly what ambient monitoring is being conducted in your area.

EPA and state environmental agencies oversee most of the point-of-contamination monitoring required by major environmental statutes. Other agencies that oversee or perform source monitoring include the Office of Surface Mining, Nuclear Regulatory Commission, Department of Defense, U.S. Department of Agriculture (especially the Soil Conservation Service), and the Bureau of Land Management. (See Figure 4.6.) Later chapters of this manual detail the federal and state monitoring requirements applicable to various sources of contamination. These sections will also detail the agencies responsible for handling monitoring reports and describe how you can use these reports.

MONITORING WATER AT THE TAP

Determining Who Is Responsible for Monitoring Your Tap Water

While ambient and point-of-contamination monitoring projects may be ongoing in your aquifer, additional monitoring is usually required for water drawn out of the ground and designated for home use. The Safe Drinking Water Act requires water utilities that service more than 15 individual year-round hookups to test regularly for 29 contaminants in drinking water. (See Chapter 11.) If you are serviced by a small utility or a private well, you unfortunately may find that testing is your responsibility.

Whether or not your home water is tested under authority of the SDWA, local or state boards of health may test for contaminants in area drinking-water wells.

Many kinds of private well-water testing may also help assess aquifer quality. For example, many banks require tests of water quality as part of loan agreements for home purchasers.

Even if you find that your drinking water has been tested, there is still no assurance it is free of contamination. For example, if you discover that a utility tests your water and complies with the procedures required under law, your water could still be contaminated; the SDWA does not require testing for many potential contaminants. The only way to know more about chemicals in your tap water is to have a more extensive analysis performed by a laboratory.

Monitoring Your Own Tap-Water Quality

If you want to know more about possible toxics in your water, or if you are not covered by any monitoring programs, you can collect samples of your own tap water and pay for laboratory testing. Both university and professional laboratories test home water samples, and some community governments offer testing programs.

How do you decide whether to choose a college or professional lab, or a government program? The most important consideration is that you find a lab with considerable experience and a staff ready to work *with* you. Other key issues include:

- How long has the lab been in business and what types of tests does it perform most frequently?
- Is the lab EPA-certified for the types of tests you need?
- Does the lab provide free EPA-approved sampling containers and transportation, and detailed directions or direct aid on how to obtain the samples?
- Will lab personnel help you understand what the results mean to you and your community rather than simply sending you a bunch of numbers?

Finally, try to find a lab nearby, because some contaminants can be destroyed in transit, reducing the accuracy of the results. The farther away the lab, the more imperative that adequate containers, transportation, and sample handling advice be provided.

College and University Laboratories

For inexpensive expertise you should consider contacting the professor in charge of a nearby college or university chemistry department and informing him or her of your efforts to document the quality of your community's groundwater. The professor might adopt your water samples as a class project or a graduate student may evaluate them for credit. If no one can adopt your project, at least the college lab may test for a lesser fee than professional labs, or may direct you to a reputable nearby lab. The biggest problem with college or university labs is that many are not EPA-certified. Even if this is the case, their inexpensive or free tests will help you determine the general severity of your problem and their expertise can help you decide whether to pay a professional lab for more detailed tests.

FIGURE 4.6

OFFICE OF TECHNOLOGY ASSESSMENT SUMMARY OF FEDERAL GROUNDWATER MONITORING PROVISIONS AND OBJECTIVES

Statutory authority	Monitoring provisions[a]	Monitoring objectives
Atomic Energy Act	Groundwater monitoring is specified in Federal regulations for low-level radioactive waste disposal sites. The facility license must specify the monitoring requirements for the source. The monitoring program must include: —Pre-operational monitoring program conducted over a 12-month period. Parameters not specified. —Monitoring during construction and operation to provide early warning of releases of radionuclides from the site. Parameters and sampling frequencies not specified. —Post-operational monitoring program to provide early warning of releases of radionuclides from the site. Parameters and sampling frequencies not specified. System design is based on operating history, closure, and stabilization of the site.	To obtain background water quality data and to evaluate whether groundwater is being contaminated.
	Groundwater monitoring related to the development of geologic repositories will be conducted. Measurements will include the rate and location of water inflow into subsurface areas and changes in groundwater conditions.	To confirm geotechnical and design parameters and to ensure that the design of the geologic repository accommodates actual field conditions.
	Groundwater monitoring may be conducted by DOE, as necessary, part of remedial action programs at storage and disposal facilities for radioactive substances.	To characterize a contamination problem and to select and evaluate the effectiveness of corrective measures.
Clean Water Act —Sections 201 and 405	Groundwater monitoring requirements are established on a case-by-case basis for the land application of wastewater and sludge from sewage treatment plants.	To evaluate whether groundwater is being contaminated.
—Section 208	No explicit requirements are established; however, groundwater monitoring studies are being conducted by SCS under the Rural Clean Water Program to evaluate the impacts of agricultural practices and to design and determine the effectiveness of Best Management Practices.	To characterize a contamination problem and to select and evaluate the effectiveness of corrective measures.

Statute	Description	Purpose
Coastal Zone Management Act	The statute does not authorize development of regulations for sources. Thus, any groundwater monitoring conducted would be the result of requirements established by a State plan (e.g, monitoring with respect to salt-water intrusion) authorized and funded by CZMA.	
Comprehensive Environmental Response, Compensation, and Liability Act	Groundwater monitoring may be conducted by EPA (or a State) as necessary to respond to releases of any hazardous substance, contaminant, or pollutant (as defined by CERCLA).	To characterize a contamination problem (e.g., to assess the impacts of the situation, to identify or verify the source(s), and to select and evaluate the effectiveness of corrective measures).
Federal Insecticide, Fungicide, and Rodenticide Act— Section 3	No monitoring requirements established for pesticide users. However, monitoring may be conducted by EPA in instances where certain pesticides are contaminating groundwater.[b]	To characterize a contamination problem.
Federal Land Policy and Management Act (and Associated Mining Laws)	Groundwater monitoring is specified in Federal regulations for geothermal recovery operations on Federal lands for a period of at least one year prior to production. Parameters and monitoring frequency are not specified. Explicit groundwater monitoring requirements for mineral operations on Federal lands are not established in Federal regulations. Monitoring may be required (as a permit condition) by BLM.	To obtain background water quality data.
Hazardous Liquid Pipeline Safety Act	Although the statute authorizes development of regulations for certain pipelines for public safety purposes, the regulatory requirements focus on design and operation and do not provide for groundwater monitoring.	
Hazardous Materials Transportation Act	Although the statute authorizes development of regulations for transportation for public safety purposes, the regulatory requirements focus on design and operation and do not provide for groundwater monitoring.	
National Environmental Policy Act	The statute does not authorize development of regulations for sources.	

Continued on next page

FIGURE 4.6 (continued)

Statutory authority	Monitoring provisions[a]	Monitoring objectives
Reclamation Act	No explicit requirements established; however, monitoring may be conducted, as necessary, as part of water supply development projects.	
Resource Conservation and Recovery Act	Groundwater monitoring is specified in Federal regulations for all hazardous waste land disposal facilities (e.g., landfills, surface impoundments, waste piles, and land treatment units)	
—Subtitle C	*Interim Status monitoring* requirements must be met until a final permit is issued. These requirements specify the installation of at least one upgradient well and three downgradient wells. Samples must be taken quarterly during the first year and analyzed for the National Interim Drinking Water Regulations, water quality indicator parameters (chloride, iron, manganese, phenols, sodium, and sulfate), and indicator parameters (pH, specific conductance, TOC, and TOX). In subsequent years, each well is sampled and analyzed quarterly for the six background water quality indicator parameters and semiannually for the four indicator parameters. *Groundwater monitoring requirements can be waived* by an owner/operator if a written determination indicating that there is low potential for waste migration via the upper-most aquifer to water supply wells or surface water is made and certified by a qualified geologist or engineer. The determination is not submitted to EPA for verification or approval. The monitoring requirements for a *fully permitted facility* are comprised of a three-part program: —*Detection Monitoring*—Implemented when a permit is issued and there is no indication of leakage from a facility. Parameters are specified in the permit. Samples must be taken and analyzed at least semiannually. Exemptions from detection monitoring program may be granted by the	To obtain background water quality data and evaluate whether groundwater is being contaminated. To obtain background water quality data or evaluate whether groundwater is being contaminated (detection monitoring), to determine whether groundwater quality standards are being met (compliance monitoring), and to evaluate the effectiveness of corrective action measures.

regulatory authority for landfills, surface impoundments, and waste piles with double liners and leak detection systems.

—*Compliance Monitoring*—Implemented when groundwater contamination is detected. Monitoring is conducted to determine whether specified concentration levels for certain parameters are being exceeded (levels are based on background concentrations, maximum contaminant levels specified by the National Drinking Water Regulations [if higher than background], or an alternative concentration limit [established on a site-specific basis]). Samples must be taken and analyzed at least quarterly for parameters specified in the permit. Samples must also be analyzed for a specific list of 375 hazardous constituents (Appendix VIII, 40 CFR 261) at least annually.

—*Corrective Action Monitoring*—Implemented if compliance monitoring indicates that specific concentration levels for specified parameters are being exceeded (and corrective measures are required). Monitoring must continue until specified concentration levels are met. Parameters and monitoring frequency not specified.

—Exemption from groundwater monitoring requirements may be granted by the regulatory authority if there is no potential for migration of liquid to the uppermost aquifer during the active life and closure and post-closure periods.

—Subtitle D

Groundwater monitoring may be required by State solid waste programs. Federal requirements for State programs *recommend* the establishment of monitoring requirements.

Safe Drinking Water
Act
—Part C—
Underground
Injection Control
Program

Groundwater monitoring requirements may be specified in a facility permit for injection wells used for in-situ or solution mining of minerals (Class III wells) where injection is into a formation containing less than 10,000 mg/l TDS. Parameters and monitoring frequency not specified except in areas subject to subsidence or collapse where monitoring is required on a quarterly basis.

To evaluate whether groundwater is being contaminated.

Continued on next page

FIGURE 4.6 (continued)

Statutory authority	Monitoring provisions[a]	Monitoring objectives
	Groundwater monitoring may also be specified in a permit for wells which inject beneath the deepest underground source of drinking water (Class I wells). Parameters and monitoring frequency not specified in Federal regulations.	
Surface Mining Control and Reclamation Act	Groundwater monitoring is specified in Federal regulations for surface and underground coal mining operations to determine the impacts on the hydrologic balance of the mining and adjacent areas. A groundwater monitoring plan must be developed for each mining operation (including reclamation). At a minimum, parameters must include total dissolved solids or specific conductance, pH, total iron, and total manganese. Samples must be taken and analyzed on a quarterly basis. Monitoring of a particular water-bearing stratum may be waived by the regulatory authority if it can be demonstrated that it is not a stratum which serves as an aquifer that significantly ensures the hydrologic balance of the cumulative impact area.	To obtain background water quality data and evaluate whether groundwater is being contaminated.
Toxic Substance Control Act— Section 6	Groundwater monitoring specified in Federal regulations requires monitoring prior to commencement of disposal operations for PCBs. Only three wells are required if underlying earth materials are homogenous, impermeable, and uniformly sloping in one direction. Parameters include (at a minimum) PCBs, pH, specific conductance, and chlorinated organics. Monitoring frequency not specified. No requirements are established for active life or after closure.	To obtain background water quality data.
Uranium Mill Tailings Radiation Control Act	Federal regulatory requirements for active mill tailings sites are, for the most part, the same as those established under Subtitle C of RCRA.	To obtain background water quality data, evaluate whether groundwater is being contaminated, determine whether groundwater quality standards are being met, and evaluate the effectiveness of corrective action measures.

Water Research and Development Act	Groundwater monitoring for inactive sites may be conducted if necessary to determine the nature of the problem and for the selection of an appropriate remedial action. The statute does not authorize the development of regulations for sources. Groundwater monitoring may be conducted as part of projects funded by the act.	To obtain background water quality data and to characterize a contamination problem.

aThe monitoring provisions in this table are either those specified by regulations for existing and new sources, or for groundwater monitoring that may be conducted as part of an investigatory study or remedial action program.
bPesticide manufacturers may be required by EPA to submit groundwater monitoring data as part of the registration requirements for a pesticide product to evaluate the potential for a pesticide to contaminate groundwater.

SOURCE: *Office of Technology Assessment, "Planning Workshop to Develop Recommendations for a Groundwater Monitoring Strategy," March 1985.*

An example of an uncertified but very helpful university laboratory is the Student Environmental Health Project's (STEHP) service at Vanderbilt University. Staffed by undergraduate and graduate chemistry and chemical engineering students, STEHP's lab is an integral part of a community outreach and direct aid project. STEHP asks donations of $5 per sample for metal screening and $12 per sample for organic chemical screening. They provide EPA-approved sampling jars for water, detailed directions on how to obtain samples, and in two to four weeks they send you the results as well as an analysis of what the numbers mean to you. The type of information STEHP's lab requests from you includes a map of your area with indications of sites where samples were taken, listings of chemicals you suspect may be in your community, and any potential health problems that may be of concern to you.

Professional Laboratories

Many professional labs follow EPA testing procedures and are EPA-certified. For a complete list of labs specializing in water testing, write to the American Council of Independent Laboratories, 1725 K Street, N.W., Suite 301, Washington, D.C. 20006 (telephone 202-887-5872), or to the American Association for Laboratory Accreditation, 656 Quince Orchard Road, Suite 704, Gaithersburg, MD 20878 (telephone 301-670-1377).

Government Programs

Local and state governments often have free testing services for those community groups requesting it. The quality and coverage of their testing varies widely, from a typical minimum of free coliform bacteria testing to a program such as Michigan's, in which any citizen may request free testing for chemical contaminants by the Michigan Department of Public Health or local health departments. When requesting testing aid through any type of program, it is important to be as specific as possible about your contamination concerns. You might also check to see if a local health agency can arrange quantity discounts with a private lab if you and several neighbors agree to test well water for the same contaminants; you may also be able to arrange quantity discounts on your own. If no inexpensive testing option currently exists and everyone in your county draws water from the same aquifer, you could work toward the establishment of a tax-supported countywide testing program.

What Do You Want the Laboratory to Look For?

Once a laboratory receives a sample, whether from a government agency, an owner of a waste facility, or from you, it generally cannot simply run the water through its testing equipment and have the results show exactly what is in the water. The laboratory needs to know which contaminants to look for. The contaminants a water supplier must look for are determined by law, but when you want a test of your home water, or when a waste facility is monitoring nearby groundwater, the selection of contaminants to test for can appear to be an overwhelming process: with thousands of potential chemical contaminants, which ones should be chosen?

Hazardous contaminants in a water sample can be completely overlooked while laboratory personnel search in vain for the contaminants selected by the party who turned in the water sample. Therefore, it is crucial that the selection of contaminants be based on the best available information. Government agencies or facility owners have to base their selection of contaminants on any historic evidence of activities that have taken place at the site along with current knowledge of the chemicals found at the site. The different laws controlling these facilities might also require that they look for specified contaminants.

For testing home tap water, you may also want to begin by gathering historic and current information on the land uses in your area, and then associating these land uses with the chemicals they are, or were, most likely to release. While this task will not be easy, it is practical. For example, there might be a facility near you that uses a dangerous chemical not included on any standard EPA or laboratory list of contaminants. Use the information on conducting land use inventories in Chapter 5 to pinpoint possible sources over your aquifer, and contact local university chemistry professors, state water quality agencies, or organizations such as Citizen's Clearinghouse for Hazardous Waste or United States Public Interest Research Group (U.S. PIRG) for help with the chemicals commonly released by those sources. Also of assistance are federal and state programs that require companies to report the types of chemicals they use at their facilities. (See Chapter 17.) These reports may give you some leads on potential groundwater contaminants to look for. (You might consider selecting a laboratory based on its willingness to help you with the contaminant-selection process.)

You can also use lists of contaminants required to be tested under various federal and state laws. For example, if your water is not monitored by a public water utility, or if you want to cross-check your utility's results, you can add to your list those chemicals required under the SDWA. Other lists of contaminants you should consider incorporating include those that EPA is required to regulate over the next several years under the 1986 SDWA amendments (see Chapter 11), chemicals covered by EPA's health advisories (see Chapters 11 and 20), and the EPA priority pollutant list developed under the Clean Water Act.

Together these lists cover hundreds of chemicals, but they include those most likely to contaminate groundwater and cause health problems. For example, the priority pollutant list is quite long, but its 129 chemicals represent a good spectrum of chemicals likely to leach from contamination sources; the EPA has approved testing procedures for each chemical; and there are "water quality criteria" documents for many of the 129 chemicals. These documents provide EPA's latest information on the contaminants' behavior in water and their effects on health and the environment.

Some contaminants may be on more than one list. You can begin finalizing the list of contaminants you want tested by eliminating these duplicates.

Next, compare the lists of potential contaminants from your inquiry into area land uses and from the regulatory agency lists with a laboratory price list. The price lists often group contaminants under chemical category headings. Different chemical categories require different testing methods, and each testing method has an individual price. If, with the help of any local experts, you can group your list of potential contaminants into chemical testing categories, you could choose which contaminants the laboratory should test first, based on your largest chemical category, i.e., the category into which the most individual contaminants on your lists fell. Once you pay the initial fee for the laboratory to adjust its equipment to detect for that category, the additional fee for testing more individual contaminants will be comparatively small; this may save some agonizing over the merits of the costs of testing for a contaminant versus your degree of concern about its presence in your groundwater.

Depending on the costs, you should also consider collecting at least two samples from the same tap on different days and different times, and then running the same tests on these samples. Because a water sample is just a "snapshot" of what is actually coming through your tap over time, running the same tests on water collected on different days will give you a better idea of the variety of contaminant levels to which you are exposed. Performing a number of tests allows you to average results and gives them greater credibility with public officials.

SCRUTINIZING WATER TEST RESULTS

Due to the inherent difficulties in selecting contaminants for testing, it is imperative that you review any monitoring reports with skepticism. For example, earlier we suggested that you contact local or state agencies regarding past testing of your water. When you do this, you might be informed that tests have been performed which revealed no evidence of

Testing Your Water

Though the tap is probably the best source for determining what is in the water you actually drink, high levels of lead in your lab results may be more an indicator of leaching lead plumbing parts than of groundwater contamination. If you compare tap water test results with utility monitoring results and find that high lead levels are indicated only on your tap water results, you could obtain water samples directly from the water utility plant, and have your laboratory check the lead levels in these samples. This will tell you whether the contamination originates in your plumbing.

---

contamination. Unfortunately, you cannot stop there. You need to find out exactly which contaminants were tested for and by what methods. Not all common water testing methods reveal possible toxic contaminants. Older tests focus more on basic water quality contamination by natural constituents such as bacteria and minerals than on toxic contamination. Even if the tests are recent, the equipment required to test chemical contaminants is expensive enough that your local or state agency may still be performing only the basic water-quality tests and those required by law.

The difficulty of selecting contaminants for testing is also frequently manifested in the performance of general screening tests for indicators of contamination rather than looking for individual contaminants. Screening tests are often suggested as a way to save money when testing water samples, but they have serious shortcomings. Such tests only measure the *combined* total of a selected group rather than the individual totals for the chemical constitutents that make up the group. The benchmark levels for groups of contaminants may be set many times higher than the minute but toxic amount of any one contaminant. Thus, for example, a seemingly reassuring low test result for total organic halides *could* mask very high levels for the one or two contaminants that may in fact be found in the sample.

Two popular screening tests are the "TOX," total halogenated organics, and "TOC," total organic carbon tests. Some other screening tests that are only general indicators of water quality are referred to as "traditional pollution parameters." These include such common tests as alkalinity, biological oxygen demand (BOD), chemical oxygen demand (COD), color, total dissolved solids (TDS), hardness, pH, total residue, total suspended solids (TSS), and turbidity.

To ensure quality and accuracy in water testing results, some laws may require the responsible monitoring party to prepare and implement some type of written "groundwater sampling and analysis plan." Such a plan requires the responsible party to record all the procedures and techniques of sample collection, preservation and shipment, lab analytical procedures, chain of custody control (documenting who did what to the sample from the time of collection to the time of analysis), as well as field and laboratory quality-control procedures used to ensure the reliability and validity of any field and laboratory data gathered. Sampling and analysis plans are not required for everyone who tests water samples, but you should consider incorporating some of the following basics when checking on the accuracy of any water test results you review, whether from your own lab's testing or from government or private industry testing:

- Are the lab and its personnel certified by EPA or the state for the tests performed?
- Were standard EPA methods of analysis used at the laboratory? If other methods were used, were they approved in advance?
- Were samples analyzed shortly after sample collection? Some chemicals will rapidly degrade over a few days.
- Were duplicate, blank, and spiked samples analyzed? These samples are used to check the adequacy of laboratory equipment.

Duplicate samples, also called "split samples," involve dividing one sample into two bottles. Because the two different bottles were collected at the same time and place by the same methods, the results from testing the two samples should be exactly the same. Requesting that the same samples be split between different labs during government or private monitoring projects is also a technique used by community groups to ensure honesty in results. Blank samples are usually samples of pure deionized water tested to ensure no residual contamination exists in the equipment. Spiked samples are those with a known quantity of a chemical added to the sample. The sample is then analyzed to see if the laboratory equipment accurately detects the chemical.

Once you obtain quality-assured results from the monitoring of a private facility, public water utility, or your home water, you need to determine their significance. All should be compared with any federal and state water standards. Unfortunately, standards may not exist for many of the contaminants on your list. However, the presence of any contaminant in your aquifer should be of concern. You need to seek out advice from experts and begin compiling information on the health and environmental effects of all the contaminants in your water. For chemicals that have federal or state standards, remember that a standard is not solely based on the contaminant's health effects. Standards are often set at levels that balance the contaminant's health impacts with the economic impacts that would result from a stricter standard for that contaminant. Therefore, any contaminant that does not exceed, but comes close to, its standard should also be of concern. Another economic/health balancing act arises with the established levels of detection for certain contaminants. The level of detection establishes a limit for the testing of chemicals. Laboratories may not test for the contaminant below this level, even though new methods and equipment may be able to detect lower concentrations of the chemical. Levels for contaminants in the water sample not detected above the level of detection are reported as "ND" (not detectable), or are not included on the list of results. This does not mean that these chemicals are not present in the water.

When comparing lab results with federal or state standards, make sure the units of measurement are the same. The four most common units of measurement are milligrams per liter (mg/l) and micrograms per liter (mcg/l), both of which measure the weight of the chemical in a volume of water; or parts per million (ppm) and parts per billion (ppb), which measure the weight of the chemical per weight of water or soil. With water samples, mg/l and ppm are practically equal and are used interchangeably, as are mcg and ppb. If you are having your own water tested, you can request that the lab report all its results in the same type of units, such as parts per million and billion, to aid your comparisons. Scrutinizing others' results may require converting the units; conversion tables can be found in any good dictionary.

This chapter has introduced you to some of the problems associated with groundwater monitoring and water quality testing, but the basic problem is not one of technology and technique. Rather, it is that groundwater monitoring and testing do not *prevent* contamination, they only *detect* it. By the time a well has detected contaminants, you may already have a serious problem. Pinpointing and fixing the cause of the contamination will not remove contaminants from the aquifer.

Though the technology exists to remove contaminants from groundwater, the techniques are expensive, usually costing tens of millions of dollars. Facing such high costs, decisionmakers may find it more cost-effective to change local water hookups to a different system, drill a new well, or simply continue monitoring the movement of the contaminant plume, only changing area hookups when it is decided the plume is getting too close to the well. None of these options restores the aquifer to its original quality. Assurances about the adequacy of groundwater monitoring and testing do not mean that your aquifer will remain uncontaminated. You need to take action to prevent contamination sources from degrading groundwater quality in the first place.

REFERENCES

Testing Groundwater Quality

American Chemical Society. *Groundwater Information Pamphlet*. Washington, D.C. 1983.

Citizen's Clearinghouse for Hazardous Wastes. *Environmental Testing*. Arlington, Va. 1985.

Conservation Foundation. *A Guide to Groundwater Pollution: Problems, Causes, and Government Responses*. Washington, D.C. 1987.

Environmental Defense Fund. *Dumpsite Cleanups: A Citizen's Guide to the Superfund Program*. Washington, D.C. 1983.

EPA, Office of Groundwater Protection. *Groundwater Monitoring Strategy*. Washington, D.C. 1985.

EPA, Office of Groundwater Protection. *Planning Workshop to Develop Recommendations for a Groundwater Monitoring Strategy*. Washington, D.C. 1985.

EPA, Office of Solid Waste and Emergency Response. *Draft RCRA Groundwater Monitoring Technical Enforcement Guidance Document*. Washington, D.C. 1985.

Gordon, Wendy. *A Citizen's Handbook on Groundwater Protection*. Natural Resources Defense Council. New York, N.Y. 1984.

King, Jonathan. *Troubled Water*. Rodale Press. Emmaus, Pa. 1985.

Moore, Andrew Owens. *Making Polluters Pay: A Citizens' Guide to Legal Action and Organizing*. Environmental Action Foundation. Washington, D.C. 1987.

O'Donnell, Arleen. *Protecting and Maintaining Private Wells*. Massachusetts Audubon Society. Lincoln, Mass. 1985.

Scalf, Marion R., James F. McNabb, William Dunlap, and Roger L. Cosby. *Manual of Groundwater Quality Sampling Procedures*. Environmental Protection Agency. Washington, D.C. 1982.

Smith, Velma. "Water, Water Everywhere But Not a Drop to Drink." *Resources*, Vol. 7, no. 1. Washington, D.C. 1987.

Student Environmental Health Project, Center for Health Services, Vanderbilt University. *Water Problems and Coal Mining: A Citizen's Handbook*. Nashville, Tenn. 1984.

Squires, Sally. "Home Filtration Offers Limited Benefits." *Washington Post*, January 22, 1986.

Squires, Sally. "When to Worry and How to Choose a Lab." *Washington Post*, January 22, 1986.

Wilson, Reid. *Testing for Toxics: A Guide to Investigating Drinking Water Quality*. U.S. Public Interest Research Group. Washington, D.C. 1986.

CHAPTER 5

Mapping Aquifers and Contamination

IF YOU WANT to find out exactly where your drinking water comes from and which sources threaten its quality, the best strategy is to gather or develop maps of your area's aquifers and potential contamination sources. This process will not be easy. Because groundwater contamination is only recently receiving the attention of lawmakers, very few areas of the country have adequate aquifer maps. More commonly, when you inquire about the status of mapping projects in your area you will find either that mapping efforts have recently been funded but finished products will not be ready for months or years, or that no aquifer mapping projects are in progress and no one has requested the funding to begin them.

Gathering existing maps or working toward the initiation of new projects are important efforts. Groundwater maps can help your group focus its efforts on those areas most vulnerable to contamination and will vividly portray this vulnerability during your community education efforts. They will be an important tool in public meetings or hearings when you are trying to prove, for example, why a facility should not locate in a certain area, or why a facility's permit should include certain stringent conditions. And, if your group decides to work on protecting groundwater through any type of zoning controls, local ordinances, or state regulations, groundwater maps will be the backbone of your campaign for appropriate legal protection.

To build a broad base of support for your mapping efforts, stress the benefits accurate maps can offer to the local community. The information such maps provide can be used to assist in early detection of contamination incidents (thus reducing liability of area businesses) and can be incorporated into emergency response plans (thus improving human safety and avoiding inappropriate response actions).

GETTING STARTED

Your first task is to determine what has already been done. Contact the local U.S. Geological Survey office, local or regional planning agency, state water board, or groundwater office to find out if any modern aquifer mapping efforts have taken place. If they have, you are very lucky. If you find that these maps are inadequate, or that minimum or no mapping efforts have taken place in your community, you will have to begin working on two different but related tasks: gathering existing data from local, state, and federal agencies and compiling this information for your own homemade aquifer maps or overlays; and getting those agencies with mapping authority and expertise to actually begin mapping the aquifers. This is a political process that could require placing the issue before local voters or pushing in the right places for the release of available funds.

COMPILING YOUR OWN WATER SUPPLY MAPS

Because groundwater cannot be seen, the home mapping process involves the compilation of a lot of clues. These clues are usually pieces of information gathered for use in situations not necessarily related to groundwater. However, with some knowledge of local hydrogeology, you can put the pieces together in an entirely new way to form a map indicating the location of your aquifer's boundaries, of its recharge areas, and of potential sources of contamination that threaten drinking-water quality. The pieces of information can be adjusted to the same scale and either put together on one map, or transferred to individual transparencies which are then placed over one another to form an "overlay" image.

It is not only citizen activists who rely on this type of "low-tech" mapmaking. For example, Massachusetts has developed overlay maps that show surface drainage basins, groundwater favorability, location of public drinking-water supplies, and potential sources of contamination.

The more clues you can gather, the greater your chances for a fairly accurate map. The agencies to contact and the information to ask for are listed below:

Who to Contact:
U. S. Geological Survey (USGS)
Local USDA Soil Conservation Service or Cooperative Extension Agents
Planning departments
Health agencies

Soil and water conservation districts
Water utilities
State water boards
College and university geology departments

What to Ask For:
Watershed boundaries
Topographic maps
Soil maps
Soil survey reports
Percolation tests
Well driller records
Public water supply quality and quantity data

Following are some ideas on how you can use the above sources to lay out possible groundwater and recharge areas. The types of information available in your area will help determine which steps are most appropriate when making your map.

Before you begin, however, we should warn you that the data discussed above will be reported on maps of many different scales. As you begin to adjust, reduce, and enlarge the various pieces of information to fit within a common set of overlays or maps, some distortion and inaccuracy will occur. To determine how to compensate for this problem, seek the help of those with experience in mapping (for some initial leads on who to contact, see this chapter's references).

Identifying Aquifers

DEFINE THE WATERSHED BOUNDARY

Topographic ("topo") maps show land elevations through contour lines. By using these maps to connect the highest elevations of land, you can get a general idea of the watershed boundaries, which in turn may be boundaries for your aquifer (aquifers may be smaller or larger than this boundary, see Chapter 1). The topographic maps of your area will be a great background map onto which to add the rest of the information. (See Figure 5.1.)

IDENTIFY KNOWN AQUIFER AREAS

Some aquifers near you may be identified in a USGS hydrologic atlas. However, these atlases detail only major aquifers, which, based on geologists' estimation of well yield, are determined to be favorable for development as a water resource. Although they will not include every aquifer in

FIGURE 5.1
WATERSHED BOUNDARIES

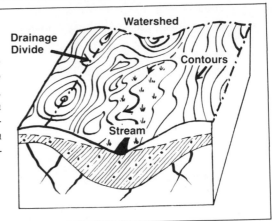

Watershed boundaries are drawn by connecting the highest elevations of land, as indicated by contours on a topographic map. Reprinted with permission from the Massachusetts Audubon Society.

your area, the atlases are a good place to start. Make sure that you check with the local USGS office to see what is already known about groundwater in your area.

Well-driller records can also be used to find where, and at what depths, drillers have reported hitting water. You can chart these well locations and depths on the topographic maps. You can also use records showing present and proposed public water-supply wells, and any land areas set aside for their protection, to find groundwater in your area. The yield of these wells may be an indication of the size of the aquifer supplying the well.

IDENTIFY PERMEABLE SOILS

Unconfined aquifers are found under relatively permeable soils. USGS surficial geologic maps or the U.S. Department of Agriculture's soil maps and soil survey reports show the types of soil near the land's surface and identify its relative permeability. Your group can look at these maps to gain a better understanding of where aquifers may be within the watershed based on the soil's permeability. However, because soil maps show only those soil types relatively close to the surface, they cannot be relied on as definitive indicators of unconfined aquifers. For example, such maps may not show an impermeable deposit just below the depth at which soil was measured.

You can also use local percolation test results to help identify the permeability of area soils. Required by many states when determining the suitability of a site for a septic system, the percolation tests estimate water's "percolation rate" in area soils, i.e., the rate of water movement downward through the pores between soil particles. Well-driller records in some states may also indicate the permeability of area soils.

Identifying Possible Recharge Areas

Topographic maps not only show land elevations, they also show wetlands, rivers, roads, some buildings, and other interesting landmarks; if you define the watershed boundary as described above, you can use the topo map to identify those rivers and wetlands that may be possible recharge areas, and you can see the types of land areas that water runs over before entering recharge areas within the watershed.

You can also use the USGS surficial geologic maps, USDA soil maps and soil survey reports, percolation reports, and well-driller records to mark areas of permeable soils that are indicative not only of aquifers, but also of recharge areas. The same cautions on potential underlying impermeable deposits apply when using these reports as possible recharge area indicators.

Identifying Possible Threats to Water Quality

This manual's chapter on contamination sources alerts you to the sources that have contaminated aquifers across the country. Do you know how many of these threats exist in your community? Identifying potential sources of groundwater contamination is essentially tantamount to preparing a land-use inventory. You might want to focus your inventory on the potential recharge areas revealed by your mapping efforts. In this way, you can document those sources that most endanger you aquifer's quality. Remember that for certain kinds of aquifers, recharge areas—and thus, the areas where inappropriate or careless land uses pose the greatest threat to groundwater quality—can be located far away from current and potential sites of drinking-water wells. (See Chapter 1.)

You can begin the inventory by writing down the sources and local practices with which you and your colleagues are familiar. Interviewing long-time residents will help locate former dumpsites, industries, and earlier spills that most people have forgotten. The local agencies suggested as resources for mapping information will also help with source information. Later chapters of this manual explain in great detail the types of source-specific information that must by law be made available to you, how you can get it, and how you can best use it. Taking this information, place the locations of potential contamination sources on your map or on the final transparency in order to give lawmakers, agency decision-makers, and the rest of the community a vivid demonstration of your group's concerns.

SEEKING THE HELP OF EXPERTS

In the process of compiling your own maps you may become astonished with how little is actually known about local hydrogeology. You can use the maps you have compiled as aids in an effort to gain more detailed professionally prepared maps of your community's groundwater resources.

Experts may be able to use the data gathered by your mapping efforts and supplement it with tests or observation borings and wells, and geophysical tests. The professionally compiled map can indicate the watershed boundaries within which your supplying aquifers are located, the aquifers' boundaries, recharge areas, and direction of groundwater flow. More detailed studies can be conducted to map the cone of depression and upgradient areas of well recharge for each existing and potential community-supply well. Mapping these areas is particularly important due to their extreme vulnerability to contamination.

Your group probably will not be able to afford to hire an expert to prepare complete groundwater maps and it would not necessarily be in the best interests of the community even if you could. More effective protection efforts may be generated if groundwater mapping involves local and regional government funds and staff. Try setting up meetings with experts from the local USGS Water Resources Division office, state geological survey office, regional planning commission, local planning board, and college or university geology/hydrogeology departments. You may find that governmental experts have simply been waiting for some indicator of public interest before spending public funds to piece a mapping effort together. Also try contacting the larger environmental organizations working in your state. Several have raised funds and established contacts earmarked specifically for community groundwater mapping.

Many states have laws requiring local governments to incorporate protection of aquifers into their comprehensive plans and zoning regulations. Such laws may help you persuade your community to actively support groundwater mapping efforts. You can point out that, to comply with the state law, the community will have to invest some time and money in producing the maps to serve as the basis for defensible zoning and master-plan protective measures. Several states have grant programs that fund or offer professional staff time for local mapping projects.

New programs in the state of New Hampshire demonstrate a nationwide trend recognizing the importance of aquifer mapping. Action on groundwater mapping has taken place on several levels, with each effort complementing the other. Two integral parts of the state's five-part protection strategy are assessing the state's groundwater resources and identifying contamination threats. The state recognizes that the key to protecting any resource is in-depth knowledge of the resource itself. To learn about its groundwater resources, the state has launched a cooperative effort with USGS to develop maps of the state's sand and gravel aquifers. These maps will depict aquifer size and shape, water storage capacity, directional flows, recharge zones, and groundwater quality. The state is also compiling all well-driller reports required since 1984, sending field personnel to new well sites so that they may accurately map coordinates for each new well site, and requiring users of more than 20,000 gallons of groundwater per day to report withdrawals and discharges in order to better measure impacts on the resource. All of this information is plugged into GRANIT, a gigantic computerized resource data base sponsored cooperatively by the University of New Hampshire and the Office of State Planning.

New Hampshire is also mandating that regional planning commissions lend their expertise to local communities in order to develop better protection for water resources in their master plans. Unfortunately, inadequate funding for the program means that much of the work must come from local and regional initiatives. The town of Nashua already has excellent maps of its water supply because its mapping efforts began in 1983. Concern over two local Superfund sites, rapid population growth, and rising incidences of contamination led the Director of the Nashua Regional Planning Commission to begin rallying for support of mapping projects. His efforts paid off, as the funding for mapping rolled in from private enterprise, towns, the state, and the federal government. A year later 24 concerned experts from private and public entities joined to form the Water Resources Action Program (WRAP), whose mission was to help other New Hampshire towns organize to address their water supply problems. WRAP's volunteer professionals spent time sharing their expertise with, and designing hands-on learning experiences for, committees established by the concerned citizens in each participating community.

The groundwater mapping ideas introduced in this chapter can be an important part of effective citizen action to combat groundwater contamination. Beginning with the next section of the manual, the discussion changes from providing basic information about groundwater and its contamination to suggesting specific steps for action. First is some general advice about how to use agency procedures, and some basic information about court actions. Next is a specific and detailed discussion of the opportunities for citizen action against specific contamination sources using federal regulatory programs. The information and suggestions, including the additional reference materials, in these first five chapters should give you a strong foundation to begin taking action.

FIGURE 5.2

PARAMETERS FOR D.R.A.S.T.I.C. SYSTEM OF AQUIFER ASSESSMENT

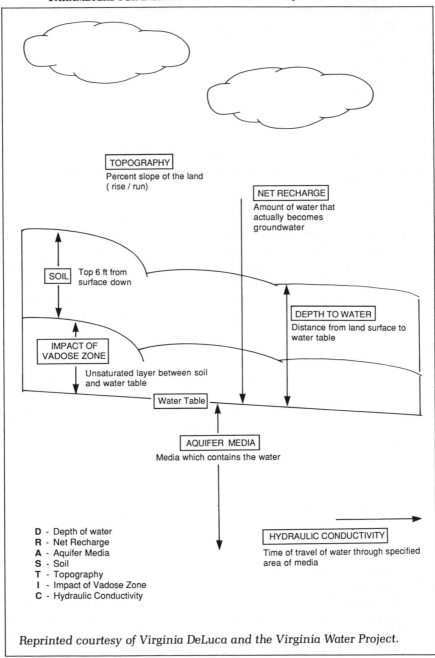

TOPOGRAPHY
Percent slope of the land
(rise / run)

NET RECHARGE
Amount of water that
actually becomes
groundwater

SOIL Top 6 ft from
surface down

DEPTH TO WATER
Distance from land surface to
water table

IMPACT OF
VADOSE ZONE
Unsaturated layer between soil
and water table

Water Table

AQUIFER MEDIA
Media which contains the water

D - Depth of water
R - Net Recharge
A - Aquifer Media
S - Soil
T - Topography
I - Impact of Vadose Zone
C - Hydraulic Conductivity

HYDRAULIC CONDUCTIVITY
Time of travel of water through specified
area of media

Reprinted courtesy of Virginia DeLuca and the Virginia Water Project.

D.R.A.S.T.I.C.

Under the sponsorship of EPA, the National Water Well Association has developed a system that uses existing data to evaluate systematically the groundwater pollution potential of specific geographic areas. The system is called D.R.A.S.T.I.C.; each letter of this acronym represents a "clue," or piece of information, relevant to groundwater vulnerability. The D.R.A.S.T.I.C. system combines all seven of these clues to produce a single numerical rating of an area's vulnerability.

For more information on D.R.A.S.T.I.C., contact the National Well Water Association at 6375 Riverside Drive, Dublin, Ohio 43017; telephone 614-761-1711.

REFERENCES

MAPPING AQUIFERS AND CONTAMINATION

Hickman, Don. *Draft Paper on Rural Community Assistance Program Work on Groundwater Reconnaisance Maps.* Winchendon, Mass. 1988.

Housatonic Valley Association, Inc. *A Final Report on the Activities of the Groundwater Action Project (GAP).* Cornwall Bridge, Conn. February, 1988.

National Water Well Association. *D.R.A.S.T.I.C.: A Standardized System for Evaluating Groundwater Pollution Potential Using Hydrogeologic Settings.* Dublin, Ohio. 1987.

O'Donnell, Arleen. *Mapping Aquifers and Recharge Areas.* Massachusetts Audubon Society. Lincoln, Mass. 1985.

Raymond, Lyle S., Jr. *Chemical Hazards in Our Groundwater: Options for Community Action.* Center for Environmental Research, Cornell University. Ithaca, N.Y. 1986.

Toppin, Kenneth W. *Hydrogeology of Stratified-Drift Aquifers and Water Quality in the Nashua Regional Planning Commission Area, South-Central New Hampshire.* U.S. Geological Survey. Bow, N.H. 1987.

PART II

CITIZEN ACTION

Being an effective citizen activist means being active in many different ways: active before regulatory agencies, active before legislatures and city councils, active in the courts and active with your neighbors. *The Poisoned Well* focuses on an important part of the citizen's overall strategy: working with administrative agencies and the courts to carry out and enforce existing laws and regulations designed to safeguard groundwater quality. Part II of this manual briefly outlines some basic procedures and principles for dealing with agencies and courts. These guidelines provide the framework for most of the specific action suggestions in Parts III and IV.

This section concludes with a short discussion of political and other tools that citizens and citizens' groups can, and generally must, use to be successful. For example, most of the suggestions for citizen action related in the following chapters would be difficult for one or two people to pursue alone without the organized support of a local citizens' group. Even though they are not covered in any detail in this manual, political and other tools must be used simultaneously with the primarily legal options that occupy the center stage of this manual. After all, agencies can ignore citizens or make arbitrary decisions, and courts can decide the law does not give you what you want; then it is up to you and your neighbors, using your combined strength and creativity, to get action.

CHAPTER 6

Freedom of Information Acts

EFFECTIVE ACTION REQUIRES information. To take action on groundwater, you need to know things like the possible sources of the contamination, the hydrogeology of the area, and the effects on the community and the environment. In many cases this information is in the hands of government agencies, and a simple request will give you access to it. Often, however, agencies will be reluctant to release important data, and you will need to use the laws that guarantee your right to public information.

The most important of these laws is the Freedom of Information Act (FOIA). With a few limited exceptions, FOIA gives you the right to request and receive any document, file, or other record in possession of any federal agency. Although FOIA does not apply to state governments, all states have similar laws governing disclosure of records held by state and local government agencies. Monitoring reports, groundwater analyses, toxicity data, and health assessments are only a sampling of the information available through FOIA.

HOW TO MAKE A REQUEST

Below is a sample FOIA request letter that you can use as a model for making your own request. Requests under state law will be similar but check with the state agency about exact procedures.

Exemptions

A FOIA request may be denied only if the requested information is specifically exempted from disclosure by the statute, and even then the agency may withhold only the exempt portions of the documents. The agency is required to excise the exempt material and release the rest of the document to you. Those exemptions that may be relevant to groundwater issues include:

FOIA Sample Request Letter

Freedom of Information Act Officer
Agency Name and Address

[Each federal agency has a FOIA officer to help you with your request. You can contact the FOIA officer directly before filing a request to get an idea of the documents that are available at the agency.]

Re: Freedom of Information Act Request

[Also write "FOIA Request Enclosed" on the outside of the envelope to speed up processing.]

Dear Sir or Madam:

This is a request for records under the Freedom of Information Act. I request a copy of [or access to] all records pertaining to groundwater quality monitoring at the Acme hazardous waste facility.

[You must reasonably describe the records you seek. A request for "all documents relating to the Acme hazardous waste facility" may be returned to you for more information. Even if answered, an overly general request may generate thousands of unhelpful pages. Be as specific as possible. You can always make a second request if you need additional documents.]

This request is made on behalf of Citizens Against Waste, a nonprofit group organized to protect groundwater quality and ensure responsible waste management in Noname County. This information will be used to evaluate facility compliance with regulatory requirements and to assess its impact on the local community. Accordingly, I request a waiver of fees for this request.

[Agencies may charge an hourly service fee plus the cost of copying the documents requested if you do not ask for a fee waiver. The fee could add up to hundreds or even thousands of dollars. The fee should be waived if your request is in the public interest, particularly if you demonstrate that the disclosure of the information will help educate or otherwise benefit the general public. If you are making the request on behalf of a citizens' group or other nonprofit organization, make this point clear as a basis for your waiver request. Fee waivers should be liberally granted.]

If a waiver is not granted [and the estimated expenses are greater than $_____], please obtain my approval before proceeding.

If this request is denied in full or in part, please identify the documents withheld and the basis for the denial.

[The agency can withhold material only if it fits within one of the Act's limited exemptions, discussed below.]

I look forward to a response within ten working days, as provided by law.

[Responses often take much longer than this, but the agency must notify you if it cannot meet this regulatory deadline. You can appeal or go to court if the agency delays its response.]

Thank you for your prompt attention to this request.

- Classified materials relating to national defense or foreign policy (possibly applicable to a federal facility);
- Material specifically exempted by the statute governing the material (such as certain toxicity data under the Toxic Substances Control Act);
- Trade secrets and commercial or financial information claimed to be privileged or confidential (such as chemical identity information);
- Materials disclosing the agency's deliberative process (portions of internal predecisional memos recommending that certain action be taken by the agency, for example);
- Certain investigatory records compiled for law enforcement purposes.

APPEAL PROCEDURES

If the agency denies your request, you can appeal the decision by sending a letter to the agency (the proper office should be indicated in the agency's letter of denial). The appeal letter should describe the original request and the denial (you can attach copies of each) and explain why you think the denial was unwarranted. You can also appeal if the agency delays its response beyond ten working days or refuses to grant a fee waiver.

EPA requires you to mail your appeal within 30 days of the date when you received the agency's initial determination. (This is true for most federal agencies, but check with each to verify the time frame.) You should receive a response within 20 working days.

Do not hesitate to use these appeal rights, particularly if a request is denied because documents contain allegedly confidential commercial information or would disclose agency deliberative processes. These exemptions should be narrowly applied, but they are easily abused. The commercial or trade secret exemption generally applies only to information that if disclosed would harm the competitive position of the company. (There are ways to obtain this information in some cases under other laws. See Chapter 17.) The deliberative process exemption applies only to agency policy and analysis, not to basic factual material.

FOIA establishes a streamlined procedure for court challenges when an agency refuses to release information. If your appeal is denied, you may want to consult an attorney about filing a lawsuit.

REFERENCES

FREEDOM OF INFORMATION ACTS

American Civil Liberties Union (ACLU). *Litigation Under the Federal Freedom of Information Act*. Washington, D.C. 1988. (Updated annually.) To order, contact the ACLU, 122 Maryland Ave., N.E., Washington, D.C. 20002. (202) 544-1681.

(The ACLU also publishes *Your Right to Government Information*, available at the same address.)

Freedom of Information Act Clearinghouse. *The Freedom of Information Act: A User's Guide*. Washington, D.C. June 1987. To order, contact the Freedom of Information Clearinghouse, P.O. Box 19367, Washington, D.C. 20036. (202) 785-3704. The Clearinghouse is also an excellent source of information for FOIA-related questions and information, and provides legal representation for FOIA challenges.

The Reporters' Committee for Freedom of the Press. *Confidential Sources and Information*. Washington, D.C. 1985. To order, contact the Reporters' Committee for Freedom of the Press, 800 18th Street, N.W, Suite 300, Washington, D.C., 20006. (202) 466-6313.

The Reporters' Committee for Freedom of the Press. *How to Use the Federal FOI Act*. Washington, D.C. 1985. To order, contact the Reporters' Committee (address and telephone number listed above.)

Washington Researchers. *How to Find Information About Companies*. Washington, D.C. 1979. To order, contact Washington Researchers, 918 16th Street, N.W., Washington, D.C. 20006. (202) 828-4800.

LAWS AND REGULATIONS

U.S. Code Citation

5 U.S.C.A. § 552 (FOIA).

Code of Federal Regulations Citation

40 C.F.R Part 2 (EPA FOIA regulations).

CHAPTER 7

Action in the Administrative Process

GROUNDWATER PROTECTION LAWS are administered by state and federal agencies (at the federal level usually the Environmental Protection Agency [EPA]). Agencies decide how to translate the general requirements of laws into specific, enforceable, obligations for particular industries and other potential polluters. Among other things, agencies promulgate general regulations, issue specific permits, and enforce both. This chapter outlines basic principles for influencing how agencies carry out their responsibilities. The advice here is generally applicable to EPA and other federal agencies involved in groundwater protection and to many state agencies. You will need to verify the specific procedures followed by state agencies, particularly for unique programs not required by federal law.

GENERAL ADVICE

Most of the specific citizen actions discussed later in this manual involve working with administrative agencies on the federal or state level. The following general pointers apply across the board:

- At least at the outset, do not treat the agency as your enemy. You may have to become more aggressive later, but begin by trying to create an ally in your struggle.
- Be clear about what you want from the agency. Try to ask for things you know it can or should do.
- Identify individuals within the agency with responsibility for the specific problems you face. Direct your requests to specific people, whenever possible, and not to offices.
- Work at different levels of the agency. Even if decisionmakers seem not to listen, you can often find lower-level staff people sympathetic to your concerns. They can be helpful sources of information. Exert pressure, political and otherwise, on higher-level decisionmakers who have authority to take action once you have developed support within the agency.

- Strive to be precise and concise. Support your requests with documentation and rational argument. Your job is to convince the agency personnel to treat your problem as their problem. You have to convince them that you are right and that it is important to take action. Evidence is more likely to be successful than emotion.
- Use political tools. Pressure from local congressmen or organized community protests may make the agency sit up and take notice.

PARTICIPATION IN THE RULEMAKING PROCESS

One way to influence agency action on a general level is to participate in the rulemaking process. This is the process agencies use to establish specific rules, generally referred to as regulations, which define and elaborate on the general requirements set out in laws. These regulations are as important as the law itself in determining the effectiveness of a program in protecting groundwater. Your participation can help ensure that regulations impose requirements that are strict enough. As important as regulations are, they generally address national or statewide problems. Unless the particular problem you face is shared by many others, it will be difficult to convince the agency to use the rulemaking process to solve your local problem. This strategy is more likely to be successful on the state and local level than with EPA and other federal agencies where the influence of local community organizations or individuals will be diluted substantially.

Commenting on Proposed Regulations

As a general matter, the process begins with public notice that the agency has prepared a draft or proposed regulation for public comment. For federal agencies, the notice is published in the *Federal Register*, the official daily publication of federal agency action, available at most public libraries. (See Chapter 10 for more about the *Federal Register.*) Some states have similar publications or use local media. You should also ask relevant agencies to put you on their mailing list for direct notice of upcoming groundwater-related rulemaking.

The agency will establish a specific time limit (usually at least 30 days) for the public to submit written comments on its proposed action. In many cases the agency also will hold a public hearing. (See page 93 for tips on giving effective oral comments.) To maximize the impact of your comments, consider the following suggestions:

- Follow any agency suggestions for organizing or presenting your comments.

- Focus comments on the issues or problems raised by the proposed regulation. You may want to suggest the agency expand the scope of its regulation but try not to bring in unrelated problems. Where you think the agency is wrong, offer suggestions for solutions or even for specific revisions of the proposed regulations.
- Use your own experience. Comments that explain how a regulation fails to address a particular problem in your community may be more persuasive than general statements. Use specific case studies or examples to illustrate your concerns.
- Work with organized community groups. Coordinate with regional, statewide, or even national groups working on the same issues to present the agency with a widely supported position.
- Seek the assistance of experts (hydrogeologists, engineers, professors, chemists, or other groundwater professionals) in developing your position and encourage them to comment on the regulation on your behalf.
- Where you can, use legal arguments to bolster your view. Agencies will be particularly sensitive to arguments that a proposed regulation is inconsistent with the statute. You may want to ask a lawyer to help focus legal arguments.

Negotiated Rulemaking

In some cases, agencies may initiate a negotiated rulemaking process, and meet with representatives from government, industry, environmental, and other groups to reach a compromise before circulating a draft regulation for public comment. You can encourage the agency to ensure that groups representing your groundwater concerns are included in such discussions.

Initiating the Rulemaking Process

You do not have to wait for the agency to decide to change or add to its regulations. You can initiate the rulemaking process to seek either a new regulation or changes in an existing one. Most agencies require that you submit a written request (often called a petition) that describes the regulatory changes or additions you propose and explains the need for them. If the agency is willing to consider the changes, it will then begin the rulemaking process described above.

The petition process is likely to be a difficult option for local citizen groups to use effectively, especially on the national level. You will need to develop compelling legal and policy requirements and probably expert evidence in support of your position, along with strong political support. As part of an overall strategy to deal with groundwater problems in your area, however, the process may be worth pursuing, particularly at the state level.

Judicial Review

In most cases, if the agency issues a final rule that you believe fails to
protect groundwater, you can file a lawsuit to challenge the decision in
court. See pages 98–99 for a full discussion of judicial review.

USING THE PERMIT PROCESS

Many laws establish a permit system to control and regulate potential
sources of groundwater pollution. Under federal law, facilities that dispose of hazardous waste, for example, must have a permit from EPA (or a
state agency) to operate lawfully. Permits usually include specific conditions on the design and operation of the facility to prevent contamination
problems. Using opportunities to influence agency decisions to issue
permits and agency decisions about the conditions of those permits is one
of the most important methods for citizen action against groundwater
contamination. The content of citizen input will depend on the problem
at hand, but the process for these agency decisions is usually much the
same. This section outlines suggestions for participating in the process
that applies to several important groundwater-related permits. See Figure 7.1 for a graphic summary of EPA's permit and appeal process.

Important Early Steps in the Process

As a general rule, the permit process begins with an application from the
facility owner to the regulatory agency. The basic steps that follow are the
preparation of a draft permit, public comment, and final agency decision
on the permit. To be most effective in the process, you should not wait
until the formal public comment period to begin working with the
agency. By the time a draft permit is written by the agency, many preliminary decisions will have been made that could be difficult to reverse.

By looking for clues in local news stories and other community
sources, you may be able to identify planned new facilities with potential
impact on groundwater quality even before an application for a permit is
submitted. Another approach is to develop a relationship with staff at the
agency who can notify you when an applicant first contacts the agency.
You can also request the agency to notify you that a permit application has
been received (most agencies have a mailing list for this purpose).

As soon as you learn of a possible new facility or permit, begin gathering information about the facility and its site (suggestions for specific
information sources are contained in Part III), identify the responsible
staff within the agency, consult with knowledgeable experts, and begin to

make your views known to the agency in informal meetings and letters. This kind of early involvement is particularly important if you flatly oppose the granting of a permit. The agency will often make a preliminary decision to issue or deny a permit before it seeks any formal public comment.

Commenting on the Draft Permit

Once it makes a preliminary decision to issue a permit, the agency will prepare a draft permit with proposed conditions and notify the public that comments on the draft permit will be accepted. The length of the public comment period will vary, depending on the type of permit, but will generally be at least 30 days (federal law requires 45 days for hazardous waste facility permits and provides for opportunity to request more time if necessary). Make sure you are on the agency's mailing list so you will receive direct mail notice that a draft permit is available for comment.

The next step is to review the draft permit and, if you have not already done so, gather background information to evaluate the permit and its conditions. The application will contain pertinent information about the facility and, sometimes, information about its expected environmental effects, including the impact on groundwater. In most cases, the agency should also prepare a brief statement (called a fact sheet or statement of basis) explaining the major issues involved in the draft permit. For hazardous waste facility permits, EPA generally establishes an information repository at a public place that includes all relevant documents and supporting materials for you to review. For all EPA-issued permits, the agency must assemble an "administrative record" consisting of the documents on which the decision to prepare a draft permit was based, and must make this record available for public review.

The following chapters of this manual provide specific suggestions concerning issues to consider carefully in preparing comments on permits for the major types of facilities or contamination sources. Some basic points to remember for all permits are:

- Organize your comments so the agency can easily understand what you think is wrong (or right) with the permit.
- Suggest specific changes in the permit where you can.
- Support your comments with your own relevant experience (particularly if you know about groundwater problems in the area).
- Use experts to help you evaluate technical aspects of the permit and provide support in the form of their own comments or a statement you can attach to yours.
- Coordinate with other local groups affected by the permit.

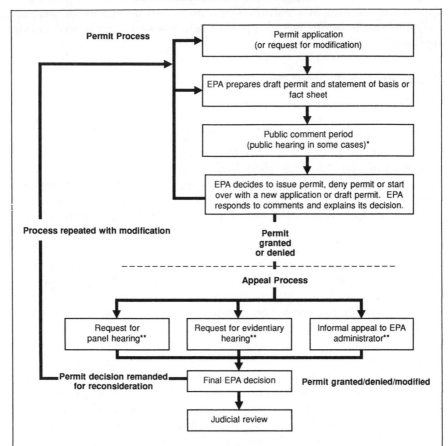

FIGURE 7.1
EPA PERMIT AND APPEAL PROCESS

* Requests for public hearings should be made at the beginning of the comment period. EPA will determine whether to hold a public hearing based on the amount of interest or controversy and the types of issues involved. For some permits, EPA may hold a more formal type of hearing called a panel hearing. In this process, a panel of agency staff not only hears statements, but can ask questions of witnesses (usually technical witnesses or experts).

** Depending on the type of permit and issues involved, appeals can take the form of an informal written process, a panel hearing process (see Note *), or a formal evidentiary hearing process before an administrative law judge (much like the trial-type procedure followed in court). Though a lawyer's assistance could be helpful in any of the three procedures, you will likely need a lawyer to participate effectively in the evidentiary hearing process.

- Where you can, respond to the arguments of your opposition.
- Be thorough—your ability to challenge the permit later may be limited to those arguments made early in the process.

You should also request the agency to hold a public hearing on the draft permit. Though not required to do so in most cases, the agency will generally hold a hearing if significant interest is generated by the permit. The agency should provide at least 30 days notice to the public prior to the hearing. You should urge the agency to hold the hearing in a location convenient to concerned citizens (not 150 miles away at the state capitol). Contact the agency before the hearing to get an idea of how it will be conducted and what time limits will be imposed on speakers. You may have to provide advance notice to the agency that you intend to speak at the hearing to ensure that you are given a time spot.

Public hearings will generally not be formal proceedings. An agency official will be present to receive comments, oral and written, and a record may be made of the presentations. There may be a brief question and answer session, but it will probably not be the best forum in which to obtain answers to anything more than simple questions. You do not usually need a lawyer—there will be no cross-examination of witnesses or other judicial procedures. (There are exceptions to this rule. For certain permits issued by EPA under the Clean Water Act and related permits, EPA may hold special, more formal hearings that involve trial-type procedures and where you may want the assistance of a lawyer.) You can solicit the help of experts for public hearings, either to help prepare your comments or to present separate testimony on more technical points.

At the hearing, try to keep your oral comments simple. Make a few basic points and leave the details to written summaries. Testify from your knowledge and experience, and get experts to testify on points that are technical. Use visual aids (photographs or maps, for example) to make your statement more memorable. If you feel comfortable speaking in public, it will usually be more effective to speak from an outline than to read a prepared statement. The hearing is your opportunity to show the agency the breadth of support in the community for your views. Persuade others who share your concerns to come, even if they do not actually speak.

Reviewing the Agency Decision

After receiving public comment, the agency will decide whether to issue the permit as proposed, issue a modified permit, start over with another draft, or maybe even to deny the permit. Look for public notice of the decision (you should get personal notice if you filed comments) and request a copy of the agency's response to public comment. The response should address all significant comments, and explain why the agency made its decision and any modifications in the permit. For EPA-issued permits, this response, together with a record of the hearing and copies of public comments, becomes part of the formal administrative record supporting the agency decision.

Challenging a Permit Decision

In most cases, you will have a right to challenge permit decisions you believe to be inadequate through an appeal to a higher level within the agency. For permits issued by state agencies, your rights are determined by state law. Most states provide for at least one layer of administrative appeal. Check with the regulatory agency to learn about your rights and the specific steps to follow.

You can initiate an appeal against permits issued by EPA by requesting the Administrator of EPA to review the permit decision within 30 days of the final decision to issue a permit. Your request, which is called a "petition for review," must explain why review is needed. Generally, you will be required to demonstrate that you brought the issues you are appealing to the attention of the agency during the comment period. If you did not participate or comment during that period, you will generally be allowed to appeal only those changes made between the draft and final permit. Your appeal should be designed to convince the Administrator that the permit is unlawful (it was issued without required conditions or without following required procedures, for example), that the permit is based on a mistaken factual premise (the agency did not accurately assess the vulnerability of groundwater in the region, for example) or that lower-level officials made an error in judgment or policy in issuing the permit (the most difficult sort of appeal to win).

If your petition is granted, the agency will set a schedule for submission of briefs, which consist of written argument and supporting material, or it may decide to hold a special hearing to develop factual disputes more fully. Depending on the complexity of the issues you raise and the type of procedure EPA follows, you may want the assistance of a lawyer. (For certain permits issued under the Clean Water Act and related permits, EPA follows special appeal procedures that are like court trials. You generally will need a lawyer to help conduct such an appeal.)

If your appeal is successful, EPA will either make changes in the permit or start the process over again with a new draft. On the other hand, EPA may reject your appeal and then your only legal remedy is to go to court. Your rights to judicial review of agency decisions are described beginning on p. 98.

Petitioning for Modifications or Termination of the Permit

Once a permit is issued and the facility is operating, situations may arise that require the permit to be changed or even terminated. In the case of most permits issued under the federal environmental laws, you can petition the agency to modify the permit (or revoke and reissue the permit if many significant changes are necessary) if you can show, among other things, new information about potential environmental problems or changes in the facility or its operation. If the facility consistently violates its permit's conditions or endangers health or the environment (including groundwater), you can request that the permit be terminated. If the agency agrees that there is a problem, it will restart the permit process to consider the suggested modifications or termination. You have the same rights to participate as you would if a new permit were being issued.

MAKING THE AGENCY ENFORCE THE LAW

What do you do when a facility is not complying with the law—when a hazardous waste landfill is violating the groundwater monitoring condition of its permit, for example, or an injection well is being used for types of waste not permitted by regulation—and groundwater quality is threatened as a result? Your first step should be to get the regulatory agency to take enforcement action. Whether enforcing general regulations or the terms of a specific permit, the agency will have authority to require the violator to correct the problem, either by issuing an agency order or by taking the violator to court. Later chapters of the manual describe specific enforcement procedures that apply in particular circumstances, but this section outlines general principles for citizen participation that will apply in most situations.

Initiating Enforcement Action

To bring a possible violation or other problem affecting groundwater to the attention of the agency, the best approach is to submit a written request for action (sometimes called a petition). Write a letter to the appropriate agency (depending on the problem, it could be a federal or state agency, or both—see Chapter 10) and include the following information:

> Your name and address and the name of any organization you represent
> Description of the problem:
> —type of facility
> —nature of the problem or violation

State-EPA Enforcement Agreements

Several federal environmental laws allow states, rather than the EPA, to take
primary responsibility for implementing the regulatory program (issue permits, conduct inspections, enforce
restrictions, and the like). But even in
states with approval to run the federal
program, the EPA generally retains
some oversight and enforcement authority. To coordinate enforcement in
these states, EPA and the state will have
entered into an agreement that spells
out the respective enforcement roles of
the agencies. The agreement also will
require regular reports from the state to
the EPA describing state enforcement
actions and listing detailed information about facilities known to be in violation of requirements. You can request
copies of these reports and review enforcement agreements at EPA or state
agency offices. They can be helpful
in citizen efforts to oversee agency enforcement.

  —effects on groundwater
  —potential health effects, etc.
A request for specific action from the agency

The action that you request will depend on the problem. Initially you
may want the agency to conduct an inspection to verify a suspected
problem. If you already have information indicating a violation of legal
requirements or other threats to groundwater quality, you will ask that the
agency issue an enforcement order or at least begin enforcement proceedings. The types of orders the agency can issue will depend on the situation and the statute involved. In an emergency you need to stress the need
for immediate action by the agency.

As a general rule, the more specific information you have to support
your request, the more likely the agency will be to respond positively.
Attach copies of inspection reports, monitoring records, photographs,
statements from nearby residents, and any other documents that verify
your concern. You need to give the agency a reason to divert its focus from
an already overloaded schedule to the problem you have discovered.
Violations of regulations or permit requirements that are likely to lead to,
or prevent the discovery of, contamination and health problems will
receive more attention than other problems at the facility.

In addition to sending in a written request, you should try to work
directly with agency staff. Organize informal meetings to educate them
about the problem. You should also make informal contact with staff at
the agency before sending in a written request for action to make sure it is
directed to the right office and person, and that it contains all the information they will need to see.

In most cases, a decision about enforcement action is within the discretion of the agency; that is, you cannot use the courts to force the agency to

act (there are exceptions to this rule, but you will need to talk about them with a lawyer.) You can, however, exert political pressure on the agency and continue to develop and submit your own evidence until the agency is persuaded to act. In some situations, you may be able to take a polluter to court yourself if the agency refuses to act. (See Chapter 8.)

Participating in Agency Enforcement Actions

If the agency decides to initiate action against the facility to correct the violation or other problem, you will in most cases have a chance to become involved. If a state agency is responsible for enforcement, your rights to participate will be determined by state law. This section summarizes citizen opportunities for EPA and other federal agency enforcement actions, but many state procedures will be similar.

Except for some situations where the agency may simply issue a warning letter to the violator, enforcement orders will be the main tool the agency will use to get action. Orders can require payment of penalties, put the violator on a deadline for compliance or revoke the facility's permit. As is true of court orders, failure to comply with an administrative order can result in penalties, even criminal sanctions, against the violator.

Some orders (notably, emergency action or "imminent hazard" orders, see Chapter 14) can be issued without formal procedures. Make sure the agency knows your views about what action is necessary and participate in any public hearings the agency may hold on the order. But in many cases, before it issues an order, the agency will go through a formal process that includes a trial-type hearing. The best way to influence the outcome of formal enforcement proceedings is to intervene, that is, become one of the formal parties in the process. As a party to the hearing, you can present evidence, make legal argument, and cross-examine witnesses. To intervene you must show that you will be harmed by the order and that none of the parties, including the government, adequately represents your interests. You will probably need a lawyer to help you intervene and to represent you in the process.

Alternatively, you may wish to consider participating in the hearing as an *amicus curiae*, or friend of the court. As an *amicus* you can file briefs setting forth your views, but you do not have the status of a full party (most notably, you do not have the right to appeal adverse decisions to a higher authority within the agency). As in the case of intervention, you must request the agency to allow you to participate as an *amicus*, and the agency is under no obligation to grant your request. Nevertheless, it is generally easier to obtain *amicus* status because you need not demonstrate harm and inadequacy of representation as an intervenor must.

At the conclusion of the formal hearing, the agency will decide whether to issue an order and what the order should include. If you have intervened, you can appeal an enforcement decision you think is too weak to a higher level in the agency (for EPA decisions, the appeal is to the EPA Administrator in Washington, D.C.) and so, of course, can the violator if it believes the decision is too strong. In the appeal, the parties can present their final arguments to the agency, usually in written form. The decision on the appeal is the final decision of the agency. At this point any one of the parties (including citizen intervenors who believe an order does not adequately address the problem) can file a lawsuit to challenge the decision. Turn to the next section for more about judicial review.

This whole process may be cut short if there is a settlement between the agency and the violator. A settlement agreement may provide for incomplete or less effective measures to correct the pollution problem at the facility. If you have intervened in the process, you may be able to influence the terms of a settlement agreement. Otherwise, you should attempt to exert political pressure on the agency not to settle for less than the full enforcement action you believe is necessary.

Agencies can also decide to skip this entire administrative process and enforce permit and regulatory restrictions by taking the violator directly to court. The most effective way to influence the outcome is to intervene in the court enforcement action to ensure your interests are protected. You will need a lawyer to represent you in such an effort.

As with administrative hearings, once you become party to a court case, you will be able to present evidence, cross-examine witnesses, and make legal arguments. You will also be in a better position to influence any settlement discussions. To be entitled to intervene in most court enforcement actions brought by EPA, you have to show that you will be affected by the outcome of the case. For some enforcement actions, you may also be required to demonstrate that you are not adequately represented by the parties in the case. (You may be able to intervene without meeting these tests, but the court may then restrict your rights to participate.)

As an alternative to intervening in a lawsuit, you can ask the court to allow you to participate as an *amicus curiae*.

REVIEW OF AGENCY ACTION BY THE COURTS

Most final agency action—whether it be a new regulation, a permit, or an enforcement order—can be challenged in court. This process is called judicial review, and can be initiated by those who are affected by the agency action. The reviewing court will determine whether the challenged action is consistent with the law and meets a minimum standard of reasonableness.

Judicial review of agency action is generally available to citizens as well as to those who are actually subject to the regulation, permit, or order. In some circumstances, however, courts may allow citizens to bring a review action only if they have participated in the agency proceedings leading up to the action (by commenting on permit or rule decisions or intervening in agency enforcement proceedings) and if they have exhausted any rights to appeal within the agency. In addition, most federal environmental statutes set out restrictions on this right to judicial review by, for instance, specifying which court the action must be brought in or requiring that a lawsuit be filed within a limited time after the agency action, such as 45 days. Procedures for judicial review vary among regulatory programs and are not the same in state and federal court, so you should seek the advice of a lawyer before you consider this step.

These are several problems with judicial review that make it a difficult remedy. First, courts generally apply a very deferential standard of review to agency decisions, that is, they will not overturn an agency decision unless the agency was arbitrary, failed to consider relevant information, or was otherwise clearly wrong. It will not be enough to simply demonstrate that there is a conflict among the experts or even that another resolution of the problem would be better, particularly when the case raises complicated or technical factual questions of the sort that are likely to arise in groundwater-related cases. Also in most cases a reviewing court will consider only the evidence that was before the agency when it made its decision. The result is that it is very difficult to get a court to reverse agency action. You are most likely to be successful if you raise claims that the agency failed to follow the proper procedure or did something inconsistent with the statute. Like any other court action, judicial review is time-consuming and expensive and should only be entered into after careful consideration.

REFERENCES

ACTION IN THE ADMINISTRATIVE PROCESS

Rulemaking

Cohen, Morris L., and Robert C. Berring. *How to Find the Law.* West Publishing Co. St. Paul, Minn. 1983.

East Michigan Environmental Action Council. *Community Care: A Guide to Local Environmental Action.* Birmingham, Mich. 1982. To order, write East Michigan Environmental Action Council, 21220 W. Fourteen Mile Road, Birmingham, Mich. 48010. (313) 258-5188.

Epstein, Samual S., Lester O. Brown, and Carl Pope. *Hazardous Waste in America*. Sierra Club Books. San Francisco, Calif. 1982.

Harwood, Judge Gerald. *Hearings Before an EPA Administrative Law Judge*, 17 Envtl. L. Rep. (Envtl. L. Inst.) 10441 (November 1987).

National Wildlife Federation. *The Toxic Substances Dilemma: A Plan for Citizen Action*. Washington, D.C. 1980.

Office of the Federal Register. *The Federal Register: What It Is and How to Use It*. National Archives and Records Administration. Washington, D.C. June 1985. To order, call the Office of the Federal Register at (202) 523-5240, or write the Superintendent of Documents, U.S. Government Printing Office, Washington, D.C. 20402. This is an excellent guide to the administrative process.

Willett, Edward F., Jr. *How Our Laws Are Made*. U.S. Government Printing Office. Washington, D.C. 1981. To order, contact the Superintendent of Documents, U.S. Government Printing Office, Washington, DC 20402.

Permitting and Enforcement References

Belfiglio, Jeff, Thomas Lippe, and Steve Franklin. *Hazardous Waste Disposal Sites: A Handbook for Public Input and Review*. Stanford Environmental Law Society. Palo Alto, Calif. 1981.

Gordon, Wendy. *A Citizen's Handbook on Groundwater Protection*. Natural Resources Defense Council. New York, N.Y. 1984.

Izaak Walton League of America. *Guide to Conservation Action*. Arlington, Va. 1983. To order, contact the Izaak Walton League of America, 1401 Wilson Blvd., Level B, Arlington, Va. 22209. (703) 528-1818.

Koch, Charles H. *Administrative Law and Practice*. West Publishing Company. St. Paul, Minn. 1985.

Loveland, David Gray, and Diane Greer. *Groundwater: A Citizen's Guide*. League of Women Voters Education Fund. Washington, D.C. 1986. To order, contact the League of Women Voters of the United States, 1730 M Street, N.W., Washington, D.C. 20036. (202) 429-1965.

National Campaign Against Toxic Hazards. *The Citizens' Toxics Protection Manual*. Boston, Mass. 1987.

LAWS AND REGULATIONS

Code of Federal Regulations Citations

40 C.F.R. Part 124 (Permitting process for several types of possible contamination sources).

40 C.F.R. Part 22 (Rules governing suspension or revocation of several types of permits, or assessments of penalties).

40 C.F.R. Part 25 (This section of the CFR outlines the opportunities for public participation that the EPA and states should provide in programs under the Resource Conservation and Recovery Act, the Safe Drinking Water Act, and the Clean Water Act).

CHAPTER 8

Taking the Polluter to Court

WHAT CAN YOU do when you cannot convince an agency to take action against a facility that is endangering groundwater quality? In some circumstances, one option is to sue the facility yourself. This section describes the two basic approaches to using the courts to address groundwater problems—the citizen suit and the common-law action—explaining in general terms when you can and cannot bring these lawsuits and what you may be able to accomplish.

A few general points apply to both types of cases. Lawsuits are often time-consuming and expensive, even if you can find a lawyer who will represent you for little or no fee. Between the filing of the action and a final decision, several years may pass and the cost may grow to tens of thousands of dollars. You should pursue a lawsuit only after considering carefully the advice of a lawyer about the prospects for success and the time and resources required to win.

The need for careful consideration is further underscored by the fact that, if you bring a suit that does not have a good legal or factual basis, the court can award penalties against you for so doing. In addition, private entities (such as companies or developers) that are sued by citizen plaintiffs have sometimes filed countersuits against the citizens seeking penalties for alleged interference with their business or property interests. Even if your suit is sound and well thought out and you ultimately defeat the countersuit, achieving that victory can be an expensive and time-consuming nuisance. If your suit is weak and hastily slapped together, however, you greatly increase the likelihood that such a countersuit will be filed and that, if filed, it will be successful against you.

Lawsuits should generally be used as part of a broader strategy that includes continued work with administrative agencies and political action. Even though they can be difficult, lawsuits can be a very effective remedy for some types of problems and many citizen groups have won major environmental victories in court. This section is intended to give you a basic understanding of some of the important questions that you will need to answer to decide whether a lawsuit is the right option.

101

One of the first important steps is finding the right lawyer. Look for someone with experience in the type of case you need to bring; not all lawyers do trials and very few do environmental cases. Contact state or local bar associations for suggestions, and inquire at nearby law schools. Many of the national and regional organizations listed in Appendix C know of experienced lawyers around the country. Remember to ask the advice of other citizens' groups that have fought similar battles.

Once you have identified a lawyer to talk to, prepare an organized presentation with documentation for your meeting. You need to convince the lawyer that your problem is serious. The more background work you have done and the more information you can provide, the more likely it is that the lawyer's response will be positive. And do not assume that once a lawyer agrees to take your case, your work is done. It will be important for you to continue to play an active role, assisting the attorney throughout the process.

CITIZEN SUITS

Citizen suits are, for the most part, a mechanism to enforce the requirements of environmental laws. Most federal environmental statutes, and a few state laws, include special provisions giving the citizen a right to sue. (See Figure 8.1.) Essentially, the statutes allow the citizen to step in and take action when the administrative agency does not do its job.

When Can You Bring a Citizen Suit?

Under most of the federal environmental statutes, citizens can bring two basic types of cases. In the first and most important category are suits against a person (a company, individual, or even government agency) who is in violation of a requirement imposed under the law (a regulation, a permit condition, or the terms of an enforcement order, for example). These are the kinds of cases that will be most useful to citizens trying to protect groundwater.

Citizen suits of this type could be used, for example, to challenge a violation of pollutant levels in a discharge permit, a failure to monitor groundwater quality around the facility, or the injection of unauthorized waste into a disposal well. Though conceivably a suit could be brought under this authority for any violation of a regulatory requirement, you should focus on significant violations related to groundwater protection. Suits for merely technical violations will rarely be worth your effort and money.

You should also consider the difficulty of proving the violation before

FIGURE 8.1

CITIZEN SUIT PROVISIONS UNDER VARIOUS STATUTES PROTECTING GROUNDWATER

Statute	What it Covers	Who Can Be Sued*
Resource Conservation and Recovery Act RCRA § 7002 (42 USC § 6972)	Hazardous and nonhazardous waste.	Anyone who transports stores, treats, or disposes of waste and violates a permit condition, order, or regulation under the Act. (Past violators are covered as well.)
Comprehensive Environmental Response, Compensation and Liability Act (CERCLA, or Superfund) § 310 (42 USC § 9659)	Hazardous substances or pollutants.	Anyone (including the government) who fails to perform cleanup responsibilities for releases of hazardous substances or pollutants (often from inactive or abandoned waste sites).
Safe Drinking Water Act (SDWA) § 1449 (42 USC § 300j–8)	Public water systems or underground sources of drinking water.	Public water systems suppliers of operators of underground injection wells who fail to maintain drinking-water standards or violate orders or regulations under the Act.
Clean Water Act (CWA) § 505 (33 USC § 1365)	Surface and groundwater.	Anyone who violates permits issued for discharge of pollutants into surface waters or any other order or limitation under the Act.
Surface Mining Control and Reclamation Act (SMCRA) § 520 (30 USC § 1270)	Coal mining activities both during active mining and restoration of the site to its premined condition.	Anyone who violates a permit condition, order, or regulation under the Act, particularly the prohibition against disturbing on- and off-site water quality.
Toxic Substances Control Act (TSCA) § 20 (15 USC § 2619)	Chemicals and other toxic substances.	Anyone who violates a regulation under the Act regarding the safe manufacture, use, disposal, and processing of chemicals, including PCBs.

*All of these citizen suit provisions also allow citizens to sue the EPA Administrator for failure to perform a "nondiscretionary" duty, such as to publish reports or regulations by a mandated deadline.

you decide to pursue an action. In some cases the facility will be required to submit regular reports to the agency that could disclose violations, and often agencies are required to conduct inspections and prepare reports describing violations at a given facility. You can generally obtain these reports from the agency to use as evidence in support of your claim. Other violations may require that you develop your own evidence—a task that can be complicated by lack of access to most facilities.

The second type of citizen suit is an action against the agency for failure to do something specifically required by the law. These suits are most often used to force agencies to issue regulations after a statutory deadline has passed. As a general rule, efforts to use this type of suit to force the agency to take enforcement action against a violator have not been successful. Though there are exceptions, most courts believe that enforcement decisions are best left to the agency.

One federal statute, the Resource Conservation and Recovery Act (RCRA), which regulates waste disposal, provides for a unique form of citizen suit. RCRA allows the citizen to sue, not just for violations of specific requirements of the law, but also whenever the handling of waste presents a significant threat to health or the environment. This type of citizen suit, called an imminent hazard action, is discussed in greater detail in Chapter 14.

What Relief Can You Get?

Once you have proven a violation of the statute or regulations, courts may issue two basic forms of relief. You can seek an injunction, which is an order from the court to the violator requiring that certain action be taken (a mandatory injunction) or that the violator stop certain actions (a prohibitory injunction). Injunctions are flexible remedies, giving the court the ability to shape the order to fit the violation. An injunction could, for example, require a facility to cease operations until it complies with a permit restriction, require that groundwater contamination be cleaned up by the company, or order an agency to promulgate regulations by a specific date.

Some statutes allow you to request the court to impose fines on the violator in addition to ordering compliance. As a general rule, these fines are paid to the government. You cannot generally use citizen suits to recover money as compensation for damage to your health or groundwater; for that kind of relief, see the discussion of common-law actions below. (The one exception for citizen suits is the federal coal mining law, see Chapter 21.)

Though you cannot get payment for contamination, you may be able to recover the costs of bringing the lawsuit, including your attorney's fees.

All the federal citizen-suit provisions give you the right to seek these costs from the violator if you are successful in the litigation. You may be able to use this right to attract a lawyer who otherwise could not take your case.

Citizen suits often result in settlements with the violator. Faced with the prospect of defending against a strong claim, the violator (a facility or an agency) may offer to take some action to correct the problem. Settlement agreements should clearly define the obligations of the violator and give you a means of enforcing those obligations if the violator fails to meet them.

What Are the Limitations on Citizen Suits?

STANDING

To bring a citizen suit you must be able to demonstrate that you have "standing." Generally, this means showing that you are adversely affected by the violation of the law. This requirement is often satisfied by showing that the violation leads to some potential environmental problem (like groundwater contamination) and that you live near or use the area affected. As a general rule, you cannot bring a citizen suit simply because you have an abstract or general interest in groundwater quality. In most cases, however, the standing requirement will not be a difficult hurdle for citizens challenging significant violations at facilities that could affect their groundwater. In addition, organizations can sue on behalf of the interests of their members (a citizens' group may have to be incorporated or register as a nonprofit association in some states to take advantage of this opportunity).

PRIOR NOTICE

You cannot bring a citizen suit until after you have notified the violator and the agency of the suspected violation and, in most cases, until a 60-day period has passed after the notice. The purpose of this requirement is to give the violator or the agency an opportunity to correct the problem. A notice can be just a letter that generally describes the violation (or violations) and refers to the statute or regulation that imposes the requirement. The specific requirements for notice vary to some degree depending on the statute. The 60-day waiting period does not apply, for example, to lawsuits challenging violations of the federal hazardous waste disposal law, the RCRA.

AGENCY PROSECUTION

You may be barred from bringing a citizen suit against a facility if the agency is already diligently prosecuting an enforcement action against the violator. Generally, to prevent parallel citizen suits, an agency must actually file a case against the facility in court, but in some cases an administrative enforcement proceeding may be enough.

COMMON-LAW ACTIONS

A common-law action is a way of using the courts to seek compensation or other relief for injury to you or your property. Unlike citizen suits, such actions are not tied to a violation of a statute or regulation. Lawsuits raising common-law claims are based on a body of law created by the courts over time based on particular facts and building on prior court decisions. Most common-law actions seeking relief for groundwater contamination are premised on the general common-law principle that gives a person the right to redress from someone who wrongfully causes him a "tort" (legal jargon for an injury).

There are many kinds of tort actions for many different kinds of injuries to person or property, but four common types are used in groundwater lawsuits: nuisance, negligence, strict liability, and trespass. Each has its own specific requirements that must be satisfied before a court will decide in your favor. Common law is developed by the state courts—not, as a general rule, by federal courts—and, as a result, the specific requirements for a particular type of tort action vary from state to state. Some of the general requirements are spelled out below, but you will need to consult a lawyer to determine if a particular theory may apply in your case. As a practical matter, however, several of these theories are likely to apply in most cases involving groundwater contamination.

Basic Common-Law Actions

Nuisance

To recover for groundwater contamination based on the nuisance theory, you must show that you have been or are likely to be injured in the use and enjoyment of your property by the unreasonable actions of another person. In one case, for example, plaintiffs became ill from fumes and hazardous substances entering their groundwater from an adjoining hazardous waste site. The site contained a huge trench where barrels of chemicals were dumped. The chemicals spilled onto the marshy soil and eventually contaminated the groundwater below. The court held that the site created a nuisance and awarded the plaintiffs compensation for their injuries and inconvenience.[1]

The main problem with a nuisance action from the citizens' viewpoint

[1] *Wood v. Picillo*, 443 A.2d 1244, 17 Env't Rep. Cases (BNA) 1386 (R.I. Sup. Ct. 1982).

is that the court will balance the value of the harmful activity against the injuries suffered by the plaintiffs to determine if the offending action is unreasonable. As a result, you may not succeed if the court determines the benefits provided by a particular facility outweigh the harms it causes. An advantage, however, is that you can win under the nuisance theory even if the activity is in compliance with a permit or other regulatory requirements.

NEGLIGENCE

To win a negligence action, you must show that the injury you have suffered was caused by defendant's failure to act with reasonable care. Negligence actions typically involve incidents like hazardous material spills from improperly loaded trucks or boats, failure to give warnings or take prompt remedial action, or misuse of chemicals or pesticides—cases where the negligent party "should have known better." Courts have awarded relief to plaintiffs for groundwater contamination based on a showing, for example, that a landfill operator failed to monitor for leakage into groundwater or operate the landfill in a safe manner.[2] You may be able to show that the defendant failed to exercise proper care if you can prove a violation of a statute or demonstrate that the company did not meet industry standards. Companies often successfully defend negligence lawsuits by claiming their activity was conducted in accordance with the state-of-the-art of the time.

STRICT LIABILITY

In many states, you may be able to get relief for groundwater contamination using the strict liability theory—that is, liability without fault—if you can show that the activity causing the problem is abnormally dangerous. For example, courts have found companies strictly liable for contamination caused by oil-drilling wastes disposed of in gravel pits, by locating underground chemical storage tanks near drinking-water wells, or by operating hazardous waste facilities.[3] The advantage to this approach is that you need not prove that defendant's conduct was negligent or otherwise unreasonable, but only that it caused you injury.

TRESPASS

Groundwater contamination that actually affects your property may give you grounds for an action under the trespass theory. The basic

---

[2] *Sterling v. Velsicol Chemical Corp.,* 647 F.Supp. 303, 316–17 (W.D. Tenn. 1986), *affirmed in part, reversed in part on other grounds,* 855 F.2d 1188 (6th Cir. 1988); *Ayers v. Jackson Township,* 525 A.2d 287, 292 (N.J. 1987).

[3] *Branch v. Western Petroleum,* 657 P.2d 267, 18 Env't Rep. Cases (BNA) 1364 (Utah 1982); *Exxon Corp. v. Yarema,* 516 A.2d 990, 1005 (Md.App. 1986); *Sterling v. Velsicol Chemical Co., supra* note 2, at 311–13.

requirement is that you show a physical intrusion onto your property. Contamination of the surface or subsurface (soil or groundwater) by leaching chemicals or other materials will satisfy this requirement.[4]

What Type of Relief Can You Get?

The most common form of relief in a common-law action is monetary compensation, called damages. You can recover money for a variety of injuries, including:

- Actual physical injury or illness
- Damage to property
- Increased risk of disease or immune system damage
- Emotional injuries

You can also obtain money for future medical surveillance. In cases where the conduct of the defendant is particularly egregious, you can recover punitive damages.

You can also seek an injunction. The court can order the defendant to clean up the contamination, provide alternate water supplies, alter its operations to prevent future harm, or even shut down altogether. The availability of this kind of relief depends in part on the nature of the injuries you suffer and the common-law theory under which the defendant is liable.

What Are the Problems with Common-Law Actions?

STATUTE OF LIMITATIONS

You must make sure that you bring a common-law action within a limited time specified by law (called the "statute of limitations"). For example, your state may require you to file an action for property damage within three years after the incident causing the injury occurred. Plaintiffs have often been prevented from successfully bringing common-law actions because by the time their injuries were discovered, the statutory period had passed (diseases caused by exposure to toxic substances, for example, may not develop until many years after exposure). In recent amendments to Superfund, however, Congress required all states to begin measuring the statute of limitations period for injuries caused by contaminants from the date when a person could reasonably *discover* his injuries, not from the time of exposure.[5]

---

[4] *Sterling v. Velsicol, Supra* note 2, at 317–19.

[5] Superfund § 309, 42 U.S.C. § 9658 (West Supp. 1988)

CAUSATION

All common-law actions discussed in this section require that you prove the defendant caused you injury. This can be troublesome in some cases involving groundwater contamination. It may be difficult to show that a particular contaminant comes from a specific facility or source, especially when there are several possible sources in the area or the hydrogeology of the area is complex. In some cases where several defendants are known to have contributed to a single, indivisible harm, you may still be able to recover from any or all of them under the "joint and several liability" rule. This means, for instance, that if only one of several culprits is available to be sued (because the others are either unknown, located outside the boundaries of the court's jurisdictional area, or financially insolvent), you can collect the full amount of your damages from that defendant.

Even if you can show a particular source caused the contamination, it may be hard to show that the contamination caused your injury. Many health problems caused by toxic chemicals are slow to develop and have a variety of possible causes; a company may argue, for example, that a plaintiff's cancer resulted from a poor diet and not from drinking contaminated water. Showing that a particular contaminant caused your injury may be easier after recent amendments to the federal Superfund Law. A new agency, the Agency for Toxic Substances and Disease Registry, was created to prepare toxicological profiles of the most common contaminants, explaining the health risks caused by such substances. This data should be very helpful to plaintiffs in common-law actions. (See Chapter 13.) Nevertheless, causation problems will likely require that you produce experts to testify in your behalf, thus making common-law actions expensive to bring.

REFERENCES

TAKING THE POLLUTER TO COURT

Citizen Suit References

Belfiglio, Jeff, Thomas Lippe, and Steve Franklin. Hazardous Waste Disposal Sites: A Handbook for Public Input and Review. Stanford Environmental Law Society. Palo Alto, Calif. 1981.

Citizen's Clearinghouse for Hazardous Wastes. User's Guide to Lawyers. Arlington, Va. 1985.

Environmental Law Institute. Citizen Suits: An Analysis of Citizen Enforcement Action Under EPA-Administered Statutes. Washington, D.C. 1984.

Jorgenson, Lisa, and Jeffrey Kimmel. Environmental Citizen Suits: Confronting the Corporation. Bureau of National Affairs. Washington, D.C. 1988. To order BNA publications, call the BNA Response Center at (800) 372-1033, or write BNA's Customer Service Center, 9435 Key West Avenue, Rockville, Md. 20850.

Miller, Jeffrey G., and Environmental Law Institute. Citizen Suits: Private Enforcement of Federal Pollution Control Laws. Wiley Law Publications. New York, N.Y. 1987.

Common Law References

Environmental Defense Fund. Dumpsite Cleanups: A Citizen's Guide to the Superfund Program. Washington, D.C. 1984.

Epstein, Samuel S., Lester O. Brown, and Carl Pope. Hazardous Waste in America. Sierra Club Books. San Francisco, Calif. 1982.

Moore, Andrew Owens. Making Polluters Pay: A Citizens' Guide to Legal Action and Organizing. Environmental Action Foundation. Washington, D.C. 1987. To order, contact Environmental Action Foundation, 1525 New Hampshire Avenue, N.W., Washington, D.C. 20036. (202) 745-4870.

National Campaign Against Toxic Hazards (NCATH). The Citizens' Toxics Protection Manual. Boston, Mass. 1987.

Prosser, William. Handbook of the Law of Torts (4th Edition). West Publishing Co. St. Paul, Minn. 1971. (This is a treatise written by a law professor on several common-law actions.)

NOTE: For thorough information on common-law actions, consult a local law library at a law school or law firm; they should have several books describing these actions.

CHAPTER 9

Grass-Roots Action

"At the heart of the environmental movement . . .
is the capacity to organize at the community level."

Lois Gibbs, Love Canal Organizer

I often feel humbled at what people can do when
they decide they CAN. The secret is . . . to focus
on building "People Power." Groups who win are
groups who pick themselves up after every fall.
And they remember the basics: strength comes from
people acting together out of mutual need and
in mutual aid.

Will Collette, Citizen's Clearinghouse
for Hazardous Waste

CHAPTER 8 TALKED ABOUT taking the polluter to court. This chapter talks about action you can take outside of court—action to persuade legislators to strengthen the laws, to encourage polluters to improve their practices, and action you can take yourself in your own home and your own neighborhood to reduce the pollution that threatens groundwater. Because knowing how to organize is important for court action and essential for direct action, this chapter begins with a short discussion about organizing.

111

ORGANIZING FOR COMMUNITY ACTION

The first question is "why organize?" After reading the preceding chapters of the book, the main reason should be clear: because it is too much work to do it all alone. Groundwater pollution is a complex problem and finding solutions requires the resources of more than one citizen, no matter how creative and energetic that individual might be.

Another reason to organize is to share your knowledge with others whose groundwater is threatened. Groundwater pollution usually affects whole communities, and the work you do can help others if they know about it.

Organizing also allows you to build a base of power, "people power." Organizing may seem like just more work at first, but the time you invest in the beginning will repay you many times over in strength and endurance and ultimately will help you achieve your goals.

The next question is "who will help?" Start with other people who are affected by the groundwater pollution you are concerned about, other people who use the groundwater, other people whose children use the groundwater. These people have as much to gain as you do. They should be willing to join together to form the core of your citizens' group.

Check with the relevant federal, state, and local agencies and find out who else has expressed concern about groundwater problems in your area. You can use the Freedom of Information Act for the federal agencies (see Chapter 6), and most states have an equivalent law you can use at the state level. You also should attend any public meetings these agencies have on your issue (or any related issues), and find out who else there might support your position. And, of course, talk to your neighbors. Many of them may not yet know that there is a problem with their groundwater.

Once you have begun finding others to join your group, you will probably find that organizing is easier than you think. Everyone has some practical experience to draw upon, and much of the rest is just common sense. To organize, you must take the people who are joining your effort, motivate them to work together, direct them towards specific goals, and learn from everything you do, your mistakes as well as your successes. You will need to define the problem and your proposed solution in nontechnical terms you can explain to your members, as well as to agency bureaucrats, legislators, industry executives, and the press. You will have to organize your meetings, set goals, allocate jobs, raise funds, and perform dozens of other tasks.

There are excellent materials available to help you learn the strategies and tactics you will need to know to be an effective organizer. While space prevents us from discussing them in this book, we have listed many references at the end of this chapter and we strongly urge you to use them.

TAKING POLITICAL ACTION

"People power" can be the promise of a group of protestors outside public offices, or floods of phone calls and letters from concerned citizens, or media coverage of your group's groundwater concerns. These are potent forces in politics. There is a saying that "all politics is local," and a strong local organization of citizens will have the best chance of getting action taken to address groundwater problems.

Political power gives you the chance to strengthen laws and regulations to protect your health—the chance to change the compromises that were made before your group was involved in protecting groundwater. In contrast, when you take a polluter to court, you pretty much have to take the law as you find it. That is why the most successful citizen efforts usually use both court action and political action.

Remember that your elected officials need you to stay in office. If they are not responsive to your concerns about groundwater, they run the risk of being voted out of office in favor of candidates who are more responsive. Elected officials control the agencies that are supposed to be protecting your groundwater. They appoint the people who head the agencies, and they appropriate the money to fund the agencies. If agency officials are not helpful, let your elected representatives know. (When officials are helpful, let your representatives know that as well.) And when the budgets of the groundwater agencies are being considered by the legislators, urge them to provide adequate funding to address the specific problems you are concerned about. Agencies generally appreciate this kind of citizen support.

You can use your "people power" with industry as well as with legislators. Many companies are beginning to see the value in responding to citizen concerns. Be mindful, however, that some community groups have reported efforts by industry, and even federal agencies, to co-opt members of their groups. Co-option may begin as innocently as an invitation to have lunch with the company executives and may progress to suggestions that you would be a better representative of the public— provided you disassociate from the citizens group. A few citizen groups have even reported threats and intimidation. But as one citizen activist said, "They have already poisoned me, my family, and my neighbors— what else can they do to us? We have to speak out."

ENCOURAGING WASTE REDUCTION AND RECYCLING

Working to reduce and recycle waste is one of the most important actions a community group can take to protect groundwater from future pollution. It also is one of the best ways to prevent new landfills and incinerators from being located in your community. It is also action your group can begin immediately, especially in your own homes and neighborhood.

Reduction is the most basic concept in the safe management of waste. By reducing the volume of materials used, the volume of waste that must be managed can be reduced. Begin in your own home: reduce your dependency on toxic chemicals and you will reduce the volume of household hazardous waste disposed in your municipal landfill. There are many combinations of things like baking soda, vinegar, lemon juice, and other food stuffs that can replace toxic drain openers, carpet stain removers, and oven cleaners. Before you buy any household products, carefully read the label and any special instructions for disposal. Some households choose to adopt more complicated reduction strategies and become absolutely toxics free, excluding everything with potentially harmful consequences from their building materials, home furnishings, household products, and food. Yet even the simplest techniques, if used by enough households, substantially reduce the volume of waste that can pollute groundwater. (See Figure 9.1.) For further discussion of household waste reduction, see the references at the end of this chapter.

Reduction

Industrial practices also can use a variety of techniques to reduce the volume of materials they use, especially of hazardous materials. (See Chapters 17 and 28.)

Recycling

Recycling is another important action a community group can take to help protect their groundwater from future pollution. Along with waste reduction, recycling is an action community groups can begin immediately. Across the country, recycling programs are as diverse as the towns and counties they serve, limited in scope only by the creativity and commitment of the community. Their scope ranges from families voluntarily bringing their recyclables to the sponsoring organization's drop-off points to mandatory separation and municipal curbside pickup. (See Chapter 27 for a discussion of recycling solid wastes; see Chapter 28 for recycling requirements for commercial and industrial facilities.) Once you reach out to active recyclers, you will find them eager to share their strategies and actually proud of what they have been able to do with their trash. Consult the references at the end of this chapter for further discussion of recycling.

SUCCESSFUL CITIZEN ACTION

Northern Alabama's Lauderdale Citizens for a Clean Environment (LCCE) provides an example of successful community action to establish a recycling program. LCCE used recycling as one of many tactics in their battle against a proposed 225-acre regional waste municipal landfill and three proposed incinerators.

The site of the proposed landfill lies in the vicinity of Zip City, an unincorporated rural farming community of 600 households. When the residents learned that their county commission had accepted an invitation to play host to thousands of tons per year of the south's wastes, concerned citizens formed the LCCE. Although the group's leaders had never been involved in waste issues or dreamed that trash would dominate already busy schedules, LCCE rapidly gained widespread support.

LCCE's victory initially appeared swift. Grandmothers and grandfathers protesting in front of the county courthouse, group testimony before the county commission, letters to the editor, TV news coverage, and all-you-can-eat breakfast fundraisers worked to apply enough pressure on local officials that the decision to play host community was reversed. But the victory did not last long. The company sued the county, and the LCCE mobilized local landowners to join as intervenors in the suit. The group learned that their community also had what it takes—rural conditions, moderate income, and a lack of political clout—to attract three incinerator companies. Leaders of the organization had no choice but to become experts in the nation's garbage crises, as everyone else's trash problems were landing in their lap.

To fight back, LCCE held more meetings, distributed more fact sheets, wrote more newspaper articles, gained more TV coverage, and held more fundraisers. In the words of an LCCE spokesperson, "Stop blundering blindly toward incineration and landfilling, and start recycling!" They hired hydrogeologists to document their dependency on shallow groundwater resources covered by permeable soils. The county commission established a solid waste task force to study their options and to placate members of LCCE, who worked to have one of their leaders appointed.

The LCCE asked an area recycling company to support a test program of household source separation. With the company's cooperation, LCCE signed up half of the Zip City households for LCCE's recycling program, and recruited volunteer drivers to pick up the recyclables once a week. Each household was encouraged to design their own method of source

FIGURE 9.1
HOUSEHOLD HAZARDOUS WASTE CHART

The following chart, adapted from one prepared by the Water Pollution Control Federation, will help you establish the most effective means of disposing of hazardous wastes used around your home or garden.

Column 1 represents products which can be poured down the drain with plenty of water. If you have a septic tank, additional caution should be exercised when dumping these items down the drain. In fact, there are certain chemical substances that cannot be used with a septic tank. Read the labels to determine if a product could damage the septic tank.

Column 2 represents materials which cannot be poured down the drain but can be safely disposed of in a sanitary landfill. Be certain the material is properly contained before it is put out for collection or carried to the landfill.

Column 3 represents hazardous wastes which should be saved for a community-wide collection day or given to a licensed hazardous wastes contractor. (Even the empty containers should be taken to a licensed contractor.)

Column 4 represents recyclable material. If there is a recycling program in your area, take the materials there. If not, encourage local officials to start such a program.

For more information on the safest way to dispose of these and other products, contact the United States Environmental Protection Agency.

	Type of Waste	1	2	3	4
Kitchen	Aerosol cans (empty)		x		
	Aluminum cleaners	x			
	Ammonia-based cleaners	x			
	Bug sprays			x	x
	Drain cleaners	x			
	Floor-care products			x	
	Furniture polish			x	
	Metal polish			x	
	Window cleaner			x	
	Oven cleaner (lye base)	x			
Bathroom	Alcohol-based lotions	x			
	(aftershaves, perfumes, etc.)				
	Bathroom cleaners	x			
	Depilatories	x			
	Disinfectants	x			
	Permanent lotions	x			
	Hair relaxers	x			
	Medicine (expired)	x			
	Nail polish		x		
	Nail polish remover			x	
	Toilet bowl cleaner	x			
	Tub and tile cleaners	x			

Type of Waste		1	2	3	4
Garage	Antifreeze			x	
	Automatic transmission fluid			x	
	Auto body repair products		x		
	Battery acid (or battery)			x	x
	Brake fluid			x	
	Car wax with solvent			x	
	Diesel fuel			x	x
	Fuel oil			x	x
	Gasoline			x	x
	Kerosene			x	x
	Metal polish with solvent	x			
	Motor oil			x	x
	Other oils			x	
Workshop	Paint brush cleaner with solvent			x	x
	Paint brush cleaner with TSP	x			
	Aerosol cans (empty)		x		
	Cutting oil			x	
	Glue (solvent based)			x	
	Glue (water based)	x			
	Paint—latex	x			
	Paint—oil based			x	
	Paint—auto			x	
	Paint—model			x	
	Paint thinner			x	x
	Paint stripper			x	
	Paint stripper (lye base)	x			
	Primer			x	
	Rust remover			x	
	Turpentine			x	x
	Varnish			x	
	Wood preservative			x	
Gardening	Fertilizer			x	
	Fungicide			x	
	Insecticide			x	
	Rat poison			x	
	Weed killer			x	
Miscellaneous	Ammunition			x	
	Artists' paints, mediums			x	
	Dry-cleaning solvents			x	x
	Fiberglass epoxy			x	
	Gun-cleaning solvents			x	x
	Lighter fluid			x	
	Mercury batteries			x	
	Moth balls			x	
	Old fire alarms			x	
	Photographic chemicals (unmixed)			x	
	Photographic chemicals (mixed and properly diluted)	x			
	Shoe polish		x		
	Swimming pool acid			x	

separation, with the group advocating the "four can method": separate trash cans for garbage, aluminum, paper, and glass. Their motto: "Source separation is not separating your garbage—it's not mixing it!"

Within a short time, LCCE recyclers found that "home recycling can be made quite simple with just a little thought and change of old ingrained habits." As families became familiar with their home systems and with which types of packaging to avoid at the grocery store, the amount of material recycled increased. Profits were put back into the LCCE budget. In a year the LCCE was recycling up to 2400 pounds of trash per week. The credo became "it isn't garbage until you declare it garbage!"

The group continues to increase the scope of its recycling project, with their goal being mandatory countywide curbside pickup. After discovering that a local waste hauler was voluntarily separating the trash collected on his route, LCCE hit upon the idea of gaining the cooperation of local waste haulers. LCCE proposed to sign up families along each driver's route if the hauler would agree to transport the separated wastes to the appropriate recycling centers. An added incentive for the hauler was the opportunity to keep all the money collected as a result of the recycling efforts. In order to reduce the myths of recycling and make it successful within their community, the group also designed unique adult, school, and door-to-door educational programs.

LCCE's fight to preserve clean water and air in their town and in their state required that they learn a good deal of science and work with scientists, that they learn law and work with lawyers, that they learn politics and work with politicians, and that they be creative. Their battle required an incredible amount of work. But the LCCE is a dedicated group with much to fight for. "We cannot sit by and watch corporate waste disposal practices poison us for profit. We must preserve our Earth—if not for ourselves, for our children."

REFERENCES

COMMUNITY ACTION

People Power Resources

Environmental Action (magazine). To order, contact: Environmental Action Foundation, 1525 New Hampshire Avenue, N.W., Washington, D.C. 20036.

Langton, Stuart, ed. *Environmental Leadership*. Lexington Books. Lexington, Mass. 1984.

Moore, Andrew Owens. *Making Polluters Pay: A Citizens' Guide to Legal Action and Organizing*. Environmental Action Foundation. Washington, D.C. 1987.

The National Campaign Against Toxic Hazards. *The Citizens' Toxics Protection Manual*. Boston, Mass. 1987.

Below are listed resources that are available through the Citizen's Clearinghouse for Hazardous Waste (CCHW) (703) 276-7070), a group dedicated to grass-roots organizing, education, and action. Founded by Lois Gibbs, the citizen activist who brought nationwide attention to the tragedy of Love Canal, the organization has helped hundreds of communities prevent hazardous waste from destroying their health and the beauty of their local environment:

Everyone's Backyard (subscription newsletter)
Action Bulletin (subscription newsletter)
Solid Waste Action Guidebook
How to Deal With a Proposed Facility
Media Means
Research Guide for Leaders
Fight to Win on Hazardous Waste: A Leader's Manual
Users' Guide to Lawyers
Love Canal: My Story

Waste Reduction and Recycling References

Center for Science in the Public Interest. Household Pollutants Guide. Anchor Press. Garden City, N.Y. 1978.

Citizen's Clearinghouse for Hazardous Waste. Recycling: The Answer to Our Garbage Problem. Arlington, Va. May 1987.

Citizen's Clearinghouse for Hazardous Waste. Reduction of Waste: The Only Serious Waste Management Option. Arlington, Va. December 1986.

Dadd, Debra Lynn. The Nontoxic Home. Tarcher. New York, N.Y. 1986.

EPA, Office of Solid Waste and Emergency Response. Survey of Household Hazardous Wastes and Related Collection Programs. Washington, D.C. October 1986.

Golden Empire Health Planning Center. Making the Switch: Alternatives to Using Toxic Chemicals in the Home. Sacramento, Calif. 1986.

Hirschhorn, Joel S., Cutting Production of Hazardous Waste, 91 Technology Review 52 (April 1988).

Sarokin, David J., et al. Cutting Chemical Wastes. Inform. New York, N.Y. 1985.

U.S. Congress, Office of Technology Assessment. From Pollution to Prevention: A Progress Report on Waste Reduction. Washington, D.C. June 1987.

U.S. Congress, Office of Technology Assessment. Serious Reduction of Hazardous Waste: For Pollution Prevention and Industrial Efficiency. Washington, D.C. September 1986.

PART III

GROUNDWATER AND FEDERAL PROGRAMS

Groundwater is affected by as many different laws as there are sources of contamination. There is, however, no single federal statute designed to protect groundwater. Instead, several different laws combine to provide a patchwork of protection under which some contaminant sources are regulated by more than one program and others are barely covered. The result is a confusing and incomplete federal system for safeguarding groundwater quality that in some circumstances creates powerful tools for citizens and in others leaves a vacuum to be filled in by state and local laws.

Part III of this manual is written to help you understand how to use the federal laws and regulations that do exist to stop or prevent the degradation of groundwater quality. Following an introductory chapter, this part begins with suggestions for using several federal programs that apply to most contamination problems, regardless of the specific source. Then, each of the major sources of groundwater contamination addressed by federal law is discussed in a separate chapter that explains how the various federal laws and regulations that may apply can be used by citizens to fight that particular source.

To make most effective use of this part, study both the general and the source-specific chapters. For example, to take action against contamination from underground storage tanks, you should, of course, review the remedies discussed in Chapter 19 that all relate directly to underground storage tanks. You should also review Chapters 11–15 to see whether any of the general remedies that are not tied to any particular contamination source may apply in your case.

CHAPTER 10

An Introduction to Federal Groundwater Protection

OVERVIEW OF THE FEDERAL STATUTORY FRAMEWORK

THE MAJOR FEDERAL statutes that affect groundwater are the Resource Conservation and Recovery Act (RCRA); the Comprehensive Environmental Response, Compensation, and Liability Act (CERCLA, or "Superfund," as it is commonly known) the Safe Drinking Water Act (SDWA); the Clean Water Act (CWA); the Toxic Substances Control Act (TSCA); the Federal Insecticide, Fungicide, and Rodenticide Act (FIFRA); and the Surface Mining Control and Reclamation Act (SMCRA). All of these laws, with the exception of SMCRA,[1] are administered and enforced by the U.S. Environmental Protection Agency (EPA), or in some cases by states with the approval of EPA. Later chapters discuss in detail how you can use the regulatory programs established by these laws to fight contamination problems. This section is intended only to give you a brief introduction to these programs.

Resource Conservation and Recovery Act (RCRA)

RCRA is designed to protect groundwater by regulating the handling and disposal of waste, especially hazardous waste. RCRA creates a system of hazardous waste management from "cradle to grave," that is, from the point of generation to ultimate disposal. RCRA establishes federal standards to be followed by generators and transporters of hazardous waste as well as by facilities that treat, store, or dispose of such waste.

Nonhazardous wastes are less strictly controlled by RCRA. The law establishes minimal federal waste control standards, which the states are required to follow in developing their own nonhazardous waste plans.

---

[1] SMCRA is administered by the Department of the Interior.

Comprehensive Environmental Response, Compensation, and Liability Act (CERCLA), or Superfund

Superfund gives EPA authority to act against groundwater contamination caused by inactive waste sites, accidental chemical releases, or other threatening situations involving hazardous substances. EPA can use the $8.5 billion trust fund created by Superfund to conduct its own cleanup at a site posing hazardous substance threats. If it can find the persons or companies responsible for creating the problem, EPA can also order them to perform the work, or can do the cleanup itself and then sue the culprits for reimbursement. Regardless of who undertakes the cleanup, it must be performed in accordance with the National Contingency Plan, a set of procedures that outline how a cleanup action should be planned and carried out. Superfund also provides for a special fund to be used for restoring and replacing natural resources that have been damaged by releases of hazardous substances.

Safe Drinking Water Act (SDWA)

The SDWA was enacted to ensure safe drinking-water supplies. The law requires EPA to set maximum levels for health-threatening contaminants in drinking water supplied by public water systems. The SDWA also authorizes EPA to protect underground sources of drinking water from contamination caused by injection of wastes and other substances into underground wells.

Under the SDWA, EPA may also designate any aquifer that serves as an area's principal source of drinking water as a "sole source aquifer." Federal agencies are barred from granting financial assistance to any project that could contaminate a sole source aquifer and create a significant health hazard.

The Wellhead Protection Program, also established under the SDWA, directs states to create a protective scheme for areas surrounding water wells or wellfields, in order to prevent contaminants from entering groundwater in these areas and affecting the public water supply.

Clean Water Act (CWA)

The primary object of the Clean Water Act is the control of the discharge of pollutants into the nation's lakes, rivers, streams, and other surface waters. Because of the important link between surface and groundwater in many areas, this law can be an effective tool to combat groundwater contamination. And many states with authority to implement this program have included groundwater directly in their definition of protected waters. One of the most important provisions of the Clean Water Act, particularly in relation to groundwater protection, is the National Pollutant Discharge Elimination System (NPDES), which prohibits the discharge of pollutants into water except in accordance with a permit issued by EPA (or a state agency). NPDES permits include specific standards that polluters must follow to protect water quality.

Toxic Substances Control Act (TSCA)

TSCA helps protect groundwater by giving EPA authority to control the manufacture, use, and disposal of toxic chemicals. Under TSCA, manufacturers of new chemicals or chemical mixtures must give a "premanufacture notice" to EPA before the substance enters commerce so EPA can determine if the use of the chemical or mixture will pose a significant threat to human health or the environment. TSCA also controls the disposal of polychlorinated biphenyls (PCBs), now being phased out for any use in this country.

Federal Insecticide, Fungicide, and Rodenticide Act (FIFRA)

FIFRA regulates pesticides in a manner similar to TSCA by requiring that manufacturers register pesticides with the EPA so the agency can, if necessary, impose restrictions on their use. EPA can also prohibit use of a pesticide altogether if it will have unreasonably adverse effects on the environment, including groundwater.

Surface Mining Control and Reclamation Act (SMCRA)

SMCRA regulates surface coal-mining activities to prevent contaminants from entering groundwater. This law requires that coal-mining operations receive a permit from the Department of the Interior (or authorized state agencies) and comply with design and operating requirements to protect groundwater from toxic mine drainage.

CITIZEN OVERSIGHT OF STATE AGENCY ENFORCEMENT OF FEDERAL LAWS

Most of the federal statutes affecting groundwater are put into action at first by EPA. EPA issues regulations to implement the general requirements of the law and is initially responsible for enforcing the law and regulations. Many of these laws are designed, however, to allow the states to take over the program from EPA. To receive EPA approval to manage a particular program, the state must enact an equivalent program and provide for enforcement authority in a state agency. Even after delegating primary enforcement responsibility to a state, however, EPA generally retains supervisory responsibility and can enforce some program requirements, especially if state enforcement lapses. In the extreme case, EPA can withdraw the state's approval. States can implement programs under RCRA, SDWA, SMCRA, and the Clean Water Act.

Participating in the State Authorization Process

You should participate in the authorization process if your state decides to seek enforcement authority for any of the federal programs that it does not already implement. State programs that meet minimum federal requirements can vary substantially. By becoming involved, citizen groups may be able to influence the effectiveness of the program's groundwater safeguards and increase the opportunities for public participation in the program. Even in states with already approved programs, citizens can be involved in the development and approval of revisions or modifications to the program, or when necessary, the withdrawal of EPA approval.

The approval process is similar for each of the statutes under discussion. The state submits an application to EPA and the public is given an opportunity to review the proposed program. EPA makes an initial decision to approve or disapprove and the public is given an opportunity to comment before EPA makes a final decision. Ideally, however, citizen groups should not delay their participation until the EPA authorization process begins but rather should be involved during the formulation of the state program.

You should focus on the following issues:

- Does the state program impose requirements at least as stringent as federal law, and, where necessary to protect groundwater, adopt even stricter regulations? As a general rule, states are free to adopt more stringent programs than required by federal law.
- Does the state enforcement program meet the minimum requirements of federal law, and are the agencies responsible for enforcement clearly identified?
- Is there full opportunity provided for citizen involvement in the enforcement and implementation of the state program? The program should provide for access to records, public participation in agency decisions, and the right to appeal adverse agency decisions.
- Does the state have adequate resources to implement the program once it is delegated?

You can file a lawsuit in federal court for review of EPA approval decisions if you believe a state program fails to meet the minimum federal requirements, but only within a limited time after the decision, generally 45 or 90 days. The Western Nebraska Resources Council, a local citizens' group, learned of this restriction the hard way. It attempted to challenge EPA's approval of Nebraska's underground injection control program under the SDWA, but the court refused to consider the petition because it was filed after the time limit had expired.

State Program Withdrawal

EPA can withdraw state approval if the state consistently fails to meet its enforcement obligations. In 1986, for example, EPA proposed to withdraw the authority of Illinois to enforce parts of the federal underground injection-well program of the SDWA because of the repeated failure of the state to correct violations by well operators. As a result, Illinois restructured its agency and EPA recently decided to allow the state to continue running the program.

You can request that EPA initiate the withdrawal process for an approved program in your state if the state agency is not doing its job. The withdrawal process is generally more formal than the approval process and involves adjudicatory hearings and the submission of legal argument and evidence. Citizens can take part in the process by formally intervening or, more informally, by submitting written or oral statements to EPA.

In some cases, however, even when state enforcement has been lax it may be to your advantage to keep the program at the state level and pressure the state agency to improve its performance. EPA does not have the resources to implement many programs effectively. State agencies may also be more receptive to local community concerns; of course, they may also be more subject to political influence from your opponents than EPA.

Be prepared to fight EPA if it proposes to withdraw a state program on the ground that it is too stringent. North Carolina amended its state hazardous waste program to impose tough conditions on hazardous waste discharges that would affect the drinking-water quality of its residents. GSX Chemical Services, Inc. and the Hazardous Waste Treatment Council petitioned EPA to withdraw authorization of the North Carolina hazardous waste program, alleging that the new provisions conflicted with RCRA (and therefore exceeded the state's regulatory powers).

Due in large part to vigorous and well-organized participation by citizen groups such as the Center for Community Action in Lumberton, North Carolina, EPA ultimately decided not to withdraw approval of North Carolina's program.

FINDING THE LAW

As a citizen activist, you will need to know how to find the federal laws and regulations that affect groundwater quality.

U.S. Code

Statutes are found in the *U.S. Code*. The *U.S. Code* (U.S.C.) is also available in privately published versions called U.S. Code Annotated (U.S.C.A.) and U.S. Code Services (U.S.C.S.). The Code is divided into 50 titles, each of which may contain several volumes. Each title covers a broad subject area; most of the statutes involving groundwater are found in Title 42, a collection of the laws relating to "The Public Health and Welfare." Each title is further divided into sections. For example, Superfund can be found in Title 42, sections 9601–9675, which is cited as 42 U.S.C. §§ 9601–9675. (You may see references to two different section numbers for what appears to be the same part of the law. Sometimes people use the section numbers of the original law rather than the section numbers where the law is contained in the *U.S. Code*. For example, that part of Superfund found in the *U.S. Code* at 42 U.S.C. § 9615 is § 115 of the original Superfund statute.)

Code of Federal Regulations

Regulations promulgated by EPA and other federal agencies to implement congressionally enacted statutes can be found in the *Code of Federal Regulations*, known as the C.F.R. Like the *U.S. Code*, the C.F.R. is divided into 50 different titles. Title 40, comprising ten volumes, contains the regulations published by EPA. Each title is divided into chapters that cover major topic areas, and each chapter is further divided into parts and subparts that cover specific regulatory issues. For example, the notation 40 C.F.R. § 144.14 refers to Title 40 of the C.F.R. (Protection of the Environment), Part 144 (Underground Injection Control Program), Paragraph 14 (requirements for wells injecting hazardous waste).

The Federal Register

Each title of the C.F.R. is brought up to date only once a year, so to find the most recent regulations you may have to look to the *Federal Register*. The

FIGURE 10.1
MAJOR FEDERAL LAWS RELEVANT TO GROUNDWATER PROTECTION

Statute	U.S. Code Citation	Regulations
RCRA*	42 U.S.C. §§ 6901–6991(i)	40 C.F.R. Parts 260–271
SDWA	42 U.S.C. §§ 300f–300j	40 C.F.R. Parts 141–147
TSCA	15 U.S.C. §§ 2601–2631	40 C.F.R. Parts 712–799
FIFRA	7 U.S.C. §§ 136–136y	40 C.F.R. Parts 162–180
SMCRA	30 U.S.C. §§ 1201–1328	30 C.F.R. Parts 700–955
CERCLA (Superfund)	42 U.S.C. §§ 9601–9675	40 C.F.R. Part 300
Emergency Planning and Right-to-Know Act	42 U.S.C. §§ 11001–11050	40 C.F.R. Part 355
CWA	33 U.S.C. §§ 1251–1387	40 C.F.R. Parts 100–140

*Subtitle I of RCRA, creating the Underground Storage Tank program, is found at 42 U.S.C. §§ 6991–6991(i). The regulations are contained in 40 C.F.R. Part 280.

Federal Register is a very important resource for citizens to know about and use. In addition to containing newly issued regulations not yet found in the C.F.R., this daily publication announces proposed regulations and other agency actions (like proposed cleanup plans and settlements with polluters). And it generally explains how citizens can comment or otherwise become involved in agency decisions. The *Federal Register* is referred to by its volume and page number; 51 Fed. Reg. 5219 (1986) means volume 51 at page 5219, published in 1986.

You should be able to find the U.S.C., the C.F.R., and the *Federal Register* at any law school library, and most law firms and public libraries. If you do not have access to these resources, ask the regional EPA office to send you a copy of the laws and regulations you need. Figure 10.1 lists the citations for most of the federal laws and regulations discussed in this manual.

EPA'S OFFICE OF GROUNDWATER PROTECTION

Recognizing the need to develop a federal strategy to protect groundwater in this country, EPA established the Office of Groundwater Protection (OGP) in 1984. For citizens, this office can be a helpful source of general information about groundwater protection. This office does not actually implement any of the regulatory programs discussed in the manual. It is responsible for: 1) identifying areas of inconsistency among programs at the local, state, and federal level, and formulating plans to facilitate comprehensive groundwater protection; 2) assessing the need for greater program coordination within the EPA; and 3) helping to strengthen states' capabilities to define and protect their groundwater resources.

FIGURE 10.2

MAJOR GROUNDWATER LEGISLATION INTRODUCED IN THE 100TH CONGRESS

An Act to Protect the Groundwater Resources of the United States

*S. 2091 (introduced by Senator David Durenberger)
  This legislation offers the most stringent groundwater protection of any of the legislation proposed. It would:

- require a strong federal presence in implementation of state groundwater management programs and regulation of groundwater contaminants;
- provide comprehensive groundwater protection regulations that would parallel the protection of surface water established by the Clean Water Act, including the requirement of permits and standards for discharges to groundwater;
- provide for strict controls on the application of pesticides and fertilizers.

Groundwater Protection Act of 1987

S. 20 (introduced by Senators Daniel Patrick Moynihan and George Mitchell)
H.R. 963 (introduced by Representative Albert Bustamante)
  This legislation would:

- establish a framework which would compel states to formulate groundwater management programs;
- require groundwater management programs to establish and monitor ambient groundwater standards;
- require the federal government to disseminate technical criteria and information to state and local governments regarding the health risks of contaminants in groundwater.

National Groundwater Contamination Research Act of 1987

H.R. 791 (introduced by several representatives)
S. 513 (introduced by Senator Durenberger as a companion bill to H.R. 791)
  These bills would:

- give the U.S. Geological Survey (USGS) specific authority for groundwater research activities;
- obligate the Secretary of the Interior to collect and analyze information on the nation's groundwater;
- require the Secretary to formulate a groundwater quality assessment program to coordinate government efforts, establish uniform data collection methods, and provide information and assistance. (Establishment of an interagency groundwater committee is contemplated.)

The Groundwater Research Act of 1987

S. 1105 (introduced by Senator Quentin Burdick)

  This bill would:

- establish a comprehensive groundwater research program at EPA;
- direct the USGS to develop a national inventory of groundwater resources;
- direct the Agricultural Research Service to study the efficacy of agricultural chemicals

---

* *"S" preceding a bill number means it was introduced in the Senate; H.R. denotes a bill originating in the House of Representatives.*

(and research techniques to control groundwater contamination from agricultural sources);
• establish a national information clearinghouse on groundwater and create research institutes.

The Groundwater Research Demonstration Act

H.R. 2253 (introduced by Representative James Scheuer)
   This legislation would accomplish many of the same objectives as Senator Burdick's bill.

Groundwater Protection through Bills Regulating Pesticides

H.R. 2463 (introduced by Representatives George Brown and Kika de la Garza)
S. 1419 (introduced by Senators David Durenberger and Patrick Leahy)
H.R. 3174 (introduced by Representative James Oberstar)
   These bills would generally:

• provide for the monitoring of pesticides and establishment of best management practices for pesticide use, and promulgation of regulations and standards based on the pesticide's potential to leach into groundwater;
• establish "groundwater residue guidance levels" (GRGL's). Additional monitoring and/or corrective action could be triggered if pesticides were found above these levels in an underground source of drinking water.

Hazardous Waste Reduction Acts

H.R. 2800 (introduced by Representative Howard Wolpe and several others. Senate version introduced by Senator Frank Lautenberg)
S. 1429 (introduced by Senator Joe Biden; companion bill introduced by Representative Thomas Carper)
   These bills would generally:

• provide for waste reduction measures at various stages of the production process, with special attention given to hazardous substances;
• create a national clearinghouse of information and EPA office responsible for coordinating these efforts, including the development of waste reduction plans.

Miscellaneous Bills

   Various other bills concentrate on state groundwater management programs (H.R. 2320, introduced by Representative George Miller); development of best management practices in agricultural production to lessen nitrogen contamination in groundwater (H.R. 3069, introduced by Rep. Arlan Strangeland, S. 1696, introduced by Senator David Karnes, and S. 1767, introduced by Senators Quentin Burdick and Patrick Leahy); and establishment of interagency groundwater committee to coordinate groundwater protection efforts (S. 1992, introduced by Sen. John Heinz).
   To find out the status of any of these bills, contact the office of the bill's sponsor, or ask your own representative what action is being taken.

To implement these goals, OGP issued its *Groundwater Protection Strategy* in 1984, which describes, among other things, the general nature and extent of groundwater contamination in the U.S., guidelines for regulators and citizens for EPA decisions regarding groundwater protection and cleanup, and coordination of the offices within EPA that deal with groundwater contamination problems.

A central focus of OGP's groundwater strategy (and a target of strong opposition from environmental groups) is OGP's proposal to create a groundwater classification system that would place all groundwater into one of three categories, depending on the importance of the aquifer as a source of water and the degree to which the aquifer is already contaminated. OGP proposes that these classes be used by the EPA and states to make decisions about the levels of protection and cleanup under existing regulations, to guide further regulations, and to establish enforcement priorities for the future. Some states have incorporated the classification system into their groundwater protection laws. (See Chapter 23.)

The OGP also publishes documents and fact sheets describing legislation and programs under the SDWA, RCRA, TSCA, FIFRA, CWA, Superfund, and other laws that affect groundwater quality. You can order these materials from OGP at no cost and ask to be put on OGP's mailing list to keep current with these publications.

NEW FEDERAL LEGISLATION

Finally, you need to keep track of new developments in the law. Most of the laws and programs described in this manual have been recently changed and improved by Congress, but more change is likely. In fact, several groundwater protection bills were introduced in the 100th Congress. To find out what bills have been offered and how far they have progressed through the legislative process, contact your senators and your congressman. Knowing what groundwater protection mechanisms are proposed will enable you not only to put pressure on legislators to enact good legislation, but also to use these proposals as models for state legislative action, encouraging your local representatives to act as sponsors. (See Figure 10.2, on pages 130–31.)

REFERENCES

An Introduction to Federal Groundwater Protection

Clean Water Action *et al. Protecting the Nation's Groundwater: A Proposal for Federal Legislation.* Washington, D.C. June 24, 1988.

Concern, Inc. *Groundwater: A Community Action Guide.* Washington, D.C. 1984.

Environmental Law Institute. *Law of Environmental Protection.* Clark Boardman Company. New York, N.Y. 1987.

EPA, Office of Groundwater Protection. *EPA Activities Related to Sources of Groundwater Contamination.* Washington, D.C. February 1987.

EPA, Office of Groundwater Protection. *Groundwater Protection Strategy.* Washington, D.C. August 1984.

EPA, Office of Groundwater Protection. *State and Territorial Use of Groundwater Strategy Grant Funds (Section 106 Clean Water Act).* Washington, D.C. May 1987.

Gordon, Wendy. *A Citizen's Handbook on Groundwater Protection.* Natural Resources Defense Council. New York, N.Y. 1984.

King, Jonathan. *Troubled Water.* Rodale Press. Emmaus, Pa. 1985.

Loveland, David Gray, and Diane Greer. *Groundwater: A Citizen's Guide.* League of Women Voters Education Fund. Washington, D.C. 1986. Order from: League of Women Voters of the United States, 1730 M Street, N.W., Washington, D.C. 20036. (202) 429-1965.

National Campaign Against Toxic Hazards. *The Citizens' Toxics Protection Manual.* Boston, Mass. 1987.

National Research Council, Committee on Ground Water Quality Protection. *Ground Water Quality Protection, State and Local Strategies.* National Academy Press. Washington, D.C. 1986.

Pye, Veronica I., Ruth Patrick, and John Quarles. *Groundwater Contamination in the United States.* University of Pennsylvania Press. Philadelphia, Pa. 1983.

Rodgers, William H., Jr. *Environmental Law.* West Publishing Co. St. Paul, Minn. 1986.

Sive, David, and Frank Friedman. *A Practical Guide to Environmental Law.* American Law Institute—American Bar Association Committee on Continuing Professional Education. Philadelphia, Pa. 1987.

U.S. Congress, Office of Technology Assessment. *Protecting the Nation's Groundwater from Contamination.* Washington, D.C. October 1984.

CHAPTER 11

The Safe Drinking Water Act Water Quality Programs

ONE OF THE steps you can take to protect yourself from contaminated groundwater, regardless of the source of the contamination, is to use the water quality program of the Safe Drinking Water Act (SDWA). Under this program, EPA and state agencies are responsible for ensuring that drinking water delivered to your tap by local water systems meets minimum quality standards. You can oversee the enforcement of these requirements by the agencies and take action against violators directly if the agencies fail to do their job.

Though action under this program will not protect groundwater quality directly, it can alert you to groundwater problems and to the need for action under other programs attacking the source of contamination. And in the meantime, enforcing the SDWA will help protect your health by preventing the water company from sending you water laden with contaminants.

Before turning to the discussion of direct action, you need to know several things about the program. First, the SDWA is limited to public water systems (defined as those that provide drinking water to at least 25 people or 15 service connections for at least 60 days a year). The suggestions for action in this chapter are, therefore, generally relevant mostly to customers of such systems. Those who get their water from a private well should nevertheless review the tables in this chapter summarizing the contaminants that are or will be regulated under the program. These lists can be used to help decide what contaminants to test for should you decide to test the water from your own private well or tap. In addition, some of the information-gathering suggestions may be useful if your private well is located near utility wells or draws from the same aquifer.

Second, the SDWA program applies not only to drinking water that comes from groundwater, but also to water that originates in reservoirs, rivers, and other surface sources. Action under this program, therefore, will be relevant to groundwater only if the local water system obtains its water from groundwater.

Finally, like several other federal pollution programs, the SDWA program can be administered by state agencies rather than EPA if the state adopts an equivalent program as part of its state law. Only Indiana, Wyoming, and the District of Columbia do not have this authority today. This chapter refers, therefore, to state agencies as the primary enforcers of this program. Residents of Indiana, Wyoming, and the District should turn to EPA in the first instance (EPA regional offices are listed in Appendix A).

CHECKING COMPLIANCE WITH MONITORING REQUIREMENTS AND CONTAMINATION LIMITS

The most important action you can take under the SDWA is to check the public water system's compliance with the national primary drinking-water regulations. These regulations, promulgated by EPA, require water systems on a regular basis to monitor the drinking water they deliver for specific contaminants and to ensure that these contaminants do not exceed specified levels. EPA is to promulgate these regulations for any contaminant that may adversely affect health and that is found in drinking-water supplies. For each contaminant, EPA is to establish a level that poses no threat to health, called the maximum contaminant level goal (MCLG), and then set the actual level that must be met by water systems, the maximum contaminant level (MCL), which must be as close to the MCLG as feasible, considering technological limits and cost. Monitoring requirements, governing the timing and number of water samples that must be taken by the water system and analyzed, vary depending on the contaminant and, for certain substances, on the number of people being served by the system. The method used to calculate whether an MCL is being violated also varies among contaminants.[1]

So far, EPA has issued standards for only 30 of the hundreds of potentially health-threatening contaminants. Figure 11.1 summarizes the regulations for these 30. Figure 11.2 lists those substances that EPA planned to regulate by June 1989; Figure 11.3 lists substances that will be regulated at later dates. Before you begin the steps outlined below, check with EPA to determine whether any of these compounds are now subject to

---

[1] EPA also sets other secondary regulations controlling contaminants that affect primarily aesthetic factors such as color, odor, and taste, but that can cause health problems if present at high levels. These regulations are enforceable only by the states, though EPA can pressure the states to take enforcement action if water utilities fail to comply. (See Figure 11.6.)

FIGURE 11.1

NATIONAL PRIMARY DRINKING WATER REGULATIONS

Contaminant	MCLG in mg/1* (milligrams per liter)	MCL in mg/1	EPA Monitoring Requirements for Systems Using Groundwater (generalizations—for specifics, see 40 C.F.R. Part 141)	Sources of Contamination
Inorganic Chemicals				
Arsenic	0.05	0.05	For inorganic chemicals, samples must be taken at three-year intervals. If a level over the MCL is found, three additional samples must be taken within a month. The average of these four samples, if over the MCL, constitutes a violation. For new systems, samples must be taken quarterly to establish the appropriate monitoring rate for that particular system.	Geological sources; pesticide residues; smelting operations; industrial waste
Barium	1.5	1.0		Variety of salts; geological sources; mining
Cadmium	0.005	0.01		Geological sources; mining and smelting; cigarettes; metal plating; paints; electronics
Chromium	0.12	0.05		Geological sources; industrial effluent; leather tanning; fungicide
Lead	0.020	0.05		Leaching from lead pipes and lead-soldered pipe joints; disposal of used storage batteries and other products
Mercury	0.003	0.002		Geological sources; food chain; mining; commercial processes; tool and dye
Nitrate	10.0	10.0		Fertilizers; sewage; feedlots; geological sources
Selenium	0.045	0.01		Geological sources; found primarily in soils of Western states

Fluoride	4.0	4.0		Geological sources; added to drinking water for dental treatment
Silver	—	0.05		
Organic Chemicals				
Endrin	—	0.0002	For organic chemicals, the frequency of monitoring is determined by the state, but is in no case less than three-year intervals. As with inorganic chemicals, once an excess is found, three additional samples must be taken within a month, and the average of these four readings determines compliance (the method of establishing a monitoring rate is also the same).	Pesticide used for control of cutworms, grasshoppers, and moles
Lindane	0.0002	0.004		Used for control of beetles; seed treatment; pharmaceutical preparations
Methoxychlor	0.34	0.10		Insecticide for mosquitos, horseflies. Used commonly in dairy barns.
Toxaphene	zero	0.005		Pesticide used to combat worms and insects affecting sheep, cattle
2,4–D	0.07	0.1		Herbicide used to control broadleaf weeds in agriculture, forestry, on range and pasture lands, and in gardens; to control aquatic weeds
2,4,5–TP Silvex	0.052	0.01		Herbicide

Continued on next page

FIGURE 11.1 (continued)

Contaminant	MCLG in mg/1* (milligrams per liter)	MCL in mg/1	EPA Monitoring Requirements for Systems Using Groundwater (generalizations—for specifics, see 40 C.F.R. Part 141)	Sources of Contamination
Volatile Organic Compounds				
Trichloroethylene	zero	0.005	To establish the sampling rate, systems must take four quarterly Carbon samples to determine vulnerability of system to VOCs. (States can reduce subsequent monitoring if no VOCs are found in first sample.) After this initial sampling, monitoring must be repeated anywhere from quarterly to once every five years, depending on whether VOCs were found in initial samples and how likely they are to reoccur. Violations are determined by average of quarterly sampling results.	Industrial effluent; waste from disposal of dry-cleaning materials and manufacture of pesticides, waxes, paints, and varnishes; metal degreasing; paint stripping
Carbon Tetrachloride	zero	0.005		Industrial wastes from manufacture of coolants, aerosol propellants, and cleaning agents
Vinyl Chloride	zero	0.002		Polyvinylchloride (PVC) pipes and solvent used to join them; industrial waste from the manufacture of plastics and synthetic rubber
1,2-Dichloroethane	zero	0.005		Production of vinyl chloride; used as solvent in paints, adhesives, soaps
Benzene	zero	0.005		Leaking fuel tanks; industrial effluents; solvent in the manufacture of pesticides, dyes, plastics, paints, and pharmaceuticals
para-Dichlorobenzene	0.075	0.075		Used in solvents, deodorizers, and insecticides

Contaminant			Monitoring	Source
1,1-Dichloroethylene	0.007	0.007		Found in industrial effluent; chemical processing by-products
1,1-Trichloroethane	0.2	0.2		Used in degreasing of metal parts; spot remover; film cleaner; additive in metal cutting oils
Bacterial Coliform bacteria bacteria/100ml**	zero	1 to 4	The monitoring requirements for bacterial contaminants vary depending on the size of the water system; anywhere from 1 to 500 samples are required monthly. Violations of the MCL are calculated in a variety of ways, depending on the number of persons served and the sampling technique.	Found in human and animal fecal matter
Total Trihalomethanes Bromoform/Chloroform Dibromochloromethane/ Dichlorobromomethane	—	0.10	Total trihalomethanes must be monitored quarterly at four different sample points in the system. The annual average (or any 12-month average) determines compliance.	By-product of public water supply disinfection system

Continued on next page

FIGURE 11.1 (continued)

Contaminant	MCLG in mg/1* (milligrams per liter)	MCL in mg/1	EPA Monitoring Requirements for Systems Using Groundwater (generalizations—for specifics, see 40 C.F.R. Part 141)	Sources of Contamination
Radionuclides Radium 226 & 228 (total)	zero	5pCi/1	Radionuclides must be monitored every four years, but much stricter requirements should be imposed by the states once any radiation level is found, or if the system is in the vicinity of mining activity. The average of four quarterly samples determines compliance.	Caused by decay of naturally occurring uranium, nuclear by-products
Gross alpha particle activity	zero	15pCi/1		
Gross beta particle activity	zero	4mrem/year		
Turbidity	0.1 NTU	1 or 5 NTU**	Turbidity must be monitored daily. A violation exists if the average of two consecutive samples is in excess of 5 NTU or if the monthly average of daily readings exceeds 1 NTU.	

*Except for Volatile Organic Compounds, for which final standards have been recently adopted, and fluoride, all of these MCLGs are proposed only. EPA may propose and adopt new MCLGs and MCLs for many of these contaminants in 1988. Make sure the standard you are using is the most recent.
**MCL depends on sampling technique.

SOURCE: 40 Code of Federal Regulations, Part 141.

FIGURE 11.2
CONTAMINANTS REGULATED SINCE JUNE 1989

Volatile Organic Chemicals

- Tetrachloroethylene
  Trichlorobenzene
- trans-1,2,Dichloroethylene

- Chlorobenzene
- Cis-1,2-Dichloroethylene
- Methylene chloride

Microbiology and Turbidity

* Total coliforms
* Giardia lambia

* Viruses
* Standard plate count
* Legionella

Inorganics

Antimony
Asbestos
Sulfate
Copper
- Nickel

- Nitrite
  Thallium
  Beryllium
- Cyanide

Organics

PAHs
PCBs
Atrazine
Aldicarb
- Chlordane
  Dalapon
  Diquat
  Endothall
  Glyphosate
- Carbofuran
  Alachlor
- Epichlorohydrin
- Toluene
- 2,3,7,8-TCDD (Dioxin)
- Aldicarb sulfoxide
- Heptachlor
- Styrene
  Dibromomethane

1,1,2-Trichloroethane
Vydate
Simazine
Phthalates
- Acrylamide
- Dibromochloropropane (DBCP)
- 1,2-Dichloropropane
- Pentachlorophenol
  Pichloram
  Dinoseb
- Ethylene dibromide (EDB)
- Xylene
  Hexachlorocyclopentadiene
- Aldicarb sulfone
- Ethylbenzene
- Heptachlor epoxide
  Adipates

Radionuclides

Uranium   Radon

See 53 Fed. Reg. 1892 (Jan. 22, 1988)
* New regulations have been proposed for these substances.
- Health advisories have been prepared for these contaminants. (Health advisories, prepared by
EPA's Office of Drinking Water, provide information on the health effects, treatment technology, and
analytical methods for each given contaminant and are updated as new information becomes
available. To order these documents, contact Health Advisory Program Coordinator, Office of
Drinking Water (WH-550), U.S. EPA, 401 M Street, S.W., Washington, D.C. 20460. (202) 382-7571.)

FIGURE 11.3
CONTAMINANTS TO BE REGULATED BY 1997*

EPA must set standards for 25 of these contaminants by January 1, 1991, an additional 25 by 1994, and a third set of 25 by 1997.

Previously Listed Contaminants
Zinc
Silver
Aluminum
Sodium
Dibromomethane
Molybdenum
Vanadium

Disinfectants and Disinfection By-Products
Chlorine
Hypochlorite
Chlorine dioxide
Chlorate
Chlorite
Chloramines
Ammonia
Ozone by-products
Bromodichloromethane
Bromoform
Dichloroacetonitrile
Dibromoacetonitrile
Bromochloroacetonitrile
Trichloroacetonitrile
Halogenated acids, alcohols, aldehydes, ketones, and other nitriles
Chloropicrin
Cyanogen chloride

Substances Commonly Found at Superfund Sites
Chloroform
Isophorone
2,4-Dinitrotoluene

---

*See 53 Fed. Reg. 1892 (January 22, 1988).

regulation.[2] Finally, EPA has issued a list of contaminants that water systems must check for at least every five years, but that are not yet subject to an MCL. (See Figure 11.4.) These contaminants may be prime subjects

[2] When investigating possible drinking-water violations, you should also check for additional standards set by the state. States with primary enforcement authority for the national drinking-water regulations must follow at least the minimum federal program, but can also regulate additional contaminants and set stricter standards for monitoring and enforcement. If the state has set additional standards, remember to determine how a violation of an MCL is calculated for each contaminant. Is it an annual average of quarterly samples or a single reading taken every three years?

Pesticides Designated by the National Pesticides Survey as Priority for Regulation

Cyanazine
Dicamba
ETU
Metolachlor
Metribuzin
Trifluralin
2,4,5-T

Volatile Organic Compounds of Concern Because of the Frequency with Which They Are Found in Drinking Water

1,1,2,2-Tetrachloroethane
1,3-Dichloropropane
Bromobenzene
Chloromethane
1,1-Dichloroethane
Bromomethane
1,2,3-Trichloropropane
1,1,1,2-Tetrachloroethane
Chloroethane
2,2-Dichloropropane
o-Chlorotoluene
p-Chlorotoluene
1,1-Dichloropropane
1,3-Dichloropropane

Other Contaminants of Concern

Boron
Cryptosporidium
Dibromochloromethane
Methyl tert-butyl ether
Strontium

for state regulation if they appear at high levels in your area (see Note 2) or for testing in a private water system.

To determine whether the public water system is complying with these monitoring requirements or whether an MCL has been violated, the first step is to examine the system's monitoring records. These records must be kept on file by the public water system (look for an address on your bill) and should be available to you upon request. Ask for records going back several years to get an historical picture of contamination problems.

If for some reason you have difficulty obtaining monitoring records from the public water system, request the information from the state

FIGURE 11.4
MONITORING FOR UNREGULATED CONTAMINANTS*

List 1: Monitoring Required for All Systems

Monitoring of these contaminants is required at least every five years, though states can monitor more frequently and/or can add contaminants to the list. All are volatile organic compounds (VOCs):

Bromobenzene
Bromodichloromethane
Bromoform
Bromomethane
Chlorobenzene
Chlorodibromomethane
Chloroethane
Chloroform
Chloromethane
o-Chlorotoluene
p-Chlorotoluene
Dibromomethane
m-Dichlorobenzene
o-Dichlorobenzene
trans-1,2-Dichloroethylene
cis-1,2-Dichloroethylene
Dichloromethane
1,1-Dichloroethane
1,1-Dichloropropene
1,2-Dichloropropane
1,3-Dichloropropane
1,3-Dichloropropene
2,2-Dichloropropane
Ethylbenzene
Styrene

---

*See 52 Fed. Reg. 25710 (July 8, 1987).

agency that implements the SDWA program. The water utility is required to submit all analyses, tests, and measurements to the state agency. If the proper monitoring has been done, that agency should have the records for your inspection.

Next, compare the records you gather with the regulations listed in Figure 11.1 (and any newly issued federal or state regulations) and check that the water supplier has tested water samples as required and, in particular, whether those tests show excessive contaminant levels.[3] Re-

---

[3] You may also be able to discover violations of the drinking-water quality regulations for your area by reviewing EPA's quarterly report of "significant noncompliers" (in 1987, 1200 utilities were on EPA's list) and annual summaries by state of regulatory violations. Both kinds of reports should be available at either the state regulatory agency or the regional EPA office.

1,1,2-Trichloroethane
1,1,1,2-Tetrachloroethane
1,1,2,2-Tetrachloroethane
Tetrachloroethylene
1,2,3-Trichloropropane
Toluene
p-Xylene
o-Xylene
m-Xylene

List 2: Monitoring Required for Vulnerable Systems
Ethylene dibromide (EDB)
1,2-Dibromo-3-Chloropropane (DBCP)

List 3: Monitoring Required at the State's Discretion
Bromochloromethane
n-Butylbenzene
Dichlorodifluoromethane
Fluorotrichloromethane
Hexachlorobutadiene
Isopropylbenzene
p-Isopropyltoluene
Naphthalene
n-Propylbenzene
sec-Butylbenzene
tert-Butylbenzene
1,2,3-Trichlorobenzene
1,2,4-Trichlorobenzene
1,2,4-Trimethylbenzene
1,3,5-Trimethylbenzene

member that a single test result showing that a particular contaminant exceeds the MCL amount does not necessarily mean the water system is in violation of the regulations. Violations are sometimes determined by average or multiple tests. Figure 11.1 summarizes the general rules here, but you will need to check the detailed requirements of the regulations or ask for assistance from the agency to interpret the records if excessive levels for any regulated contaminant are shown.

If you find that your water supplier is violating monitoring requirements or distributing water that does not meet MCL standards, you need to take action to have the problem corrected. Specific suggestions for initiating agency enforcement action are given in the section "Participating in Enforcement."

You should also check to see that the public is being notified properly. Anytime the supplier fails to meet SDWA requirements, it is supposed to notify its customers. Unfortunately, compliance with this requirement is poor. In 1985, 146,000 SDWA violations were committed (by EPA's best estimate) but the public was properly notified in only 16,000 cases. How the supplier must notify the public depends on the severity of the problem. The notice should generally contain an explanation of the problem, any potential health risk, the steps being taken to correct the problem, and any measures the public should take in the interim, including seeking alternative water supplies. By ensuring proper notice of violations is given, you can substantially increase the numbers of people in the community who are aware of the problem and who may be willing to join in an organized effort to address groundwater contamination in the area. Figure 11.5 summarizes the public notice requirements.

POLICING VARIANCES AND EXEMPTIONS

In some circumstances, a public water system can legally exceed an MCL if a variance or exemption is issued by the agency. You can take steps to prevent the agency from abusing this loophole or, if a variance or exemption is granted, to enforce the conditions that accompany it.

The agency may grant a variance if, despite application of the best technology and treatment methods, a system cannot meet an MCL because of the poor quality of its water resources. Exemptions from specific standards may be granted to water systems if compelling reasons (such as poor economic conditions) prevent them from achieving immediate compliance. Exemptions and variances may be granted only if they will not pose an unreasonable health risk, and must include a schedule that will require the system to reach full compliance by a specified date.

For variances or exemptions not yet granted, you should take advantage of required public notice and hearing opportunities to make sure the process is not misused. Ask the agency for copies of the application and any other relevant documents to determine if the variance or exemption request is justified. If the application is granted, make sure that the water system is put on a schedule to bring it into compliance and monitor the system's progress to make sure it stays on schedule.

The state may provide a means for challenging variances or exemptions that you believe were improperly issued. In addition to using any state appeal process, you should notify the EPA regional office any time the state issues a variance or exemption that you believe to be improper. If EPA finds a pattern of abuses in variances or exemptions issued by a state, it can step in and modify those that need change. Take advantage of the public comment and hearing opportunities that EPA should provide in this process.

FIGURE 11.5

NOTIFICATION REQUIREMENTS FOR PUBLIC WATER SYSTEMS*

	Type of Notice Given			Timing of Notice	
	Publish newspaper notices or post if no newspaper available?	*Mail notice to customers?*	*Provide electronic media notice?*	*Initial notice within what period?*	*Repeat notice required?*
Notice Requirements for Tier I Violations	yes	yes†	yes (for variances and exemptions only)	Newspaper = 14 days Mail = 45 days Elec. = 72 hours Post = 14 days	Newspaper = none Mail = quarterly Elec. = none Post = continuous
Notice Requirements for Tier II Violations	yes		no	Mail = none Post = 3 months Newspaper = 3 months	Mail = quarterly Hand delivery = quarterly Post = continuous Newspaper = none

Notes:

Tier I Violations consist of:
1. MCL exceeded
2. Treatment technique inadequate
3. Variance or exemption schedule

Tier II Violations consist of:
1. Monitoring violations
2. Testing procedure violations
3. Variance or exemption issued improperly

* Effective April 28, 1989. See 40 C.F.R. §§ 141.32, 142, and 143.5. See 52 Fed. Reg. 10974 (April 6, 1987).

† Requirement may be waived. See 40 C.F.R. § 142.32(a)(1)(ii).

PARTICIPATING IN ENFORCEMENT

State Agency Enforcement Action

To take action against violations of SDWA requirements—inadequate monitoring, MCL violations, missing public notice, failure to comply with conditions attached to variances or exemptions—first notify the state regulatory agency responsible for the SDWA program.[4] Document the violation as best you can with copies of monitoring records or other information. You may be able to generate more action from the agency if you can establish a prior history of compliance problems or consumer complaints about water quality. If the water system is on EPA's list of "significant noncompliers" or similar state lists, use this fact to ask for more aggressive enforcement.

Ask the agency to make a formal determination of a system's compliance by conducting a thorough review of the system's reports and an inspection, if necessary. If violations by the system are infrequent or minor, the agency may initiate enforcement with a notice of violation or warning letter to the water system. Follow up on the utility's response to such an approach and insist on more aggressive action, as described below, if the problem is not corrected.

The agency may also initiate negotiations with a public water system to bring it into compliance. Ask for public notice and an opportunity to participate or comment on any agreement before it comes into effect. A negotiated remedial action should include a specific schedule of compliance that is legally enforceable.

If the water supplier has consistently violated SDWA regulations or if violations may cause health problems (if contaminant levels exceed an MCL, for instance), request the agency to take formal enforcement action by issuing an administrative order to force the public water system into compliance. These orders may be issued without any formal procedure, but if the supplier requests a hearing, citizens should be given the opportunity to participate. You can also request a hearing if the water supplier chooses not to, but the agency may not hold a hearing unless enough public interest is generated. Even if there is not a hearing, you can submit comments to the agency about the violation and the adequacy of the agency's order.

In any of these procedures, the agency has a variety of enforcement options. It may require the water supplier to adopt new treatment, disinfection, or filtration techniques, including advanced methods like

---

[4] The state agency may delegate compliance and enforcement authority to local health departments for certain systems within its borders.

granular-activated carbon beds or aeration systems to remove organic chemicals. It may also provide training programs to educate suppliers about monitoring and treatment techniques. You may also want to request that, as an interim measure, the supplier be required to provide treatment devices for private homes or even bottled water. Where the water system is unable to meet drinking-water standards with existing supplies, enforcement may mean finding a new source of water, including drilling new wells. For example, in Duluth, Minnesota, new wells were required when asbestos was found in drinking water and could not be removed with treatment systems.

In the extreme case or when the supplier fails to comply with agency orders, the agency may file a civil action against the supplier in court. You may consider intervening in such actions to ensure your interests are protected by the court's remedial orders. (See Chapter 7.)

EPA Enforcement Action

If the state agency does not enforce SDWA requirements, ask the regional EPA office to take action. Once aware of a violation, EPA is required to notify the state agency and the violator of the problem and to provide advice and technical assistance to bring about compliance. EPA may hold a public hearing as part of this process to find a way to reach compliance. (You can petition the EPA regional office for such a hearing if you think it will be helpful.) If 30 days pass, and the state agency has not initiated any enforcement action, EPA is required to either issue an administrative order or file an action in court to force compliance. Before issuing an administrative order, EPA must give the violator and the public an opportunity to request an informal public hearing. Make sure you receive notice of any hearing and participate to ensure the remedy meets your concerns. (Where the agency seeks to impose a fine on the violator, this hearing will be a formal, trial-type hearing. See Chapter 7 for more about participating in this process).

If the EPA initiates a court action to force compliance, you may want to consider intervening to protect your interests. (See Chapter 7.)

Taking the Case to Court

If you do not think the state agency or EPA is taking adequate enforcement action, you can sue the public water system directly for violating the SDWA. Generally, you will want to consider a citizen suit only for significant noncompliance with the SDWA, such as persistent violations of MCLs, regular failure to monitor water quality, or failure to take corrective action once enforcement orders are issued against a system. (See Chapter 8 for general discussion of citizen suits.)

FIGURE 11.6
NATIONAL SECONDARY DRINKING WATER REGULATIONS*

Contaminant	Level
Chloride	250 mg/1
Color	15 color units
Copper	1 mg/1
Corrosivity	Noncorrosive
Fluoride	2.0 mg/1
Foaming Agents	0.5 mg/1
Iron	0.3 mg/1
Manganese	0.05 mg/1
Odor	3 threshold odor number
pH	6.5–8.5
Sulfate	250 mg/1
Total dissolved solids (TDS)	500 mg/1
Zinc	5 mg/1

* See 40 C.F.R. § 143.3.

IMMINENT HAZARD SITUATIONS

If you are confronted with a contamination problem that presents an imminent and substantial threat to human health, request EPA to use its emergency authority under the SDWA to issue an order or commence a lawsuit for immediate action. These actions can be taken without any showing of a violation of the specific requirements of the SDWA and even if only a private well is threatened, as long as the threat to human health can be demonstrated. (See the general discussion of imminent hazard actions at Chapter 14.)

REFERENCES

THE SAFE DRINKING WATER ACT WATER QUALITY PROGRAMS

Camp, Dresser, and McKee, Inc. Summary of Revisions of the Drinking Water Regulations and Amendments to the Safe Drinking Water Act. Boston, Mass. June 1986.
"Drinking Water." Buyer's Market, Vol. 3, no. 1. March 1987.
Concern, Inc. Drinking Water: A Community Action Guide. Washington, D.C. December 1986. To order, contact Concern, Inc., 1794 Columbia Road, N.W., Washington, D.C. 20009. (202) 328-8160.
EPA, Office of Groundwater Protection. Groundwater Provisions of the SDWA Amendments of 1986 (Fact Sheet). Washington, D.C. June 1986.

EPA, Office of Groundwater Protection. *Groundwater Protection Update* (Fact Sheet). Washington, D.C. December 1986.

EPA, Office of Drinking Water. *Is Your Drinking Water Safe?* Washington, D.C. March 1985.

Getches, David H. *Controlling Groundwater Use and Quality: A Fragmented System,* 17 Natural Resources Lawyer 623 (1985).

U.S. General Accounting Office. *States' Compliance Lacking in Meeting Safe Drinking Water Regulations.* Washington, D.C. March 3, 1982.

Wilson, Reid. *Testing for Toxics: A Guide to Investigating Drinking Water Quality.* U.S. Public Interest Research Group. Washington, D.C. March 1986.

LAWS AND REGULATIONS

U.S. Code Citation

Safe Drinking Water Act. 42 U.S.C.A. §§ 300f-300j (West 1982 & Supp. 1988).

Code of Federal Regulations Citations

40 C.F.R. Parts 141, 142 (National Primary Drinking Water Regulations).
40 C.F.R. Part 143 (National Secondary Drinking Water Regulations).

Federal Register Citations

52 Fed. Reg. 20672 (June 2, 1987) (Incorporating the changes of the 1986 SDWA Amendments into the C.F.R.).

52 Fed. Reg. 25690 (July 8, 1987) (National Primary Drinking Water Regulations).

52 Fed. Reg. 41534 (October 28, 1987) (Announcing the new public notification requirements for PWS violations).

53 Fed. Reg. 565 (Jan. 8, 1988) (Availability of new health advisories for 50 pesticides).

53 Fed. Reg. 1892 (January 22, 1988) (Priority list of contaminants for future regulation).

CHAPTER 12

Aquifer Protection Programs

THE MOST EFFECTIVE way to protect groundwater quality may be to impose restrictions on the use of land in areas where spills, leaks, or other contamination are most likely to affect aquifers, wellhead areas of influence and recharge areas in particular. Because protection such as this generally involves land use planning on a local level, most regulatory action of this kind comes from state and local programs. We introduce some of these in Chapter 24. However, two federal programs established by the Safe Drinking Water Act (SDWA) do provide an opportunity for this sort of aquifer protection for areas around public water supply wells and, on a more limited scale, for aquifers that serve as the main source of drinking water for an area or community. This chapter describes how to initiate and participate in action under these programs.

WELLHEAD PROTECTION PROGRAM

You should participate in the formulation and implementation of state plans to protect aquifers under the federal Wellhead Protection Program. By June 1989, all states must develop programs to prevent contaminants from entering public water supply wells. The heart of the program is a requirement that each state protect the area around public water supply wells—called the wellhead protection area.

Though the SDWA does specify a few general requirements for these programs, states are given a relatively free hand to decide what areas to protect, what contaminants must be addressed, and what kind of regulatory measures are necessary. Citizen action, therefore, can make the difference between a perfunctory effort and a successful program.

Initiating State Action

The first step is to contact the state agency responsible for drinking water or groundwater regulation to find out if a wellhead protection program is in place or if one is being developed. Generally, there will be a lead agency coordinating the program's components, with local and regional agencies providing site-specific data on wellheads around the state. At this writing, approximately 20 states have initiated program development efforts with the EPA regional offices. No programs have yet been approved by EPA.

If your state has not already begun drafting a program for submission to the EPA regional office, bring political pressure to bear on the responsible agencies. Persuading the state to act is critical because, unlike other federal programs, SDWA does not authorize EPA to undertake the program in lieu of the states.

Participating in Program Development

Once planning is underway, there are several ways to influence the outcome. Program development will be likely to take place in two stages. First, the state agency, with input from EPA and citizens, will draft a program description defining in general terms how the program will work and which agencies will be responsible for its components. Next, the state will begin to implement the general program by applying its general principles to specific areas. For example, individual wellhead areas may be delineated in this second phase and land use restrictions appropriate for each area may be imposed.

To influence the development of the general state program, take advantage of the public participation opportunities the state should provide. Ask for public notice of any program meetings and begin working with the agencies at the earliest possible opportunity. Try to get members of your group on to the technical and citizens' advisory committees the state is required to establish. At the very least, make sure you comment on the draft program plan the agency issues for public review. In all of these processes, the two main points to focus on are: the size of the wellhead protection area and the program developed to protect the area.

DEFINITION OF THE WELLHEAD PROTECTION AREA

The first step in making sure the state's program provides comprehensive protection for wellheads within the state is to ensure that the wellhead protection areas will be big enough. The SDWA defines a wellhead protection area as the area surrounding a water well or wellfield supplying a public water system, through which contaminants are reasonably likely to move and eventually enter a drinking-water supply. It is designed to include the groundwater and surface water supplies that are likely to be drawn into the well system. In general, a wellhead protection area will include the area of influence or contribution around a pumping well and surrounding recharge areas. (See Figure 12.1.)

EPA has decided that state programs need not actually identify each wellhead area in the state by 1989, but programs should provide specific guidance for later decisions that delineate protection areas for each wellhead. (Some groups believe the decision to allow states to submit programs even though they do not actually designate wellhead protection areas is inconsistent with the SDWA. This situation may offer grounds for challenging EPA approval of a state program that you believe is inadequate.) Among the factors that determine how large an area needs to be protected and that should be considered under the state program are:

- *Nature of the potential contaminants.* Some contaminants move for long distances in aquifers; others tend to stay put. Some contaminants survive for long periods without breaking down; others may be short-lived, particularly some biological contaminants.
- *Location of existing and potential contamination sources in relation to water supply wells.*
- *Aquifer characteristics and local hydrogeology.* Unconfined aquifers (those not protected by impermeable barriers) are more susceptible to contamination from sources near the wellhead and more likely to benefit from wellhead area protection. The groundwater flow pattern, aquifer boundaries, local weather and recharge patterns, and the aquifer's ability to assimilate contaminants may all influence the proper delineation of protection boundaries.
- *Water withdrawal rates.* Higher water use volumes and associated high pumping rates will increase the rate at which contaminants may move from their source through the aquifer to the water supply well.

These factors can be analyzed in a variety of different ways to calculate the proper protection area. For example, the data can be used to determine the time of travel for various contaminants, that is, the amount of time it will take contaminants to move from their source to the water well. The area to be protected can then be selected based on a travel time determined to be acceptable. (Acceptable travel times may be measured in days or many years depending on the contaminants involved.)

States may also vary widely in the actual method used to establish boundaries. Some programs may permit protection areas to be determined simply by drawing a circle around a well, its size dependent on a general consideration of the factors listed above. Others may require complex analytical modeling or even extensive hydrogeologic mapping to determine the size and shape of the protection area best for each wellhead. You should advocate the use of methods that will ensure the most careful consideration of site-specific factors. Keep in mind, however, that such methods are likely to be more costly and time-consuming, and may delay the implementation of protection for the wellhead area.

FIGURE 12.1
ZONES OF INFLUENCE AND CONTRIBUTION

Below is an example of where the Zone of Influence and Zone of Contribution might be in relation to a pumping well. The Zone of Transport is the distance contaminants will move over a given period. SOURCE: EPA, Guidelines for Delineation of Wellhead Protection Areas, 1987.

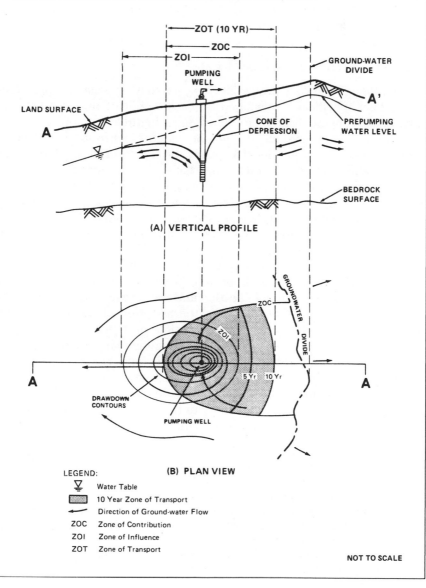

The proper size of the protection area will vary with local conditions. In those communities or states that already have programs in place (see Chapter 24 for more about these), wellhead protection areas as small as a few hundred feet have been designated. In most cases, however, it is generally agreed that protection areas must have a radius of at least 1/4 mile if they are to be effective, and in some cases as large as several miles across. For EPA's best advice on protection area size, ask for a copy of their *Guidelines for Delineation of Wellhead Protection Areas*. It contains detailed information about the factors that should be considered and the methods that can be used to establish protection areas.

DEFINING A STRONG REGULATORY PROGRAM

Even though the state protection program may be generally described rather than specifically developed for each wellhead area, you can still take steps to ensure that the program is meaningful and enforceable. Your two primary concerns should be the careful delineation of responsibilities among state and local agencies and the description of the actual use restrictions and other protections that will be imposed.

The obligations of each agency responsible for implementing and enforcing the program must be clearly defined. The program should be incorporated into existing state laws, or new laws should be enacted that permit citizens to sue to enforce the program and compel the responsible agencies to do their jobs.

The critical component of any program will be, of course, the actual procedures and restrictions imposed to protect the aquifer. The program could range from providing education and training to implementing control measures. You should advocate the adoption of control measures such as restrictions on building construction, pesticide use, agricultural practices, waste disposal, industrial or commercial development, and transportation of hazardous materials in the zone. For examples of the states' programs that have been the most successful, and for information useful in formulating these programs, request EPA's *Annotated Bibliography on Wellhead Protection Areas* from the EPA regional office.

In addition to your focus on these central components, make sure the state program does the following:

- Identifies all human-generated sources of contamination that may have an adverse impact on human health, including potential sources as well as actual sources surrounding wellheads. EPA has particularly emphasized pesticide use as a source that should be considered. A proper identification of contaminants may affect both the wellhead protection area and the kinds of restrictions deemed necessary.
- Provides contingency plans for alternative water supplies for each public water system in the event of contamination. Not only should these sup-

plies be identified, but plans must be made to ensure that the water supply will not be interrupted.

- Requires that before new wells are drilled, all potential sources of contamination that may affect the well are evaluated. Without this review, new wells might escape the protection offered by the program.
- Requires public notice, a comment period, and opportunity for hearing concerning any changes made to the program, including modification of protection areas and controls used to prevent contamination from entering the well. The plan should require such reconsideration whenever water use changes or new contamination sources will alter the effectiveness of the program.

EPA's advice on wellhead programs is summarized in *Guidance for Applicants for State Wellhead Protection Program Assistance Funds*, available from EPA regional offices.

If you find the state unresponsive to concerns you have about any of these major issues, you can turn to the EPA. To be eligible for federal funding, states must secure EPA's approval for their programs. By contacting the EPA regional office directly you may be able to generate some extra scrutiny of inadequate provisions, and perhaps even convince EPA to ask the state to make changes in its proposal.

Participating in the Implementation and Enforcement of the State Program

Once a statewide program is approved and in place, there will still remain a great deal of implementation to complete. The general principles in the program must be applied by the responsible agencies to each public wellfield in the state. It is this stage that will determine whether in fact your aquifer and water supply are protected. It is too early to tell what form this process will take in the states, but you should make use of every public participation opportunity that the state program offers.

You also need to ensure that your system gets proper priority in this process. The states are not likely to implement the program for all wellfields at once. Priorities are likely to be established among wellfields based on such factors as the susceptibility of the wellfield to contamination and the size of the water system. Given EPA's estimate that there are over 104,000 wells serving public water systems in this country, this implementation process will obviously take some time. Nevertheless, the SDWA requires each state to try to implement its program at the local level within two years of writing the general plan.

After the program is put into place for particular wellfields, it must then be enforced. You should police the responsible agencies to see that they are enforcing the program requirements and, if the program allows

FIGURE 12.2

DESIGNATED SOLE SOURCE AQUIFERS—NATIONALLY (STATUS AS OF 12/31/88)

State	Aquifer and/or Location	Federal Register Citation
Arizona	Bisbee-Naco Aquifer, Cochise County	53 FR 38337(1988)
	Upper Santa Cruz & Avra Altar Basin Aquifers	49 FR 2948(1984)
California	Santa Margarita Aquifer, Scotts Valley, Santa Cruz County	50 FR 2023(1985)
	Fresno County	44 FR 57251(1979)
Florida	Biscayne Aquifer	44 FR 58797(1979)
	Volusia-Floridan Aquifer; Volusia, Flager & Putnam Counties	52 FR 44221(1987)
Hawaii	Southern Oahu Basalt Aquifer	52 FR 45496(1987)
Indiana	St. Joseph Aquifer System, Elkart County	53 FR 23682(1988)
Louisiana	Chicot Aquifer	53 FR 20893(1988)
Louisiana/ Mississippi	Southern Hills Aquifer System	53 FR 25538(1988)
Maine	Monhegan Island	53 FR 24496(1988)
Maryland	Maryland Piedmont Aquifer; Montgomery, Frederick, Howard & Carroll Counties	45 FR 57165(1980)
Massachusetts	Cape Cod Aquifer	47 FR 30282(1982)
	Nantucket Island Aquifer	49 FR 2952(1984)
	Martha's Vineyard Regional Aquifer	52 FR 3451(1988)
	Head of Neponset Aquifer Area	53 FR 49920(1988)
Montana	Missoula Valley Aquifer	53 FR 20895(1988)
New Jersey	Buried Valley Aquifer System	45 FR 30537(1980)
	Rockaway River Basin Area	49 FR 2946(1984)
	N.J. Coastal Plain Aquifer System	53 FR 23791(1988)
New Jersey/ New York	N.J. Fifteen Basin Aquifer Systems	53 FR 23685(1988)
	Brunswick Shale and Sandstone Aquifer, Ridgewood Area	49 FR 2943(1984)
	Highlands Aquifer System	52 FR 37213(1987)
New York	Nassau/Suffolk Counties, Long Island	43 FR 26611(1978)
	Clinton Street-Ballpark Valley Aquifer System; Broome and Tioga Counties	50 FR 2025(1985)
	Cortland-Homer-Preble Aquifer System	53 FR 22045(1988)
	Cattaraugus Creek Basin Aquifer System (CCBA)	52 FR 36100(1987)
	Schenectady/Niskayuna; Schenectady, Saratoga and Albany Counties	50 FR 2022(1985)
	Kings/Queens Counties	49 FR 2950(1984)

SOURCE: EPA. Continued on next page

FIGURE 12.2 (continued)

State	Aquifer and/or Location	Federal Register Citation
Ohio	Buried Valley Aquifer System (BVAS)	53 FR 15876(1988)
	Bass Island Aquifer, Catawba Island	52 FR 37009(1987)
	Pleasant City Aquifer, Guernsey County	52 FR 32342(1987)
	OKI-Miami Buried Valley Aquifer	53 FR 25670(1988)
Oregon	North Florence-Dunal Aquifer, Lane County	52 FR 37519(1987)
Pennsylvania	Seven Valleys Aquifer, York County	50 FR 9126(1985)
Rhode Island	Block Island Aquifer	49 FR 2952(1984)
	Hunt-Annaquatucket-Pettaquamscutt Aquifer System (HAP)	53 FR 19026(1988)
Rhode Island/ Connecticut	Pawcatuck Basin Aquifer System	53 FR 17108(1988)
Texas	Edwards Aquifer, San Antonio Area	40 FR 58344(1975)
	Edwards Aquifer, Austin Area	53 FR 20897(1988)
Virginia	Prospect Hill Aquifer, Clark County	52 FR 21733(1987)
Washington	Cedar Valley Aquifer, King County	53 FR 38779(1988)
	Cross Valley Aquifer, Snohomish & King Counties	52 FR 18606(1987)
	Camano Island Aquifer	47 FR 14779(1982)
	Whidbey Island Aquifer	47 FR 14779(1982)
	Newberg Area Aquifer, Snohomish County	52 FR 37215(1987)
Washington/ Idaho	Rathdrum Prairie Aquifer, Spokane Valley	43 FR 5566(1978)
	Lewistown Basin Aquifer	53 FR 38782(1988)
Guam	Northern Guam	53 FR 17868(1978)

it, bring enforcement action yourself when necessary. Look for changes in the land use or water use patterns of the area that may require amendments to the protection scheme.

SOLE SOURCE AQUIFER PROGRAM

If groundwater is the primary source of drinking water in your area, you can take advantage of another, more limited federal program—the Sole Source Aquifer Program. The first step is to petition the regional EPA office to have the aquifer designated as a "sole or principal source" of drinking water. Once an aquifer is designated a sole source aquifer, federal agencies are barred from granting financial assistance to projects that could, in the opinion of EPA, contaminate the aquifer. Highway projects, housing developments, and sewage water treatment works are the major activities affected by sole source aquifer designations. You may also find that a sole-source aquifer designation will assist in efforts to develop state and local protection programs for the aquifer. (See Chapter 23.) As of December 31, 1988, 49 aquifers were designated by the EPA as "sole source." (See Figure 12.2, on pages 158–59.)

A sole-source aquifer petition must include sufficient technical information to:

(1) Verify that the aquifer is the sole or principal source of drinking water (more than 50 percent);
(2) Substantiate the boundaries of the aquifer and its recharge area; and
(3) Show that contamination of the aquifer would create a significant health problem.

For assistance in preparation of the petition, present your ideas to local or state government officials or planners who have expertise in the technical and hydrogeological information necessary for the designation. Information required in these petitions is also available from the USGS, the state geological survey or water resources agency, soil conservation service, state and local health and environmental agencies, and local public water suppliers.

Once EPA has received a complete petition, it will conduct a detailed review to verify the technical information you present. Request a hearing during this phase of the review to support the petition and defuse any doubts EPA has concerning its validity or need. Encourage others in your community to participate in the hearing or file written comments, focusing on the lack of alternative water supplies and the importance of preserving the aquifer's quality. Generally, the designation decision process takes a minimum of six months from the submission of a petition to a determination by EPA.

If EPA denies the petition, you can modify and resubmit it. Request a meeting with EPA officials to discuss what must be amended in the petition to meet sole source aquifer criteria.

Sole Source Aquifer Demonstration Program

For aquifers designated as sole source aquifers before June 1988, EPA grant money may be available for the development of special state or local protection programs for especially important portions of the aquifer called Critical Aquifer Protection Areas. Enacted in 1986 as part of the Safe Drinking Water Act, this program, called the Sole Source Aquifer Demonstration Program, has not yet been funded by Congress, and EPA is not yet even accepting applications. Ask the EPA regional office about the status of the program and for more information about this source of funding.

Citizen Success

Four sole source aquifer designations were made for the Buried Valley Aquifers, Rockaway Quarternary Systems, Highland Aquifer System, and Ridgewood Aquifers in New Jersey, thanks in large part to citizen efforts. According to Ella Filipone of the Passaic River Coalition (PRC), this success was achieved by working with municipal officials to make them partners in the petition process. These officials in turn put the PRC in touch with city planners, water engineers, or attorneys within each municipality, who could quickly provide them with population statistics and hydrogeologic data necessary to complete the petition. Once the petition had been drafted, the PRC, municipal officials, and other interested persons held a public meeting to refine its contents and eliminate possible conflict between local governments and environmental groups. In the end, the coalition was able to submit to EPA a petition agreed to by the representatives of all local interests.

REFERENCES

AQUIFER PROTECTION PROGRAMS

EPA, Office of Groundwater Protection. *An Annotated Bibliography of Wellhead Protection Programs.* Washington, D.C. August 1987.
EPA, Office of Groundwater Protection. *Guidance for Applicants for State Wellhead Protection Program Assistance Funds Under the Safe Drinking Water Act.* Washington, D.C. June 1987.
EPA, Office of Groundwater Protection. *Sole Source Aquifer Petitioner Guidance.* Washington, D.C. February 1987. (Environmental groups object to some provisions in this guidance but it provides some information about this process.)
EPA, Office of Groundwater Protection. *Wellhead Protection: A Decision-Makers' Guide.* Washington, D.C. May 1987.
Feliciano, Donald V. *Wellhead Protection: The New Federal Role in Groundwater Protection.* Environment and Natural Resources Policy Division, U.S. Congressional Research Service. Washington, D.C. June 20, 1986.

LAWS AND REGULATIONS

U.S. Code Citation

Safe Drinking Water Act. 42 U.S.C.A. §§ 300f–300j (West 1982 & Supp. 1988).

Code of Federal Regulations Citation

40 C.F.R. Subpart 149 (Sole Source Aquifer Criteria).

Federal Register Citation

52 Fed. Reg. 23982 (June 26, 1987) (Criteria for identifying critical aquifer protection areas).

CHAPTER 13

Action Under Superfund

COLUMBIA, MISSISSIPPI WAS for many years the home of the Reichhold Chemical plant. The company abandoned its facility there after a fire and explosion destroyed most of the buildings at the site. Tragically, thousands of barrels of benzene, an industrial solvent believed to be a cause of leukemia, were left behind, attracting children who used the area as a playground and eventually leaking into the town's aquifer and contaminating its drinking wells. Faced with serious health problems, the residents banded together to form STOP (Stop Toxic On-site Pollution) and demanded cleanup action from EPA. After months of pressure from STOP, EPA responded under its Superfund program, evacuating citizens living near the site and removing the threatening barrels.

The fundamental goals of Superfund[1] are: (1) to provide authority and money for EPA to act quickly against hazardous materials spills and to conduct longer term cleanup actions at sites contaminated by hazardous substances; and (2) to make those who create or contribute to contamination problems undertake cleanup or pay for the costs of EPA cleanup actions. This chapter will tell you about how to initiate agency cleanup action under Superfund and how to influence the agency in the right direction once it begins to use its authority.

HOW TO IDENTIFY SPILLS AND CONTAMINATED SITES SUBJECT TO SUPERFUND ACTION

The first step toward getting EPA action under Superfund is to identify the problem areas where Superfund might apply. Under Superfund, EPA can move against nearly any kind of facility or contamination source, be it an abandoned waste dump, leaking injection well, bursting pipeline, or

[1] Superfund may also be referred to by its original formal name, the Comprehensive Environmental Response, Compensation, and Liability Act (CERCLA), or by the name of recent significant amendments to the law, Superfund Amendments and Reauthorization Act (SARA).

overturned tanker truck. Superfund's only limitation is that there be a release or threatened release of hazardous substances (or other pollutants that may cause health problems).[2] These categories are broad enough that most chemical releases or contaminated sites are likely to fall within Superfund's reach.

Beginning with Chapter 16, we discuss in detail various ways to gather information about particular types of potential contamination sources. Many of these techniques will enable you to discover not only the locations of potential problem sites, but also the types of substances that may be involved. For example, one particularly important strategy is to use the new federal toxics Right-to-Know law, which requires industrial and commercial facilities to report to local committees the types of hazardous substances they handle and any releases of such compounds. (See Chapter 17.)

To identify sites that the EPA has decided require Superfund action, ask the Regional EPA office for the latest National Priorities List (more about this later). To help you locate sites identified by the EPA as possible hazards, ask the EPA regional office for a copy of the CERCLIS (Comprehensive Environmental Response Liability Information Service) inventory. CERCLIS lists over 26,000 sites that may require Superfund action (though a recent government report suggests that there might be more than 15 times this number of contaminated sites).[3] EPA also has Superfund Comprehensive Accomplishments Planning (SCAP) reports that describe, by region, Superfund cleanup action, goals, and candidate sites.

You can also request reports of hazardous substance releases from the National Response Center (NRC). Established to coordinate emergency responses to oil and other hazardous materials spills, the NRC receives reports of most potentially dangerous releases from around the country. Write a letter to the NRC, describing the area and the time period you are concerned about and request spill information from the NRC computer files. Further investigation of the reports you receive may turn up contaminated sites that deserve Superfund action.

---

[2] EPA maintains and updates a list of all hazardous substances in 40 CFR Part 302. There are currently 717 substances on the list. Oil and petroleum products (including natural gas) are excluded from the list (except where they are mixed with other listed substances or break down into hazardous components).

"Release" is defined very broadly to include any type of spill or leak of a substance into the environment. Some releases are specifically excluded, including certain radioactive contamination covered by other statutes, normal workplace releases, and federally permitted releases, such as those allowed under the Clean Water Act in compliance with NPDES permits.

[3] U.S. Government Accounting Office. *Superfund: Extent of Nation's Potential Hazardous Waste Problem Still Unknown.* GAO/RCED-88-44, p. 21. December 1987.

Do not forget to research historical information. Interview community residents about the site. Employees at the facility, or employees of waste-hauling businesses serving the site, government officials, newspaper staff, and neighbors all may have helpful information. Library research can fill in gaps; film clips and old newspapers may recount past events at the site and identify the people involved.[4]

Keep a watch on the site even after it has been identified. Illegal activities, such as "midnight dumping" of hazardous wastes at uncontrolled or closed dumps, may be continuing.

HOW TO GET ACTION TO CLEAN UP SPILLS AND OTHER EMERGENCIES

If you discover evidence of a recent or continuing spill or other release of what may be a hazardous substance or dangerous pollutant (including oil) from any kind of facility, pipeline, vehicle or other source, you can initiate emergency cleanup by reporting the spill to the National Response Center (NRC). The NRC receives spill reports from around the country and relays the information to federal EPA or Coast Guard officials in the area, called On-Scene Coordinators (OSC). The OSC is then supposed to ensure that the spill is cleaned up, either by the person responsible for the release or by federal or state emergency teams.

To report a spill, call 800-424-8802 (or 202-426-2675). Be prepared to give as much information as you can about the release, including the date, time, location, source and cause of the spill, the type and amount of spilled material, any injuries, and the amount of potential damage.

As a practical matter, since many emergency response actions are completed within hours or days, there is not likely to be much time for you to influence emergency actions before they are taken. You can, however, review the completed emergency action to determine whether further steps should be taken. Obtain from the EPA a copy of the report prepared by the OSC at the conclusion of the emergency action. It will indicate whether any further long term cleanup is to be done by the agency. If not, and if you believe problems remain at the site, you have two options. One is to pursue an informal appeal of the decision to the EPA administrator explaining why the action is inadequate and seeking further cleanup. If an appeal is not successful, you can in some circumstances file a citizen suit against the agency. (See page 189 for more detailed advice on these

---

[4] For this and other good advice about investigation, see Dean, Lillian and Elizabeth Harris, *Groundwater Contamination Sites: A Citizens's Guide to Fact Finding and Follow-up*, East Michigan Environmental Action Council. Birmingham, Michigan (1983).

procedures.) The other option is to attempt to initiate a more thorough investigation of the potential problems at the site, by filing with the EPA a petition for a preliminary assessment of the site. This option is discussed in the next section.

PARTICIPATING IN SUPERFUND CLEANUP ACTIONS

This section describes how you can initiate and participate in cleanup actions under Superfund.[5] EPA follows two basic procedures to clean up contaminated sites. A *removal action* is a short-term, limited response to prevent emergency situations and remove immediate threats to health and the environment. A *remedial action* is a longer-term, more expensive, and permanent solution for more complex contamination problems. At some sites, it may be necessary to undertake both actions: for example, removal of waste on the surface of a site to abate further contamination, and remedial action to eradicate the pollution of soil and groundwater. To ensure that the best possible cleanup is completed, you can take steps to influence how the agency carries out its responsibilities.

Before you begin the steps outlined below, check into the possibility that a state agency, rather than the EPA, may be doing the Superfund cleanup work. States are encouraged to enter into cooperative agreements with EPA under which the state agency becomes the lead agency and carries out the cleanup actions. The same opportunities for public participation that EPA is required to provide, and that are described below, should be provided by any state agency if it is conducting Superfund work under a cooperative agreement.[6] So, though we will generally refer to EPA in this section, a state agency may be filling EPA's shoes in your state for some or all of the steps in the cleanup process.

[5] EPA's regulations governing Superfund cleanup actions are called the National Contingency Plan (NCP) and can be found at 40 C.F.R. § 300. As this book was going to press, EPA published proposed changes to the NCP to incorporate the Superfund amendments passed by Congress in 1986. See 53 Fed. REg. 51394 (Dec. 21, 1988). Though the proposed changes would not significantly alter the Superfund procedures outlined in this chapter, particularly as they relate to citizen participation, you should ask EPA about the status of these changes before you begin active participation in the cleanup process. It is possible that the final rules could include important modifications of the process. The proposed revisions to the NCP and the issues raised by the changes are discussed in a recent article by an EPA attorney. See Freedman, Joseph, *Proposed Amendments to the National Contingency Plan: Explanation and Analysis.* 19 Envtl. L. Rep. (Envtl. L. Inst.) News and Analysis, 10103 (1989).

[6] Don't confuse this with separate state mini-Superfund laws. Many states have enacted such laws on their own that may authorize state response entirely independent from the federal program. See the discussion of these laws at Chapter 24.

GETTING POTENTIAL SUPERFUND SITES
EVALUATED BY EPA

Once you have information that a particular site is or may be contaminated with hazardous substances or dangerous pollutants, you can prompt EPA action at the site, assuming that the agency has not already begun, by initiating the *site evaluation* process. This is the procedure that EPA uses to determine whether to take any cleanup action at the site and what type of action is necessary.

Preliminary Assessment

You should start by simply presenting the information you have gathered to the EPA regional office and asking that a *preliminary assessment* be done. The first step in the evaluation process, the preliminary assessment is generally limited to desktop reviews of site records, permits, and other documents that may disclose information about the site and its potential hazards. The agency may also contact individuals and agencies with knowledge of the site and its history.

You can influence the agency's decision to conduct a preliminary assessment by bringing together as much information as possible about the site and submitting it for review when you make a request. Information answering any of the following questions will be helpful:

- What kinds of activities are or have been going on at the site?
- Who owned and operated the site?
- What types of materials were handled at the site and in what quantities?
- How were materials disposed?
- Is there evidence of contamination of surface water or groundwater (see Chapters 4 and 5)?
- Does the geologic structure of the area make groundwater particularly susceptible (see Chapter 1)?
- Do local emergency records (e.g., fire department) show past problems at the site?
- How close to the site do people live, work, and play?
- Are there natural areas nearby that might be affected?

If EPA does not respond to your informal requests, you can file a formal petition with the agency requesting a preliminary assessment. The agency must then complete an assessment within one year or explain why it is not needed.

In some cases, the preliminary assessment alone will convince the agency that immediate removal action is necessary (see the next section). In most cases, however, if the preliminary assessment discloses that a suspected release may threaten human health or the environment, EPA will progress to the next investigative step and conduct a *site inspection*. Pressure from affected citizens to conduct a thorough investigation is important because EPA or the states have decided not to take action after preliminary assessments in two-thirds of the sites examined to date.

If EPA does an assessment and wrongly decides not to undertake any further investigation or action, you can file an informal appeal with the Administrator, and in some circumstances bring a citizen suit challenging the agency decision. (See page 189 for more about these options.) Political pressure and continued fact-gathering about the site may be more effective, however, given the deference the courts are likely to give EPA on this threshold decision. Try to persuade the agency to explain itself at a public meeting. Take your information to the press and publicly demand further action.

Site Inspections

Taking part in the site inspection means ensuring that the agency gathers all the necessary data from the site and then properly evaluates the data. The site inspection involves an actual visit to the site and surrounding area and may include the taking of photographs and samples of air, soil, water (including groundwater), and materials found at the site. Request from the on-scene coordinator (OSC) or EPA community relations personnel a copy of the proposed sampling plan and ask to meet with them to discuss their inspection plans. You should consider seeking some expert assistance at this stage to help evaluate the site and the investigations to be done by the agency. Take advantage of the opportunities to attend public meetings and review fact sheets the OSC should be providing.[7]

Some of the important issues to review are:[8]

- How many samples were taken and where? Do they accurately reflect the potential contamination at the site? Were background samples taken? Were samples taken offsite?

[7] EPA's public relations responsibilities and policies are outlined in detail in *Community Relations in Superfund: A Handbook.*

[8] For good detailed advice on the more technical aspects of evaluating site inspections, see EDF, *Dumpsite Cleanups: A Citizen's Guide to the Superfund Program*, Washington, DC, EDF, 1984.

- Were the samples handled properly to prevent deterioration of the contaminants? (EPA guidelines are at 40 C.F.R. Part. 261, Appendix I.)
- Have the samples been analyzed according to EPA standards?

At the conclusion of the site inspection, EPA will prepare a site investigation report detailing the information obtained by the agency and how it was gathered. Ask for a copy of this report and request to meet with the OSC to raise any inadequacies with the agency. Request a supplemental inquiry if you believe it is necessary. If the site inspection discloses problems requiring immediate attention, ask EPA to initiate a removal action. (See the next section.)

Health Assessments

Another step you can take at this stage to discover more about the health threats created by the site and give you more evidence to support a request for action, is to request the Agency for Toxic Substances and Disease Registry (ATSDR) for a health risk assessment. Your request should include the name of the site and a description of the potential problem with as much information about the site and contamination as you can gather.[9] The ATSDR received about 50 petitions for health assessments in 1987, and expects many more in future years. Priority for these assessments is based on several factors, including potential risk to human health, adequacy of existing data, and EPA's schedule for Superfund cleanups or site assessments. Health assessments may lead to pilot health effects studies, full-scale epidemiological studies, development of a registry of exposed persons, and health surveillance programs, all of which will add to the bank of data you need to force a cleanup.

You should also request copies of toxicological profiles from ATSDR for hazardous substances found at the site of concern to you.[10] ATSDR is required to prepare profiles explaining the available data about health risks associated with those chemicals most commonly found at Superfund sites. ATSDR has developed a list of 100 common contaminants and has prepared draft toxicological profiles for 25 of those compounds as of July 1988. (See Figure 13.1.)

[9] The complete procedure you should follow to request a health assessment is described in the November 24, 1987 Federal Register (52 Fed. Reg. 45018 (1987)). If ATSDR chooses not to conduct an assessment for the site, it must explain its reasons to you in writing.

[10] To obtain copies of toxicological profiles from ATSDR, make a request in writing to Ms. Georgi A. Jones, Director, Office of External Affairs, ATSDR, Chamblee Building, 1600 Clifton Road, Atlanta, GA, 30333. If they are unavailable from ATSDR, contact Superfund Reports at (800) 424-9068.

FIGURE 13.1

THE ATSDR/EPA LIST OF THE 100 MOST HAZARDOUS CONTAMINANTS

Group 1*	Group 2*	Group 3	Group 4
Arsenic	Benzidine	Acrolein	Aniline
Benzene	BHC-1,2,3,4	Acrylonitrile	Benzoic acid
Benzo(a)anthracene	Bis(chloromethyl)ether	Ammonia	Bromomethane
Benzo(b)fluoranthene	Bis(2-chloroethyl)ether	Bromoform	2-Butanone
Benzo(a)pyrene	Bromodichloromethane	Chlorobenzene	Carbondisulfide
Beryllium	Carbon tetrachloride	Chlorodibromomethane	1,2-Dichlorobenzene
Bis(2-ethylhexyl)phthalate	Chlordane	Chloromethane	1,3-Dichlorobenzene
Cadmium	Chloroethane	Copper	Dichlorodifluoromethane
Chloroform	4,4-DDE, DDT, DDD	Di-N-butyl phthalate	2,4-Dichlorophenol
Chromium	3,3-Dichlorobenzidine	1,1-Dichloroethane	Diethyl phthalate
Chrysene	1,2-Dichloroethane	2,6-Dinitrotoluene	2,4-Diitrophenol
Cyanide	1,1-Dichloroethene	1,2-Diphenylhydrazine	Dimethyl phthalate
Dibenzo(a,h)anthracene	1,2-Dichloropropane	Endrin aldehyde/endrin	2,4-Dimethylphenol
1,4-Dichlorobenzene	2,4-Dinitrotoluene	Ethylbenzene	4,6-Dinitro-2-methylphenol
Dieldrin/aldrin	Isophorone	Hexachlorobenzene	1,4-Dioxane
p-Dioxin	Mercury	Indeno(1,2,3-cd)pyrene	Fluoranthene
Heptachlor/heptachlor epoxide	N-nitrosodimethylamine	Naphthalene	Fluorotrichloromethane
Lead	N-nitrosodi-n-proplamine	Nitrobenzene	Hexachlorobutadiene
Methylene chloride	Pentachlorophenol	Oxirane	Hexachloroethane
Nickel	Phenol	Silver	P-Chloro-m-cresol
N-nitrosodiphenylamine	Selenium	Total xylenes	2-Pentanone, 4-Methyl
PCB-1260,54,48,42,32,21,1016	1,1,2,2-Tetrachloroethane	Toxaphene	Phenanthrene
Tetrachloroethene	Toluene	1,2-Trans-dichloroethene	Phenol,2-methyl
Trichloroethylene	1,1,2-Trichloroethane	1,1,1-Trichloroethane	Thallium
Vinyl chloride	Zinc	2,4,6-Trichlorophenol	1,2,4-Trichlorobenzene

PARTICIPATING IN ACTIONS AGAINST
IMMEDIATE HAZARDS: REMOVAL ACTIONS

You need to be active not only during EPA's assessment of the hazards posed by a contaminated site, but also in the cleanup process itself. Any time a preliminary assessment, site inspection, or other information (such as a report to the NRC) discloses an immediate hazard situation, EPA should begin a removal action, a short-term step designed to remove any imminent hazard at the site or stabilize contamination pending further cleanup. Removal actions can range from emergency actions that may begin within a few hours of the discovery of the problem and last only days or weeks, to more extensive waste removal and containment projects that may last up to a year or longer in some circumstances. In some cases, the polluter may do the removal under agency supervision. The opportunities you have to participate and influence cleanup actions vary with the urgency of the action.

Removal Actions Taken by Agencies

Whether the removal action is short or long term, the first step is to find out quickly what EPA plans to do and try to influence the agency to address your concerns. Even for very short-term removal actions, EPA will designate an on-scene coordinator (OSC) to oversee the response and a community relations spokesperson who will provide information about the problems at the site and removal actions and act as a conduit for community concerns. Request copies of the pollution reports written by the OSC, which, depending on how long the response lasts, may be prepared on a daily basis. Ask to meet with the spokesperson and the OSC or technical staff for explanations of the removal action and press for improvements in their response plan if you believe it fails to address specific problems.

Removal action may include any of the following, depending on the circumstances:

- Fencing of the facility
- Measures to control surface water drainage onto or from the site
- Repair of a design malfunction, such as a leak from a waste pile liner
- Restraint of chemical releases with physical or biochemical barriers or construction of caps over the waste
- Removal of the waste or contaminated soils off-site
- Provision of alternative water supplies
- Evacuation of threatened individuals

When a removal action will last longer than 45 days, you can also learn about the site and the action by checking the information repository that EPA will establish near the site at a public library, town hall, or other public building. Review the documents in the repository describing EPA actions at the site, and ask for meetings with the spokesperson or the OSC to clarify anything you do not understand or to raise objections to the actions.

The repository will also have a community relations plan. Among other things, the community relations plan will describe the ways in which the agency will keep citizens informed of new developments and solicit citizen input. Make sure the agency follows these minimum procedures and ask for improvements in the plan if it fails to provide for enough citizen involvement. In situations generating considerable citizen concern, the plan may include an on-scene information office staffed by agency personnel that can give you immediate access to information as it develops.

You can take advantage of requirements for more formal public participation procedures that are triggered when action can be delayed for at least six months (called "non-time-critical removal actions" in agency jargon). In these circumstances, make sure you become involved in the preparation of EPA's Engineering Evaluation/Cost Analysis. It is your best opportunity to influence the removal action. This analysis consists of a review of alternatives for the proposed action, a description of the selected plan, project costs, technology to be used, and a full disclosure of the environmental effects.

In cases where the agency will be doing this analysis, contact the staff as early as possible to raise questions about the removal action, preferably well before the agency actually writes a draft analysis. If you miss this opportunity, look for public notices that a draft of the analysis is ready for public comment. You can also find the draft analysis in the information repository. You will have three weeks to review and make comments on the draft before the agency makes a final decision on how to proceed with the removal action. Request that a hearing be held during this period to allow you to present oral and written comments, new data, and testimony from experts.

After the hearing and comment period, the agency will issue a final decision. Review the agency's explanation of its decision (called the responsiveness summary). This should answer each significant comment and provide an explanation of how and why the agency made its final removal decision.

Points to focus on in your review of any removal action regardless of its duration or urgency include:

- Will the action actually remove significant contamination from the site or will wastes only be contained so that further action will be needed?
- Will groundwater be monitored to detect leaching of contaminants?
- Will there be a schedule made available so that you can monitor progress at the site?
- Have predicted health effects been taken into account? Has there been any assessment of health effects on the community?
- Is the action broad enough to address migration of contaminants to adjacent areas?

Removal Actions and Responsible Parties

In some circumstances, removal actions may be carried out by persons or companies responsible for the pollution at the site (EPA calls them potentially responsible parties or PRPs). If the responsible party is identifiable and has the resources to take effective action, EPA may use its enforcement authority to force that party to take whatever actions are necessary to abate the hazard. You will still have opportunities to be involved in the action. Agency personnel will supervise the private removal through the OSC and should follow the same public participation procedures that apply to agency actions and have been described in the previous section.

In emergency situations where a response must begin immediately, the agency is likely to take the action itself unless the responsible party does so voluntarily. (If a company does take emergency action, you can request the OSC to step in and finish the job, if necessary.) Where action can be postponed or may last many months, the agency is more likely to try to get those responsible to do the work. EPA may have to go to court to force action, or may reach a settlement with those responsible that specifies the removal action that will be done. You should take advantage of rights to intervene in such court actions or get involved in pretrial settlement discussions because their outcome may affect how thoroughly a contaminated site is cleaned up. See page 189 for more advice about how to participate in lawsuits and settlement discussions.

Challenging Inadequate Removal Action

What do you do if the agency says the removal is complete, but you believe the site still presents an imminent health hazard? Your best option is to continue to work with and prod the agency. Exert political pressure on the agency by educating the public about the problems that remain and by using the media.

You can also file an informal appeal to the EPA Administrator asking for additional removal action and, if necessary, bring a citizen suit in court. (See page 189 for more about these steps.)

Challenging Inadequate Removal Action

Residents of Fayette County, West Virginia, can tell you about removal action that does not go far enough. They are plagued by a mound of PCB-contaminated soil created when electrical transformers were drained near the Shaffer Equipment Company site. Seventy families live nearby; many have traces of PCBs in their bodies and in the land surrounding their homes. PCBs have migrated into groundwater and into creeks and rivers.

EPA spent $2.3 million in an emergency response action to extract the PCBs from the soil and, when that failed, simply consolidated the contaminated soil. But that did not solve the problem. Levels of PCBs at the site reached 280,000 ppm, 5,600 times the level at which PCB-contaminated material is so toxic that it must be disposed of by incineration. Medical histories of 325 Minden, West Virginia, residents showed 8 percent had cancer, 30 percent had recurrent infections, 30 percent had kidney problems, and 8 percent had liver problems.

Concerned Citizens to Save Fayette County (CCSFC) formed to force action in the community. Appalachian Student Health Coalition helped to document health problems, and interns from the Student Environmental Health Project at Vanderbilt University helped identify the extent of the PCB contamination in the soil and water at the site. CCSFC enlisted the support of local representatives, including West Virginia Senator Rockefeller and Congressman Harley Staggers, Jr.

CCSFC is still fighting to get action. The state has so far rejected proposals to buy out the affected families and EPA is still evaluating the site to decide whether additional cleanup will be done under Superfund. This is not a story of citizen success (at least not yet) but a reminder of how important it is not to assume your problems are solved once EPA arrives on the scene.

REMEDIAL ACTIONS:
PARTICIPATING IN PERMANENT CLEANUPS

The real focus of citizen action in the Superfund cleanup process should be the remedial action phase. This is when EPA decides what steps it will take, or force the polluter to take, to clean up permanently a contaminated site. Unfortunately, not all contaminated sites will necessarily reach this stage of the process. But for those that do, citizen participation may make the difference between effective, thorough action and an incomplete cleanup that will haunt the community in the years ahead.

As is true for most of the opportunities for citizen participation, the key to effectiveness is early action. Once you know a site qualifies for Superfund cleanup, begin working with experts and the agency to make sure your concerns are resolved in the remedial action. Recent changes in

Superfund also create large incentives for those who contributed to the contamination problem at a given site, and who may be liable for the cleanup costs, to become involved early in the process and agree to do certain cleanup work in exchange for some limitation on their financial liability. You must try to be an active player in the discussions leading to a settlement agreement, primarily because such early agreements may affect how thorough the cleanup is as well as who will pay for it. We discuss these procedures in more detail at page 188.

Participating in EPA's Hazard Ranking Process

The first step is to make sure the site qualifies for Superfund remedial action. EPA cannot take remedial action at a site unless it presents a serious enough problem in comparison to other problem areas to make it onto the National Priorities List (NPL), EPA's list of the worst sites in the country. To decide which sites to put on the NPL, EPA analyzes the data it gathers during the site evaluation process (see page 167) with a model called the Hazard Ranking System (HRS). Although EPA is currently in the process of revising the HRS to conform to the 1986 Superfund Amendments, a discussion of the current system provides a useful starting point. The model assesses various factors (waste characteristics, human exposure routes, and proximity to groundwater are just a few) to produce a relative ranking of the potential for releases and health hazards at waste sites around the country. Only those sites that generate a high enough score to be included on the NPL are eligible for long-term, permanent cleanup with Superfund money, and only those near the top of the list are likely to receive quick action.[11] In April of 1988, there were 802 sites on the list, with 149 more proposed. (See Figure 13.2.)

If you have not already, review the agency's preliminary assessment and site inspection reports to ensure that they give as complete a picture as possible of the site and the surrounding area. The agency will rely on these reports for the data required by the HRS model and any inadequacies in the reports may lead to an inappropriately low hazard ranking. By studying the HRS model, which is described in its current form at 40 C.F.R. Part 300, Appendix A, you can get a good sense of the kind of information that is important. (See Figure 13.3.) Make sure the reports are accurate and complete on these points and submit additional information if you can to the agency.[12]

Once the agency has completed the HRS ranking, request copies of the

[11] EPA does have the option of using its enforcement authority to force polluters to conduct cleanups at a site even though it does not rank high enough to qualify for EPA cleanup with Superfund money. (See Chapter 14.)

[12] At this writing, EPA is revising the HRS. The proposed regulations were issued in December 1988. Check with the regional EPA office to learn about changes to the system.

FIGURE 13.2

CERCLIS-NATIONAL G.M.1 NATIONAL PRIORITIES LIST (NPL) SITES

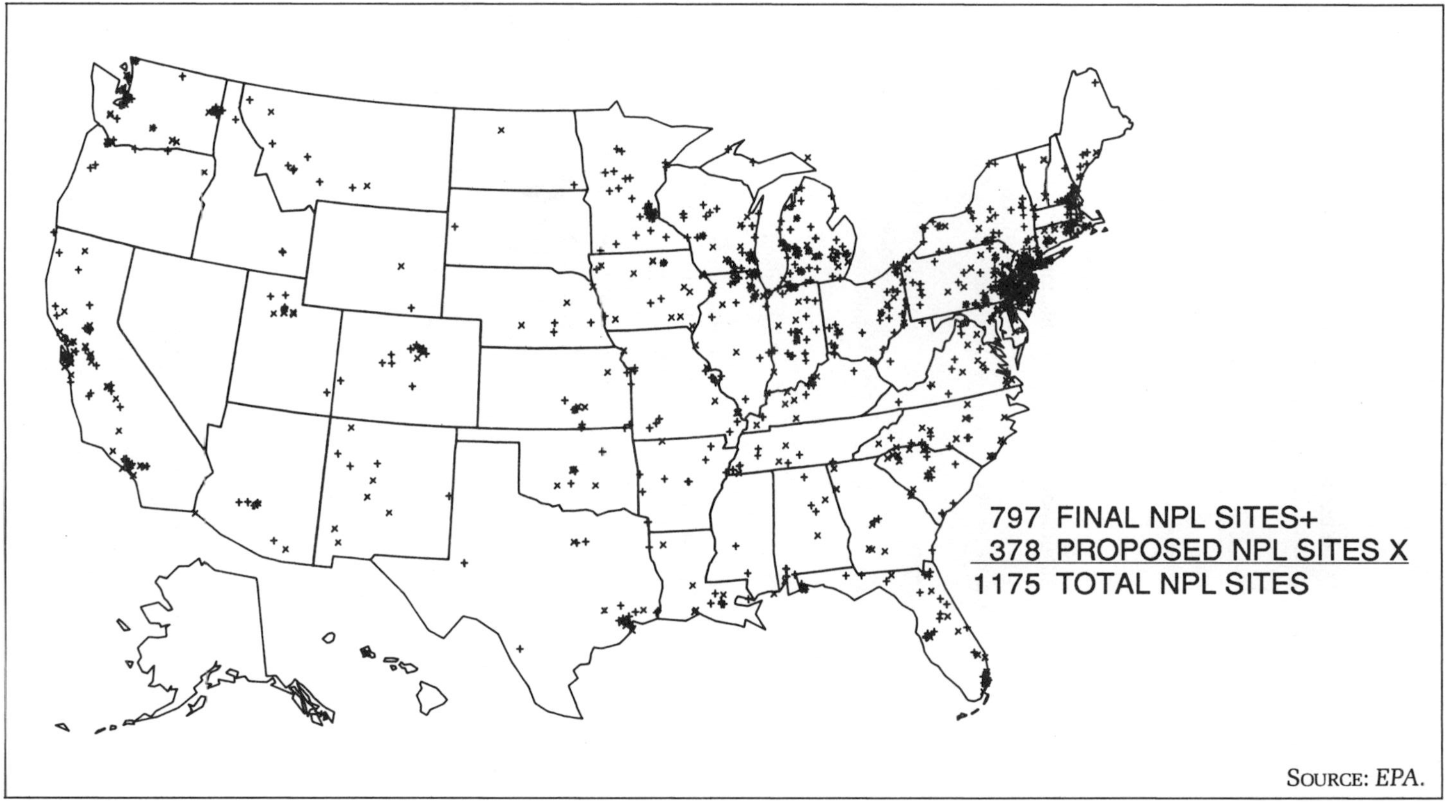

797 FINAL NPL SITES+
378 PROPOSED NPL SITES X
1175 TOTAL NPL SITES

SOURCE: EPA.

FIGURE 13.3
HAZARDOUS WASTE RANKING SYSTEM FACTORS

Hazard Mode	Factor Category	Ground Water Route
Migration	Route Characteristics	• Depth to Aquifer of Concern • Net Precipitation • Permeability of Unsaturated Zone • Physical State
	Containment	• Containment
	Waste Characteristics	• Toxicity/Persistence • Hazardous Waste Quantity
	Targets	• Ground Water Use • Distance to Nearest Well/ Population Served
Fire and Explosion	Containment	• Containment
	Waste Characteristics	• Direct Evidence • Ignitability • Reactivity • Incompatibility • Hazardous Waste Quantity
	Targets	• Distance to Nearest Population • Distance to Nearest Building • Distance to Nearest Sensitive Environment • Land Use • Population Within 2-Mile Radius • Number of Buildings Within 2-Mile Radius
Direct Contact	Observed Incident	• Observed Incident
	Accessibility	• Accessibility of Hazardous Substances
	Containment	• Containment
	Toxicity	• Toxicity
	Targets	• Population Within 1-Mile Radius • Distance to Critical Habitat

SOURCE: 40 C.F.R. Part 300, Appendix A.

Factors	
Surface Water Route	*Air Route*
• Facility Slope and Intervening Terrain	
• One-Year 24-Hour Rainfall	
• Distance to Nearest Surface Water	
• Physical State	
• Containment	
• Toxicity/Persistence	• Reactivity/Incompatibililty
• Hazardous Waste Quantity	• Toxicity
	• Hazardous Waste Quantity
• Surface Water Use	• Land Use
• Distance to Sensitive Environment	• Population Within 4-Mile Radius
• Population Served/Distance to Water Intake Downstream	• Distance to Sensitive Environment

scoring sheets and other materials (called the HRS package documents) used by EPA to arrive at the ranking. With the help of experts, if possible, review EPA's application of the model to the site and submit comments to EPA challenging any factual innacuracies or inappropriate scorings.

Citizen groups have been successful with this strategy. We Are the Endangered Residents (WATER), a citizen's group in Rhode Island, conducted its own ranking of a landfill in its community where there had been illegal open dumping. WATER used the HRS to conduct its own evaluation of the site, and found it ranked a score of 40, well above the 28.5 HRS cutoff for inclusion on the NPL. EPA's ranking of the site was 12.

FIGURE 13.4
SUPERFUND SCORECARD

Action	Description	Number
Sites in CERCLIS	Inventory of abandoned or uncontrolled hazardous waste sites	27,797
Preliminary Assessments (PA)	Initial assessment of potential hazards	25,034
Site Inspections (SI)	On-site investigation of hazards	8,075
National Priorities List (NPL)	List of hazardous waste sites eligible for Superfund monies	951
Removal Actions	Short-term actions to address immediate threats	
• NPL		264
• Non-NPL		818
Remedial Investigations/ Feasibility Studies (RI/FS)	Engineering studies at NPL sites to examine contamination and identify possible remedies	
• In Progress		472
• Cumulative Starts		578
Remedial Designs (RD)	Detailed design plans and specifications for imple-	
• In Progress	menting chosen remedy	106
• Cumulative Starts		181
Remedial Actions (RA)	Construction or implementation of chosen remedy	
• In Progress		65
• Cumulative Starts		168
NPL Site Cleanups	Site cleanup actions complete	
• Site Work Completed		22
• Deleted from NPL		13
Settlements	Sites where potentially responsible parties (PRPs)	
• Number	are conducting cleanup work	444
• Value		$750M

SOURCE: *EPA Office of Solid Waste and Emergency Response, Superfund Advisory, January 1988*

WATER reviewed the docket of the scoring conducted by the agency, and submitted comments to the EPA regional office. Explaining why the score should be higher, WATER convinced EPA and reached a settlement, giving the group the right to choose a consultant to formulate a remedy and oversee the cleanup plan.

You can also influence the HRS score for a particular site by working with state agencies. EPA must currently review state suggestions for sites suitable for the NPL and select top priority sites based in part on state submissions. Ask the state environmental agency when it plans to make its submissions to EPA so that you can comment on its selections and encourage it to include sites that concern you. (See Figure 13.4.)

How to Get Federal Money to Help Understand the Problems

To help you understand the threats posed by a contaminated site and the adequacy of cleanup measures EPA proposes for the site, you can apply for a grant from the EPA for up to $50,000 to hire hydrogeologists, toxicologists, and other experts to help you evaluate the problem. These grants are limited to cases where EPA is pursuing remedial action (that is, sites listed on the NPL). Recipients must generally contribute 35 percent of the cost of technical assistance.[13]

Ocean County Citizens for Clean Water was the first citizen's group in the nation to receive a grant, even though the program was not yet fully implemented by the EPA. They circumvented this problem by receiving the money directly from the polluter. Because the technical assistance program is designed to have responsible parties reimburse the EPA for issuance of these grants, this approach achieved the same bottom line, and gave the group the benefit of expert assistance. The group hired hydrogeologists, pathologists, and toxicologists to guide them toward an effective remedy.

Although the application procedures are overly confusing and demanding, citizens should be aggressive in applying for grants. One national activist suggests that citizens should demand that EPA walk them through the process. If EPA fails to provide sufficient help, you should contact an environmental group or your Congressperson for assistance.

Obtaining Health Effects Information

For information about the potential health risks posed by a site, ask the regional EPA office for a copy of the health assessment that should be prepared by the ATSDR. For every site listed on the NPL, the ATSDR is

---

[13] Interim application procedures for technical assistance grants are contained in the *Federal Register* of March 24, 1988 (53 Fed. Reg. 9736). These procedures may change, so contact the regional EPA office for the latest guidance.

required to conduct a health assessment within one year of the date the
site is first proposed for listing. The assessment should be completed
before the agency initiates the detailed study and analysis of potential
remedial options discussed below. The health assessment may lead to
more detailed health studies of the community or to emergency action to
reduce immediate health threats. If the ATSDR has not completed the
health assessment within the required period, you can bring a citizen suit
to force the ATSDR to do its job.

Remedial Investigation

PARTICIPATING IN THE REMEDIAL INVESTIGATION

Once EPA decides to begin a remedial action, the most important part
of the remedial process for citizen groups to follow closely is the remedial
investigation and feasibility study (referred to as the RI/FS by EPA). The
first part of this process, the remedial investigation, is designed to gather
detailed and extensive information about the site and surrounding areas.
Similar in some respects to the less involved site inspections conducted
prior to removal actions, the remedial investigation may last as long as
several years. Look for and take advantage of the same sorts of public
participation opportunities we described for removal actions. There will
be an agency staff person in charge of the process, who may be referred to
as the RPM (remedial project manager) rather than the OSC. The RPM and
agency community relations personnel should conduct community inter-
views, prepare a community relations plan, create an information repository for important documents concerning the site, and hold public
meetings to describe the agency's plans and activities.

You should focus first on the remedial investigation work plan (a copy
should be kept in the information repository) to ensure that the agency's
information about the site will be complete before it begins to plan the
cleanup action. Critique the investigation work plan and raise any inadequacies with the agency. The plan should include extensive soil, groundwater, and surface water sampling, and detailed geophysical surveys of
the area to locate and identify all contamination sources.

Submit all the information you have and can gather about the site and
its effects: well-water quality test reports, reports of unusual or severe
health problems, evidence of contamination that has migrated from the
site, historical information and the like. (See page 168 for further suggestions.) Use experts to bolster your position, and request to meet with
agency staff regularly to keep current with new developments at the site.
Remember also that any time the investigation discloses an emergency,
the agency should take immediate removal action. At the conclusion of

the investigation, review the agency's report and press for further investigation if necessary.

INFLUENCING THE FEASIBILITY STUDY

Influencing the choice of the remedial actions to be taken at a site is the most critical aspect of citizen participation in the cleanup process. EPA uses what is called a *feasibility study* to make these decisions. The final study will describe the alternatives for remedial action at the site and which ones were selected by the agency. Though the agency is required to issue a draft study, give the public a three-week comment period, and hold public hearings before making a final decision, you should begin to bring key issues to the agency's attention well before the draft study is available.

CRITICAL ISSUES IN THE FEASIBILITY STUDY ANALYSIS

Regardless of the particular remedies that may be proposed for a site, there are several general issues that you should focus on during the early phases of the feasibility study process. The first is to advocate permanent cleanup remedies. EPA has tended to select strategies that attempt to contain wastes at the site rather than remove, treat, or destroy the wastes, primarily because containment was generally much cheaper. Unfortunately, selecting this strategy almost always means the problem will resurface when the containment fails. Superfund now requires, therefore, that EPA give preference to permanent measures that treat wastes to reduce their toxicity, volume, and mobility, and strongly disfavors simply moving wastes from the contaminated site to another disposal facility or containing the wastes at the site.

Recent reports show, however, that EPA is ignoring these new requirements at the large majority of Superfund sites and still choosing "cleanup" plans that leave untreated wastes in the ground, where they are certain to cause continuing contamination problems.[14] Aggressive and organized citizen pressure on EPA to select permanent treatment remedies is essential to prevent Superfund cleanups from becoming merely temporary measures.

There may be several alternatives for remedial measures, including waste treatment, that offer relatively permanent solutions for a given site. Deciding which alternatives to support is a difficult task and depends on

[14] U.S. Congress, Office of Technology Assessment, *Are We Cleaning Up? 10 Superfund Case Studies—Special Report*, OTA-ITE-362, U.S. Government Printing Office, June 1988; Environmental Defense Fund *et al., Right Train, Wrong Track: Failed Leadership in the Superfund Cleanup Program*, Washington, D.C., June 1988.

the specific circumstances at the site. In the next section we review briefly several types of remedial action and suggest resources to use to help you make choices about which measures will work best in your case. Many remedial techniques, especially waste treatment methods, are relatively new and untested, and when proposed for a site near you deserve special scrutiny to ensure there will be adequate supervision and monitoring of the process so that unexpected problems can be identified and remedied quickly.

The second major issue that deserves close attention is the question of how clean the site should be when the action is complete. Potentially, one of the most stringent requirements of Superfund is that site cleanups meet all "applicable, relevant, and appropriate" pollution standards established under federal or state law (EPA refers to such standards as ARAR's). Deciding what standards will apply to a particular site will involve legal issues and discretionary judgment calls by EPA. These decisions deserve your closest scrutiny because they may determine, in the end, what remedial actions EPA or private parties take and how thorough and permanent the cleanup really is. EPA is so far refusing to apply protective standards at most Superfund sites. For example, even though Superfund says EPA should generally apply the stringent maximum contaminant level goals (MCLG's) set under the Safe Drinking Water Act and water quality criteria set under the Clean Water Act (see Chapters 11 and 17), EPA has actually adopted a policy that prevents their use as standards at most sites. Here again, citizen pressure must be applied to force EPA to adopt the stringent cleanup standards that the law requires.[15]

A final issue to pay close attention to in every case is the cost-effectiveness of the remedial measure. As part of its decision-making process, EPA can consider whether a particular measure is cost effective. Companies responsible for contaminated sites are likely to argue for stringent application of the cost effectiveness test to minimize their potential liability. Your goal should be to ensure that EPA first properly determines what standards the cleanup should meet to protect health, and applies the cost-balancing test only to the selection of the best

---

[15] In a few limited situations, Superfund allows EPA to select a cleanup plan that will not meet these pollution standards (or ARAR's). See 42 U.S.C. § 9621. Take a careful look any time EPA suggests one of these exemptions applies, particularly if EPA says the method selected will provide "equivalent" protection or that to meet the standards would be too expensive and prevent other needed cleanups using Superfund money. Of the several exceptions, these two are most easily misused by the EPA.

State groundwater classifications assigned to a particular aquifer (drinking water, non-drinking-water, etc.) may influence significantly whether certain pollution standards will apply to a site affecting the aquifer. (See Chapter 14 for our discussion of these classification systems.)

method to achieve that level of cleanup. EPA should not use the cost test to decide how clean to leave a site.

OTHER IMPORTANT CONSIDERATIONS

In addition to these three major points, review of any feasibility study should consider the following issues:

- Did the feasibility study evaluate a wide range of alternative remedial measures? This is particularly important where private parties are actually doing the feasibility study, because prior agreements may artificially limit the types of alternatives that will be considered.
- Will response action to be completed by private parties be properly supervised?
- Does the plan contain safety measures so that the cleanup construction work does not create new health threats?
- Is there a well-defined schedule and monitoring system in place so that you can follow the progress at the site?
- Does the plan provide for long-term operation and maintenance of the site (groundwater monitoring and pumping, waste treatment, etc.) after major construction work is finished? And in particular, since those costs are born by the state, not Superfund, does the state have adequate resources to fund the maintenance?

CHOOSING AMONG REMEDIAL MEASURES

Selecting the proper remedial measures for a given site depends largely on the characteristics of the site and the nature of the wastes involved. A wide variety of techniques are used to handle the waste itself and to control surface water, soil and groundwater contamination, and even air pollution at a site. These options are summarized in the following box. Not all of these measures provide the permanent remedies you should be advocating. To evaluate the suitability of any of these measures, you should seek expert guidance. Several of the resources listed at the end of this chapter provide more detailed advice or information about remedial measures.

REMEDIAL OPTIONS

The remedial methods explained below are matched to a particular resource, but most can be used to control a variety of waste problems, and are often used in combination with one another.

Groundwater

1. *Impermeable barriers,* including *slurry walls* and *grout curtains,* are installed beneath the ground to contain, capture, or redirect groundwater flow in the vicinity of a site. *Slurry walls,* usually made of

minerals and water, are built into vertical trenches to contain a plume, often for later pumping or treatment. (See Figure 13.5.) *Grout curtains* are created by injecting fluids into soil or rock to reduce water flow and strengthen the formation.

2. *Permeable treatment beds* are trenches placed perpendicular to groundwater flow and filled with an appropriate material to treat the plume as it flows through the material. Depending on the nature of the wastes, limestone, crushed shells, and activated carbon are used for plume treatment.
3. *Groundwater pumping* is used to remove contaminated groundwater or to isolate a plume of contamination by changing groundwater flow patterns. It is often used with slurry walls and other methods to contain contamination.
4. *Subsurface drains* are often used in conjunction with groundwater pumping. These drains include any type of buried conduit that draws in discharges by gravity flow, and collects them into a subsurface well for treatment.
5. *Leachate control* systems prevent seepage from wastes on the surface from forming or entering groundwater below. Immobilization is a common leachate control method, where contaminants are made insoluble to prevent leachate from forming or moving off or below site.

Soil
1. *Incineration* is the on- or off-site burning of soils at high temperatures with proper controls to prevent the resulting toxic ash from creating another waste problem.
2. *On-site heating* of soil is done to destroy or remove organic contamination in the subsurface through decomposition, vaporization, and distillation.
3. *Soil flushing* is a process of extracting contaminants from soil for detoxification.
4. *Surface capping* is used to cover buried waste to prevent contact with water and to prevent leachate formation. Caps can be single or multilayered, made of synthetic materials or clay. The proper structure and design of a cap is site-specific.
5. *Air stripping* is a process in which volatile chemicals in water or soil are transferred to gas. (See controls on air pollution.)
6. *Bioreclamation* is the process of injecting nutrients and sometimes air into the subsurface of a site to enhance biodegradation in contaminated zones.
7. *Microbiological degradation* involves injecting bacteria into soil or wastes to detoxify contaminants.
8. *Dredging or excavation* removes contaminated soil from the site. Cau-

FIGURE 13.5
EXAMPLES OF SLURRY WALLS

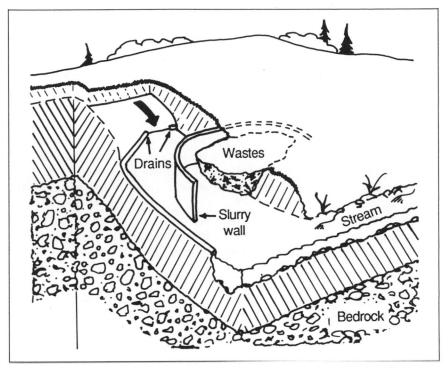

tion must be taken not to spread contamination by stirring up wastes at the site.

Surface Water

1. *Floating covers* are synthetic covers placed over surface impoundments with proper anchoring at the edges to prevent run-off into surface water.
2. *Grading* is a general term for the techniques used to reshape soil or wastes left on-site to help prevent surface water infiltration and run-off.
3. *Revegetation* is the growing of a vegetative cover (grass, etc.) over waste disposal sites to decrease erosion by wind and water.
4. *Dikes and berms* are well-compacted ridges or ledges to divert run-off from sites to manmade drainageways or outlets to prevent waste migration.
5. *Channels and waterways* are manmade tunnels used to intercept run-off. *Seepage basins and ditches* are also used for this purpose.

Sedimentation basins and ditches are used to collect the sediment suspended in the liquid run-off.

Contained Wastes

1. *Activated carbon treatment* involves the application of carbon directly to the wastes to absorb hazardous constituents.
2. *Filtration* is a process of removing suspended solids from leachate by forcing the liquid through a drainage system.
3. *Precipitation* is a process of transforming a liquid waste into a solid waste for easier containment (and possibly removal) at a site.
4. *Biological methods,* including *wastewater treatment technology, anaerobic lagoons,* and *biological reactors* are used to treat the waste on-site by using air or nutrients to decontaminate the waste.
5. *Chemical methods* to treat waste on-site include *chlorination* and *oxidation*.
6. *Physical methods,* many of which are discussed as treatment for soil or groundwater contamination, include *incineration, dredging, excavation, carbon absorption* or *permeable bed treatments.*

Air

1. *Dust suppression devices,* made from a wide range of natural and synthetic materials, bond soil to prevent contaminants from blowing off-site.
2. *Spraying of water* on wastes is the most commonly used method to control dust emissions, especially at excavation areas.
3. *Gas collection systems* alter the path of gas migration by mechanical means. This system typically consists of gas extraction wells, gas collection tanks, vacuum blowers to pull in gases from the wastes, and a treatment system.
4. *Liners and caps* prevent fumes from escaping wastes contained on-site.
5. *Wind fences* are porous screens that deflect wind to prevent erosion or dispersion at a site.
6. *Pipe vents* are used to channel gases from a site into collection centers for treatment.

REVIEWING THE FINAL DECISION

To evaluate the agency's final remedial decision, review the agency's Record of Decision, a document that announces the selected remedial action and explains why alternative plans were not selected, and the responsiveness summary, which gives the agency response to citizen comments and explains any changes to the final remedial plan made as a result of public participation. EPA should provide public notice when the

Remedial Plan Success

In Massachusetts, Clean Water Action (CWA), a national environmental organization, and People United to Restore the Environment (PURE), a local citizens' group, worked together to convince EPA to adopt a remedial plan considered by many to be a model for future cleanups.

Together, PURE and CWA spread awareness of the threat created by an abandoned chemical and pesticide plant, organizing a door-to-door canvassing drive, appearing on cable television, and running petition drives to generate support. Contamination spread 60 feet below the site, causing the town to close three of its wells. A nearby river was polluted, threatening a local reservoir, the community's alternative source of drinking water. A citizens' task force formed of representatives from PURE, CWA, elected officials, and citizens-turned-activists united to establish goals for cleanup at the site, including a plan to restore contaminated groundwater in the area, destroy toxic-laden soil at the site, and a guarantee of citizen oversight of the remedial action conducted.

These demands were met by the EPA in a $67 million cleanup plan that includes a permanent remedy instead of merely containing wastes on-site and gives local citizens substantial opportunity to monitor the cleanup.

decision is available and will place a copy in the information repository. If the decision is inadequate, see page 189 for our suggestions on challenges to EPA decisions.

Following Design and Implementation of the Remedial Action

The last step in ensuring effective Superfund cleanup is becoming involved in the actual design and implementation phases of remedial action. During the design phase, final engineering and construction decisions are made to implement the selected remedial plan. Request information about the design process and give the agency your comments. Stay involved so that the design decisions are consistent with the remedial plan.

Once construction and cleanup begin, you should continue to follow the progress at the site to ensure that schedules are adhered to and all necessary work is completed. Ask for regular meetings with the remedial project manager and encourage the agency to provide site tours so you can see the work yourself. When the cleanup is done, make sure that any long-term operation and maintenance at the site (most common when wastes are left at the site) is done properly (generally by state agencies). Ask for groundwater monitoring reports and maintenance records, and inspect the area yourself to look for obvious surface problems (erosion of

Federal Facilities

What do you do if the site is on a military installation or other federal facility? Nothing much different. Federal facilities are subject to the same basic process as any other Superfund site. At military installations, EPA and the Defense Department share responsibility for cleanup decisions, so you will have to work with both agencies. Otherwise the process is the same. You can get information about contamination at federal areas from the Federal Agency Hazardous Waste Compliance Docket at the EPA regional office.

Do not rely on federal agencies to discover all the problems on their property. The Army's Toxic and Hazardous Materials Agency conducted a records search of the Twin Cities Ammunitions Plant in New Brighton, Minnesota in 1978, and claimed the study showed no evidence of significant potential for groundwater contamination. However, a few years later the

Minnesota Pollution Control Agency discovered trichloroethylene (TCE) in the groundwater beneath and surrounding the plant. Some of New Brighton's municipal wells were later determined by the Minnesota Health Department to have up to 180 parts per billion of TCE in 1981, more than 6 times the state limit at the time. The city spent $5 million on deepening the existing wells and drilling three new ones. An additional $10 million was spent to clean the existing wells for use.

One additional set of requirements that you can use are the deadlines for evaluating and cleaning up federal sites. By April 1989, EPA must complete a preliminary assessment and decide whether to conduct cleanup for all contaminated facilities on the Compliance Docket list. You can bring a citizen suit to force EPA to make necessary decisions for listed sites.

surface caps, for example). Look for and participate in EPA's review of the effectiveness of the remedial action; such a review must be done every five years if wastes are left on the site.

PARTICIPATING IN CLEANUPS INVOLVING THE POLLUTER

You need to pay particular attention to the possibility that those responsible for the contamination (called potentially responsible parties, or PRPs, in agency jargon) will be involved in and influence the cleanup at a site. Responsible parties can include any company connected with the site—generators and transporters of the hazardous material and owners of the site. Such companies may become involved either through settlements with EPA before cleanup action begins or through EPA orders or court actions brought by EPA to force cleanup.

EPA may reach settlements with responsible parties before it begins any cleanup action through new special notice procedures. Once EPA is aware of a site that needs action and knows the identities of at least some of those involved with the site, it has the option of initiating settlement by notifying all the parties of the problem and their potential liability. Discussions are then likely to proceed quickly (EPA guidance says within 90 to 120 days) and may result in an agreement that outlines not only liability, but the cleanup options that will be evaluated. As a result of such an agreement the PRP may take responsibility for the remedial investigation and feasibility study, recommend cleanup actions, and even do the cleanup work.

As soon as you become aware of a potential Superfund site, ask EPA to be notified of the initiation of any such discussions so that you can sit in on discussions or at least follow their progress to ensure that no shortcuts are included in the agreement that limits the types of cleanup options that will be considered.[16] Though the public will have an opportunity to comment on any agreement before it becomes final, you are likely to be too late to achieve major changes in the agreement if you wait for the public comment period before getting involved.

EPA may also use its enforcement authority to issue an order or to go to court to force responsible parties to clean up a site (even if it is not listed in the NPL but still presents an imminent hazard). (See Chapter 13 for more about this EPA authority under Superfund and other laws.)

CHALLENGING AGENCY REMOVAL AND REMEDIAL DECISIONS

What can you do if EPA's final removal or remedial decisions or actions stop short of what you believe is necessary for effective cleanup? Aside from the political pressure techniques we have already mentioned, you have several legal options for action.

First, you can challenge an inadequate decision through an informal administrative appeal. The appeal can be simply a letter to the EPA Administrator in Washington, D.C., requesting a review of the decision. The request should explain why the action is inadequate and include evidence to support the points of the appeal.

If an appeal is unsuccessful, you can generally bring a citizen suit against the agency, but there are several limitations. First, you cannot

---

[16] In addition to normal cleanup costs, the Superfund statute authorizes the collection of damages from persons responsible for injury to or loss of natural resources caused by the release of a hazardous substance. See 42 U.S.C. § 9607. The amounts collected are to be paid into a special trust fund for use in restoring or replacing natural resources. Remember to keep these provisions in mind when evaluating the options available to EPA.

bring a lawsuit to challenge a removal action if it is going to be followed with a permanent remedial action cleanup. Second, whether challenging a removal or remedial action, you cannot bring a lawsuit before the action is begun, but must wait until a "distinct and separate phase" of the action is completed. This unique rule is Congress's attempt to balance the need to prevent litigation from delaying important cleanup against the need to permit timely challenges so that truly inadequate removal or remedial measures can be quickly corrected before it is too late or becomes too expensive to do so. Under this rule you could not challenge in court, for example, an EPA decision to remove only some wastes from a site and merely contain others until after it has completed the waste removal phase of the action, but you need not wait until the entire remedial action is complete, the site capped and revegetated, with the wastes you want removed still in the ground. EPA may indicate in the record of decision or other documents what it believes are the proper points during the cleanup to bring legal challenges, but you can argue that a smaller piece of the cleanup is a "distinct phase" and attempt to bring litigation sooner.

Issues to raise in an administrative appeal or a citizen suit include:

Contaminants left on-site have not been treated to a level required by standards or regulations under the SDWA, CWA, FIFRA, TSCA, or RCRA or stricter, applicable state standards.

Any wastes moved off-site have not been transferred to a facility operating in substantial compliance with state and federal regulations.

Waivers from applicable cleanup standards have been granted inappropriately.

For removal actions, an imminent hazard still exists that must be addressed with more removal action, or, for remedial challenges, further measures must be taken to remedy permanently the threat of health or environmental problems.

Relevant information was not considered by the agency.

To win a lawsuit, you will be required to show that the agency's decision was arbitrary and capricious. This is a difficult standard to meet, particularly because the evidence you can rely on will be limited to the record before the agency at the time of its decision and because the case will be likely to turn on difficult scientific or engineering questions that generally lead courts to defer to agency decisions.

As a last resort, you could consider a lawsuit directly against the responsible party. Depending on the nature of the materials at the site, you may be able to bring such a lawsuit under the imminent hazard citizen suit provision of RCRA. (See Chapter 14.) Finally, do not forget options under state law, particularly state agency mini-superfund-type cleanups (see Chapter 25), or nuisance lawsuits against the polluter (see Chapter 8).

TAKING YOUR OWN ACTION

In some cases, you may decide as a group to take actions to protect yourself from threats posed by a contaminated site when the agency does not act quickly or does not do enough. For instance, you may want to:

- Find alternate water supplies
- Monitor wells and water supplies
- Conduct soil tests
- Undergo medical tests and treatment, if necessary
- Permanently or temporarily relocate homes and businesses

You can seek to recover the costs of these types of actions from either those responsible for the contamination or the Superfund. To recover these expenses, your action must be consistent with the National Contingency Plan, which generally outlines when certain types of response are appropriate. (See 40 C.F.R. Part 300 and 40 C.F.R. §§ 300.25 and 300.71, in particular.) Recovering costs from those responsible may only require the filing of a claim with the company or individual, but in some cases you may also have to file a lawsuit. To get reimbursement from the Superfund for actions taken, you must first have submitted your claim to the private party responsible for the contamination, and have failed to get relief from this source. For some types of response action, you may need prior approval from a federal official before reimbursement can be awarded. Check with EPA for forms to apply for response costs and more detailed instructions on these procedures.

REFERENCES

SUPERFUND CLEANUPS

Environmental Defense Fund. *Dumpsite Cleanups: A Citizen's Guide to the Superfund Program.* Washington, D.C. 1984.
Environmental Defense Fund et al. *Right Train, Wrong Track: Failed Leadership in the Superfund Cleanup Program.* Washington, D.C. June 1988.
Environmental Law Reporter. *Superfund Deskbook.* Environmental Law Institute. Washington, D.C. 1986.
EPA, Office of Emergency and Remedial Response. *National Priorities List Fact Book.* Washington, D.C. June 1986.

EPA, Office of Solid Waste and Emergency Response. *Superfund: A Six-Year Perspective.* Washington, D.C. October 1986.

Freedman, Joseph. *Proposed Amendments to the National Contingency Plan: Explanation and Analysis.* News and Analysis, 19 Envtl. L. Rep. (Envtl. L. Inst.) 10103 (1989).

Gaba, Jeffrey M. *Recovering Hazardous Waste Cleanup Costs: The Private Cause of Action Under CERCLA.* 13 Ecology Law Quarterly 181 (1986).

ICF, Incorporated. *Community Relations in Superfund: A Handbook.* Washington, D.C. March 1986. Prepared for Office of Emergency and Remedial Response, EPA. OSWER Directive No. 9230.0-3A.

Johnson, Barry L. *Health Effects of Hazardous Waste: The Expanding Functions of the Agency for Toxic Substances and Disease Registry* [18 News & Analysis] Envtl. L. Rep. 10132 (April 1988).

Lipsett, Brian. "STOP Foils EPA." *Everyone's Backyard*, Spring 1988.

U.S. Congress, Office of Technology Assessment. *Are We Cleaning Up? 10 Superfund Case Studies—Special Report.* Washington, D.C. June 1988.

U.S. General Accounting Office. *Hazardous Waste: Corrective Action Cleanups Will Take Years to Complete.* Washington, D.C. December 1987.

U.S. General Accounting Office. *Hazardous Waste: Environmental Safeguards Jeopardized When Facilities Cease Operating.* Washington, D.C. February 1986.

U.S. General Accounting Office. *Hazardous Waste: EPA's Consideration of Permanent Cleanup Remedies.* Washington, D.C. July 1986.

U.S. General Accounting Office. *Superfund, Extent of Nation's Potential Hazardous Waste Problem Still Unknown.* Washington, D.C. December 1987.

LAWS AND REGULATIONS

U.S. Code Citation

CERCLA, or Superfund, 42 U.S.C.A. §§ 9601–9675 (West 1983 & Supp. 1988).

Code of Federal Regulations Citation

40 C.F.R. Part 300 (National Contingency Plan).

CHAPTER 14

Imminent Hazard Actions

AN UNDERGROUND STORAGE tank is leaking toxic chemicals into the ground at a nearby chemical company. An abandoned waste dump sitting above a shallow drinking-water aquifer contains rusted and deteriorating drums full of hazardous wastes. An injection well used to dispose of sewage develops a leak that may threaten a deep aquifer. Some of these problems may be addressed by EPA using Superfund, as discussed in the previous chapter. But where Superfund does not apply or EPA decides not to take immediate agency action with Superfund money, what else can be done?

Where groundwater is contaminated or threatened by an industrial facility, waste site, or nearly any contamination source, you may be able to get action under "imminent hazard" authorities contained in most federal pollution laws. These imminent hazard provisions give EPA (and in one case, citizens themselves) authority to force persons or companies associated with the contamination to take immediate action to prevent or remove the threat.

WHEN IS AN IMMINENT
HAZARD ACTION APPROPRIATE?

The first step is to identify whether the situation you face may be subject to an imminent hazard action. Figure 14.1 sets out briefly and generally the laws that give EPA authority to act against imminent hazards. As you can see, there is considerable overlap among these statutes. In most instances where contamination threatens groundwater and health, EPA will have authority to act under several of these laws.

These laws share the general requirement that EPA must be able to demonstrate a substantial and imminent threat to health or the environment in order to force responsible persons to act. What does that mean? The courts or the agency will decide what presents an imminent hazard depending on the particular facts involved in a specific case, but some general rules are developing. The kinds of evidence that have been used to

FIGURE 14.1
EPA IMMINENT HAZARD AUTHORITIES

Statute	What Substances Trigger Action?	What Must Be Threatened for Action to Be Triggered?
CERCLA 42 USC § 9606	Hazardous substances	Public health, welfare, or the environment
RCRA 42 USC § 6973	Hazardous and nonhazardous waste	Human health and the environment
RCRA Subtitle I (Underground Storage Tanks) 42 USC § 6991b(h)	Petroleum product from underground storage tanks	Human health and the environment
SDWA 42 USC § 300i	Any contaminant likely to enter a public water supply or underground source of drinking water	Human health
TSCA 15 USC § 2606	Hazardous chemicals (or products containing them) that pose an unreasonable risk to health or the environment	Human health or the environment
CWA 33 USC § 1364	Pollution from source or combination of sources discharging to water or its surrounding area	Health or welfare (such as a threat to livelihood, i.e., the ability to market shellfish)
CWA 33 USC § 1321(e)	Oil or hazardous substance discharge, actual or threatened, into water	Public health or welfare (such as fish, shellfish, beaches, shorelines, etc.)

Against Whom Is Action Taken?	Process to Compel Responsible Parties to Clean Up Hazardous Conditions
Past or present owners or operators of the waste site, generators, those who arrange for disposal, and transporters responsible for creating the hazard	Administrative orders and court action
Any person who contributed to the condition of the site, including past or present waste site owner/operators, generators, and/or transporters who contributed to the problem at the site	Administrative orders and court action
Underground storage tank owner/operator	Administrative orders
Persons causing or contributing to the contamination	Administrative orders and court action
Persons who manufacture, process, distribute, use, or dispose of the imminently hazardous chemical or product	Court action
Persons causing or contributing to pollution	Court action
Persons responsible for the discharge	Court action

show that an imminent hazard may exist include proof that the substances at a site are highly toxic, or, if less toxic, are present in high concentrations, and that exposure to the substances is likely (if, for example, groundwater is already contaminated or is likely to be, given the area's geologic conditions). There need be no showing of an actual harm to people or the environment, but only that there is a probability of harm if corrective action is not taken.

One example of a successful action concerned a site in Kingston, New Hampshire used for many years by several waste processing companies, where EPA found that a variety of toxic chemicals, including benzene, chloroform, and volatile organic compounds, had contaminated the soil and groundwater. Even though drinking-water supplies had not yet been affected, EPA sued the responsible companies using its imminent hazard authority. The court ordered the companies to cease operations and conduct a cleanup to prevent the contamination problem from becoming a health problem.

These laws authorize broad forms of relief. EPA or a court, depending on the law, can issue orders that require responsible persons or companies to halt an ongoing activity, erect barriers to stop the release of contaminants, monitor groundwater quality, plan and carry out cleanup of the contaminants at the site, provide alternative water supplies, and the like. As a general rule, orders can be structured to fit the circumstances of the case, requiring whatever is necessary to remove the threat.

HOW TO PRODUCE ACTION

Once you are convinced that an imminent hazard may exist, you should begin efforts to spur action by requesting EPA to use its authority, since most of the imminent hazard laws authorize only EPA to take action. Direct a request for action to the EPA regional office. Identify the site, explain why it might create an imminent hazard in your view, and ask for specific action. Attach as much evidence as possible to demonstrate the seriousness of the problem: photographs showing chemical leaks or corroded containers, groundwater sample tests indicating contamination, anecdotal information about suspicious health problems in the area, and the like. Many times, evidence to support a request can be found in regulatory agency files. The following chapters explain methods for gathering helpful information about specific contamination sources from the EPA, state regulatory agencies and other public offices.

As a practical matter, however, convincing EPA to take action under the imminent hazard authorities will be difficult. You will have to demonstrate that the problem is more serious than others faced by the agency.

Requests for imminent hazard action should be carefully coordinated with your strategy to obtain action under the Superfund program. Rather than do the cleanup with Superfund money, as described in the previous chapter, EPA can sometimes require responsible persons to respond to a contamination problem. This authority comes from the Superfund imminent hazard provision. Sometimes effective cleanup may be more likely if EPA does the work under Superfund, but in other cases cleanup may be faster and equally thorough if private cleanup action is compelled through the use of imminent hazard orders. And of course, EPA can use this imminent hazard authority only if it can identify responsible parties who have the money and resources to comply with any orders that might be issued.

If EPA does initiate action, you have a few limited opportunities to participate and to try to ensure that the remedy will remove the threat to groundwater. In circumstances where EPA decides to issue its own order to the responsible parties, there is generally no formal procedure. You can attempt to influence the content of the order informally by meeting with responsible officials and using political tools.

When EPA decides instead to go directly to court to force action by the responsible parties, you may be able to intervene in the case and present your arguments directly to the court. (See Chapter 7.) In addition, in some circumstances if EPA reaches a settlement with the responsible party after a lawsuit is filed, EPA must publish a description of the settlement in the *Federal Register* and solicit public comment before the settlement can be approved by the court. You can use this opportunity to ask for improvements in the agreement even if you have not intervened in the case.

BRINGING A CITIZEN SUIT

When EPA refuses to initiate an imminent hazard action, you can, in some circumstances, bring your own. The Resource Conservation and Recovery Act (RCRA) gives affected citizens the opportunity to bring a court action against anyone responsible for a contamination threat that may present an imminent hazard. There are two significant limitations. First, a waste must be involved (as opposed to a chemical product or other substance not yet considered a waste). Second, citizens cannot file such a lawsuit when EPA or other agencies have begun one of several specific types of actions to address the problem, such as initiating a cleanup action under Superfund. Otherwise citizen imminent hazard lawsuits are similar to EPA actions—the same general rules about what constitutes an imminent hazard apply and the citizen can seek the same types of broad relief. In addition, the same sort of evidence that you would use to persuade the agency to initiate action can be used to convince the court of the need for relief.

Imminent hazard citizen suits will often be very complex and expensive to bring and will require the help of a lawyer and scientific experts. In many ways, these suits will be similar to the common-law actions in state court described in Chapter 8.

Not many citizens have filed lawsuits under this imminent hazard authority and fewer yet have achieved success. One example of at least partial success comes from Gettysburg, Pennsylvania. Neighbors of property used for disposal of chemical wastes in open lagoons and other uncontrolled open dumps sued the owner under the imminent hazard provision of RCRA. Though the plaintiffs were not able to develop enough evidence of harm to get a preliminary injunction from the court, their lawsuit forced the owner into negotiations. Eventually, they reached an agreement settling their claims in the case.

REFERENCES

IMMINENT HAZARD ACTIONS

EPA, Office of Solid Waste and Emergency Response. *Final Revised Guidance Memorandum on the Use and Issuance of Administrative Orders under §7003 of the Resource Conservation and Recovery Act (RCRA)*. Washington, D.C. September 26, 1984.

EPA, Office of Solid Waste and Emergency Response. *Issuance of Administrative Orders for Immediate Removal Actions*. OSWER Directive. Washington, D.C. February 21, 1984.

Jorgenson, Lisa, and Jeffrey J. Kimmel. *Environmental Citizen Suits: Confronting the Corporation*. Bureau of National Affairs, Inc. Washington, D.C. 1988.

LAWS AND REGULATIONS

Federal Register Citations

50 Fed. Reg. 45933 (Nov. 5, 1985) (Notice of Procedures for Planning and Implementing Response Actions).

CHAPTER 15

The National Environmental Policy Act

The National Environmental Policy Act of 1969 (NEPA) requires federal agencies to study the environmental impacts of their proposed actions, including the impacts on groundwater. Where the proposed federal action is "major" and where the impacts may be "significant," the agency must prepare a document referred to as an Environmental Impact Statement (EIS). An EIS is done in two stages: a draft EIS, which is circulated to the public, as well as to federal, state, and local government agencies, for comments, and a final EIS, which incorporates or otherwise responds to the comments. When it is not clear whether impacts of the proposed action will be significant, a document referred to as an Environmental Assessment (EA) generally must be prepared.

FEDERAL ACTION

To determine whether a project that may affect groundwater is covered by NEPA, you must first find out whether the project is federal: Federal actions include projects entirely or partially financed, assisted, conducted, regulated, permitted, or approved by federal agencies. Also included are new or revised agency rules, regulations, plans, policies, or procedures of federal agencies, as well as legislative proposals. Examples include roads and airports that are financed with federal funds and private construction that requires federal permits, such as a shopping center that needs a permit from the Corps of Engineers to fill in a wetland. State actions and occasionally even private actions may be covered by state NEPAs. (See Chapter 26.)

MAJOR ACTIONS WITH SIGNIFICANT
ENVIRONMENTAL EFFECTS

Where a federal project threatens your groundwater, an EIS will be required if the project is major and has potentially significant effects on the environment. ("Major" reinforces but does not have a separate meaning from "significant.") The significance of environmental impacts is determined by looking first at the *context* of the action. A hazardous waste site will have potentially more significant impacts when it is located in an area with porous soil, near a drinking-water supply, than when it is in an area with non-porous soil far away from people and their drinking water. In addition to context, significance depends on the *intensity* of the impacts. Bigger impacts are more significant, as are those that affect public health or safety.

If it is not clear whether a federal action will have significant environmental impacts, an EA generally will be required. An EA is used to determine whether a full EIS must be prepared. If the EA concludes that no EIS is needed, the agency then issues a "finding of no significant impact" (FONSI).

ALTERNATIVES TO THE ACTION

Probably the most important part of an EIS or EA is the search for alternative ways to carry out the project without causing as much impact on groundwater or other aspects of the environment. The discussion of alternatives must present the environmental effects of the proposed action and all reasonable alternatives in comparative form, so that the decisionmaker and public will have a clear choice among options.

NEPA places on the federal agency the duty to undertake the search for better alternatives. Even so, you must alert the agency to any alternatives you think would help protect groundwater, especially if the alternatives are unconventional. The less conventional your alternative is, the more evidence you will have to present to convince the agency it should be studied. Make your suggestions as early as possible and include them as part of your comments on the draft EIS. Focus on how your alternative will better protect groundwater. Alternatives must be discussed in both EISs and EAs.

MITIGATION MEASURES

Measures for mitigating potential harm to groundwater must be studied by the federal agency preparing the EIS or EA. Again, you should alert the agency as early as possible to any mitigation measures you think will better protect groundwater. Mitigation includes avoiding the adverse impact by scaling back on the project as well as measures to minimize impacts, to rectify impacts by restoring the environment, and to compensate for the impacts by replacing or providing substitute resources. Keep in mind that mitigation measures can be made mandatory conditions for going forward with the project. Urge the agency to make your suggested conditions mandatory, even when only an EA is being prepared.

EXEMPTIONS FOR EPA ACTIONS

Actions by EPA under the Clean Air Act are exempt from NEPA, as are most actions under the Clean Water Act (only the construction of public sewage treatment works and permits for new pollution sources are not exempt). Other federal actions undertaken to protect the environment also may be exempt, but these will be few and far between.

HOW TO USE NEPA

NEPA gives you the opportunity to work directly with the federal agencies that make the decisions to build, fund, or permit projects that threaten your groundwater. Each agency has its own NEPA regulations, which you should study carefully.

Two federal agencies are involved in all NEPA reviews: the Council on Environmental Quality (CEQ) and the Environmental Protection Agency (EPA). CEQ ensures that all other federal agencies fully implement NEPA. The CEQ NEPA regulations have the force of law, and all other agencies must ensure that their NEPA regulations are consistent. Study CEQ's regulations.

EPA reviews all federal actions, including actions that require an EIS, and submits comments on the environmental impacts. If EPA determines that the federal project is "unsatisfactory from the standpoint of public health or welfare or environmental quality," the matter is referred to CEQ for mediation, and if that is not successful, to the President. The public has the right to submit comments to CEQ on such referrals.

Under NEPA, the mission of both CEQ and EPA is to protect the environment, and by and large these agencies should be sympathetic to your concerns about groundwater. For example, if you have a less damaging alternative to suggest, or specific mitigation measures, try to convince EPA to submit the alternative or mitigation in its comments. (Do this in addition to submitting the comments yourself.) Comments from EPA will carry more weight. So will comments from other federal agencies, as well as state and local agencies. Search carefully for sympathetic allies. Generally these will be agencies responsible for protecting the environment, such as the federal Fish and Wildlife Service, and at the state level, the Department of Environmental Quality and the Fish and Game Department. Ask the agency that is preparing the EIS for the list of agencies that will be receiving the draft. Get to know your allies early and work with them closely.

Ultimately, however, the federal agency that is building the project (or funding or permitting it) has broad discretion to go ahead once the NEPA studies have been done—even when the public and other agencies have submitted comments against the project, and even when the project clearly will harm your groundwater. NEPA requires that environmental impacts be considered by the agencies, but it does not require that environmental factors be elevated over other factors, such as the need for housing, or roads, or whatever else is being proposed.

NEPA is designed in part to give you and the public and other agencies the opportunity to argue for other alternatives and mitigation measures. But often NEPA's greatest value is to give you the time to work on a political solution to protect your groundwater. An EIS takes more than a year in most cases, and considerably longer for large or controversial projects, where there are many comments from concerned citizens. During this period, the project cannot go forward. This gives you time to work with your state and local elected representatives as well as those in Washington, D.C. Convince them that new laws or regulations are needed to protect your groundwater. Convince them that your alternative to the project or your mitigation measures should be incorporated into a new law. And continue working with the proponents of the project to convince them that they should adopt your proposals to protect groundwater.

REFERENCES

THE NATIONAL ENVIRONMENTAL POLICY ACT

Rodgers, William H., Jr. *Handbook on Environmental Law*. West Publishing Co. St. Paul, Minn. 1977.

The following publications are available from the Council on Environmental Quality:

Appendices I, II, and III to NEPA Implementing Regulations. Includes information about contacts for further NEPA information in each agency, EIS reviewers, agencies' jurisdiction by law and special expertise.

Guidance Regarding NEPA Regulations, 1983. Covers scoping, adoption, tiering, alternatives analysis, categorical exclusions, contracting provisions.

Memorandum for General Counsels, NEPA Liaisons and Participants in Scoping: Scoping Guidance, 1981. A useful compilation of experience regarding techniques for successful scoping practices.

Preamble and final rule to amendment of 40 C.F.R. Part 1502.22, May 27, 1986. Amendment to "worst case analysis" regulation.

Recommendations of the Council on Environmental Quality Regarding the Proposed Amendments to the Army Corps of Engineers' Procedures Implementing the National Environmental Policy Act, Reprint of 52 Fed. Reg. 22,517 (June 12, 1987). Discusses Corps of Engineers' approach to the NEPA process, particularly with respect to § 404 permits under Clean Water Act.

Regulations for Implementing the Procedural Provisions of the National Environmental Policy Act. Includes NEPA, regulations, relevant Executive Orders.

LAWS AND REGULATIONS

U.S. Code Citation

The National Environmental Policy Act of 1969, 42 U.S.C.A. §§ 4321-4347 (West 1977 & Supp. 1988).

Code of Federal Regulations Citation

40 C.F.R. Part 1500-1508 (regulations for implementing NEPA).

Federal Register Citation

46 Fed. Reg. 18,026 (March 23, 1981), as amended 51 Fed. Reg. 15,618 (April 25, 1986) ("Forty Most Asked Questions Concerning CEQ's National Environmental Policy Act Regulations").

CHAPTER 16

Waste Disposal Facilities

EACH YEAR AMERICAN industry produces over 264 million metric tons of hazardous waste—enough to fill the New Orleans Superdome almost 1,500 times over.[1] Ensuring that this waste is properly disposed of so as not to contaminate groundwater poses a tremendous challenge for industry, for government regulators, and for citizen activists. And the problem is not limited to hazardous waste facilities. EPA recently estimated that 37,000 nonhazardous waste facilities are receiving small quantities of hazardous waste from households and small businesses.[2] These small quantities can add up to big problems—in 1986, 20 percent of the nation's Superfund sites were found among supposedly "nonhazardous" municipal landfills.

To stop waste disposal from becoming an environmental nightmare in years ahead, Congress created a comprehensive regulatory program, called the Resource Conservation and Recovery Act (RCRA), which requires safe and environmentally sound treatment of wastes. For hazardous wastes in particular, Congress established a "cradle to grave" system that controls such waste from initial generation to final disposal. This chapter will focus on the requirements of RCRA that apply to hazardous waste. There are, however, some federal restrictions that govern nonhazardous waste dumps (often called solid waste facilities). Suggestions for using the solid waste program are outlined at the end of the chapter. Beginning in 1989, Congress will be reexamining federal control of nonhazardous waste disposal and is expected to pass new, stricter legislation regulating the problem. Citizen groups should follow and get involved in this process, because disposal of solid waste is a growing problem that presents a real threat to groundwater quality around the country.

[1] *National Geographic*, March 1985, page 325.

[2] Christopher Harris, William Want, and Morris Ward, *Hazardous Waste: Confronting the Challenge*, ELI (Quorum Books, 1987), page 221.

Treatment, Storage, and Disposal Facilities

RCRA's permit program applies to hazardous waste treatment, storage, and disposal facilities. An explanation of these concepts is in order.

A disposal facility is defined in EPA's regulations as "a facility or part of a facility at which hazardous waste is intentionally placed into or on any land or water, and at which waste will remain after closure." (See 40 C.F.R. § 260.10.) The most common kinds of disposal facilities are surface impoundments (ponds or pits containing liquid wastes, either alone or mixed with solids), waste piles (accumulations on the surface of the land of noncontainerized solid waste), and landfills (facilities where wastes are placed in or on the land. Note that EPA's definition of "landfill" specifically excludes surface impoundments, waste piles, and several other kinds of disposal facilities).

Storage facilities are those where hazardous wastes are held "for a temporary period, at the end of which the hazardous waste is treated, disposed of, or stored elsewhere." (See 40 C.F.R. § 260.10.) The key characteristic here is the *temporary* presence of the wastes. A tank or a group of containers (such as metal drums) are examples of facilities that could be used for purposes of storage.

Treatment facilities are those which alter wastes physically, chemically, or biologically to make them safer or more manageable, or to extract recyclable matter or energy from them. (See 40 C.F.R. § 260.10.) Treatment includes such activities as burning wastes in incinerators or applying chemicals to change their physical or chemical composition.

Although RCRA places important regulatory obligations on those who produce (see Chapter 17) and transport (see Chapter 22) hazardous waste ("generators" and "transporters"), the heart of the program is the extensive set of requirements that must be followed by those who treat, store, or dispose of hazardous waste.

Each treatment, storage, or disposal facility (TSDF) must apply for and obtain a regulatory permit. To obtain such a permit, the applicant must agree to design, operate, and maintain its facility in an environmentally sound manner; to monitor the groundwater around its facility for chemical contamination (if it is a land disposal facility); to remedy any groundwater contamination it has caused; to remove any remaining wastes when the facility closes, or if such removal is unfeasible, to provide 30 years of post-closure monitoring and supervision; to provide assurances concerning its ability to pay for the costs of closure and postclosure care and to compensate for any accidents resulting in release of its wastes; and to report regularly to regulatory authorities concerning its operations.

Preexisting TSDFs that have applied for permits are allowed to continue operating temporarily as "interim status" facilities, provided that they follow a slightly less demanding version of the program for permitted facilities.

State governments can receive authority to regulate TSDFs by enacting a regulatory program that is "equivalent" to—in other words, at least as stringent as—EPA's program.[3] (See 42 U.S.C. § 6926 (b).) An authorized state essentially steps into EPA's shoes, and may issue permits and conduct enforcement activities. Nearly all states have received authorization to administer that portion of the RCRA program that was in existence prior to 1984, but few have assumed responsibility for the sweeping new regulatory requirements added by the 1984 RCRA amendments. Thus in many states there is a confusing dual jurisdiction, and a TSDF must obtain a permit from both EPA and the state.

In this chapter the term "agency" is used to refer to the appropriate RCRA regulatory authority: either EPA (in a completely nonauthorized state), the state agency (in a completely authorized state), or both (in a partially authorized state).

GATHERING INFORMATION ON FACILITIES

To identify which treatment, storage, and disposal facilities are present in your area, consult the agency. All active TSDFs were required to notify the agency of their existence several years ago. (See 42 U.S.C. § 6930(a).) In addition, the states were required to compile and submit to EPA an inventory "describing the location of each site . . . at which hazardous waste has at any time been stored or disposed of." (See 42 U.S.C. § 6933(a).) Because of these requirements, the agency's database on TSDFs should be fairly comprehensive. Generally, state agencies will have far more extensive files than EPA.

Once you have focused your attention on a specific TSDF that concerns you, visit the agency and ask to see the facility file (this file is likely to be voluminous, so you probably will not want to ask the agency to copy it for you until you are sure just what you need). Several helpful documents should be present in the file:

- *Permit applications.* Each facility must submit first a Part A and then a Part B application. While the Part A is fairly rudimentary (see 40 C.F.R. § 270.13), the Part B contains an extensive description of every important

---

[3] While states are not allowed to relax the minimum requirements of federal law, they are free to adopt hazardous waste programs that go beyond those requirements. See 42 U.S.C. § 6929. You should urge your state to adopt any rules that you consider necessary for proper management and disposal of hazardous wastes, even if such rules are not required by RCRA or by EPA's regulations.

What Is a Hazardous Waste?

To be regulated as a "hazardous waste" under RCRA a substance must meet three criteria:

First, it must be a *waste*, that is, a "discarded material." A material is discarded if it is abandoned, recycled, or considered "inherently waste-like." (See 42 U.S.C. § 6903(27); 40 C.F.R.) § 261.2(a).

Second, it must be a *solid waste*. Congress has counterintuitively defined as "solid" all "solid, liquid, semisolid, or contained gaseous material." (See 42 U.S.C. § 6903(27).) Certain enumerated items are excluded from regulation, specifically (1) domestic sewage; (2) industrial discharges regulated under the Clean Water Act; (3) irrigation return flows; (4) nuclear wastes; (5) certain mining materials; (5) pulping liquors; (6) spent sulfuric acid; and (7) secondary materials that are reclaimed and returned to the original process from which they were generated. (See 40 C.F.R. § 261.4(a).)

Third, it must be a *hazardous waste*. EPA has compiled regulatory lists of wastes which it has determined to be hazardous. (See 40 C.F.R. §§ 261.30 to 261.33.) Even if a waste is not on EPA's lists, however, it is hazardous if it exhibits one of four dangerous characteristics: ignitability (a propensity to catch on fire); corrosivity (a propensity to eat through metals or other materials); reactivity (a propensity to explode or generate toxic fumes); and EP toxicity (a propensity to leach toxic substances into groundwater). (See 40 C.F.R. §§ 261.20 to 261.24.)

Excluded from the definition of "hazardous wastes" are: (1) household and hotel wastes; (2) fertilizers; (3) mining overburden; (4) waste residue generated by the combustion of fossil fuels; (5) wastes produced in conjunction with exploration for and production of oil, gas, or geothermal energy; (6) certain trivalent chromium wastes; (7) mining wastes; (8) cement kiln dust waste; and (9) discarded wood or wood products. (See 40 C.F.R. § 261.4(b).)

You should be aware that if a waste included on one of EPA's lists is combined with another substance, the resulting mixture is (with some exceptions) considered to be a hazardous waste. But if a waste that is considered hazardous solely because it demonstrates one of the four key characteristics (ignitability, corrosivity, reactivity, or toxicity) is combined with another substance, the resulting mixture will be considered hazardous only if it continues to exhibit that characteristic. (See 40 C.F.R. § 261.3(a)(2).)

---

aspect of a facility's operations (see 40 C.F.R. §§ 270.14 to 270.23). Of particular interest to citizen activists, it includes a description of all past releases of hazardous wastes into the environment (see 40 C.F.R. § 270.14(c)), as well as (in the case of landfills and surface impoundments) an assessment of the potential for the public to be exposed to releases in the future. (See 42 U.S.C. § 6939a(a).)

- *Notices of Deficiency.* If the agency noticed any substantial omission from a facility's application, it may have sent a notice of deficiency detailing what additional information was needed.

- *Facility Assessments.* As part of the permit review process, the agency prepares a "facility assessment" to determine whether the facility has released wastes into the environment.
- *Groundwater Monitoring Reports.* Land disposal facilities must drill test wells to monitor for contaminants in groundwater. The results of this monitoring must be reported to the agency. (See 40 C.F.R. §§ 264.97 to 264.99, 270.30(j)(4), 265.90 to 265.94.)
- *Manifests.* Under RCRA, each shipment of hazardous wastes is accompanied by a "manifest," a document that describes the wastes being shipped and their destination. (See 42 U.S.C. §§ 6922(a)(5), 6923(a)(3), 6924(a)(2); 40 C.F.R. §§ 264.71 to 264.72, 265.71 to 265.72.) Upon receiving the wastes the TSDF must compare them against the manifest and note any discrepancies. (See 40 C.F.R. §§ 264.71(b)(2), 265.71(b)(2).) In some states, the state agency may have copies of the manifests for the facility that concerns you.

  At a minimum, the agency will have any manifest discrepancy reports and unmanifested waste reports the facility has filed. The facility must submit a manifest discrepancy report if it has been unable within 15 days to reconcile any discrepancy between the actual content of a waste shipment and the manifest's description of that shipment. (See 40 C.F.R. §§ 264.72(b), 265.72(b).) An unmanifested waste report must be submitted whenever a facility accepts wastes that are unaccompanied by a manifest. (See 40 C.F.R. §§ 264.76, 265.76.)
- *Inspection Reports.* When regulatory authorities go onsite to inspect a TSDF, they generally prepare a written report summarizing their findings. Such reports are frequently of great informational value because they represent the eyewitness observations of someone who is not employed by the facility owner.
- *Corrective-Action Programs.* If the facility has previously released wastes into the environment, the file may contain descriptions of any programs that were designed or undertaken to clean up the resulting contamination.
- *Biennial Reports.* By March 1 of each even-numbered year, each TSDF must submit to the agency a report describing the activities of the facility over the previous year. (See 40 C.F.R. § 264.75, 265.75.)
- *Administrative Orders.* If the TSDF has broken the law, regulatory authorities may have issued an administrative order directing the facility to come into compliance. The order may contain valuable information on the problems the authorities found at the facility.

If you examine the facility file and determine that necessary information has not been submitted to the agency, urge the agency to request it from the facility. RCRA gives the agency broad investigative powers, including the right to require the facility to answer written inquiries, to go onsite to conduct inspections, and take samples while doing so, and to require the facility to monitor groundwater. (See 42 U.S.C. §§ 6927, 6934.)

As a substitute for or supplement to data in the facility file, you may want to do some independent fact-gathering on the facility. Talk to local fire, police, and health officials and former employees of the facility to learn about any problems that the facility may have experienced. Visit the perimeters of the facility and look for evidence of improper waste management (strange liquids or fumes emanating from the facility, careless waste-handling techniques by personnel, waste shipments arriving outside normal business hours or at closed facilities). Have water samples taken at drinking wells or surface waters downgradient from the facility, and have them analyzed for contamination. (See Chapter 4.)

REMEDYING RELEASES INTO GROUNDWATER

If your investigations indicate that the facility has caused releases of hazardous waste into groundwater, make sure that appropriate corrective action is undertaken.

If the facility already has a permit (this includes operating facilities and closed facilities with post-closure permits), its permit will require that the owner initiate special intensified groundwater monitoring (called "compliance monitoring") to see if the contaminants exceed the maximum concentration levels prescribed by the permit. (See 40 C.F.R. § 264.98(h)(4).) If compliance monitoring reveals that the maximum contaminant levels have been exceeded, the permit will require the owner to design and implement a corrective action program. (See 40 C.F.R. § 264.99(h)(2).) Where necessary to protect human health and the environment, such a program must include provision for corrective action *outside* the facility boundary. (See 42 U.S.C. § 6924(v).)

Neither a compliance monitoring program under § 264.98(h)(4) nor a corrective action program under § 264.99(h)(2) can be implemented by the facility until it first submits the program to the agency for approval as a permit modification. You have the right to comment on the proposed modification and to participate in any public hearing.[4] (See Chapter 7.)

If the facility has failed to request necessary permit modifications, contact it directly and ask it to do so. If the facility refuses, ask the agency to take enforcement action. If the agency fails to act, you may want to consider bringing a citizen suit directly against the facility. (See 42 U.S.C.

---

[4] Regulations recently issued by EPA (53 *Fed. Reg.* 37912, Sept. 28, 1988, *codified at* 40 C.F.R. § 270.42) cut back sharply on public involvement in many kinds of permit modifications, but the compliance monitoring and corrective-action modifications discussed in the text remain subject to full public participation requirements. See 40 C.F.R. § 270.42, Appendix I, ¶¶ C.7.a, C.8.a. Procedures governing modifications in authorized states may vary from state to state.

§ 6972.) Such a suit will be difficult and time-consuming, and should only be initiated if you have strong and compelling evidence of a violation.

Interim Status Facilities

If the facility is still in interim status, ask the agency to issue an "interim status corrective action" order requiring appropriate cleanup. (See 42 U.S.C. § 6928(h).) If EPA decides to proceed with such an order, it will first allow an opportunity for public comment on the proposed corrective measures.[5] It is important that you participate at this stage, because once the agency actually issues an initial order, only the facility has the right to request and participate in an additional hearing. (See 40 C.F.R. § 24.05(a); 53 Fed. Reg. 12262 col. 1 (Apr. 13, 1988).) If EPA refuses to act against an interim status facility, you unfortunately have no right to bring a federal citizen suit to obtain an interim status corrective order; your only federal remedy is an imminent endangerment suit (see below). You may, however, have a citizen suit remedy under state law.

Continue to monitor the interim status facility as it moves into the permitting process, and make sure that no permit is issued without a requirement that the facility undertake corrective action.

Imminent Hazards

If a release from either a permitted or interim status facility is so serious that it creates an immediate danger, ask the agency to either issue an imminent hazard administrative order or commence an imminent hazard lawsuit in the appropriate court. (See 42 U.S.C. § 6973(a).) You also have the right to bring a federal imminent hazard suit if EPA and the state fail to act. (See 42 U.S.C. § 6972(a)(1)(B).) (See Chapter 14.)

INVESTIGATING AND CORRECTING VIOLATIONS OF REGULATORY REQUIREMENTS

Investigation

Even if you do not have evidence that a facility is leaking, you can take action to enforce operating requirements. RCRA requires TSDFs to follow

---

[5] "Guidance for Public Involvement In RCRA Section 3008(h) Actions," Memorandum from Assistant EPA Administrator J. Winston Porter, Office of Solid Waste and Emergency Response (May 5, 1987), at page 2. Unfortunately, EPA's practice has been to allow comment only on permanent corrective measures, not on interim ones. If you learn that EPA is considering interim measures, offer your views to the agency as soon as possible, because a formal opportunity for comment is unlikely. As for authorized states, public involvement procedures vary from state to state.

a wide variety of specific regulatory requirements designed to ensure that hazardous wastes do not escape into groundwater. Your investigations should attempt to identify any serious violations of these requirements. (See page 206 for suggestions about where to get this information.) Some of the key requirements are listed below.

Groundwater Monitoring. Both permitted and interim-status land disposal facilities must monitor their groundwater to detect any contamination resulting from their operations. Initially, a facility must monitor for certain indicator parameters (such as specific conductance, pH, and total organic carbon) in order to determine whether a release has occurred. (See 40 C.F.R. §§ 264.98(a), 265.92(b).) A permitted facility must also monitor for specific hazardous waste constituents and reaction products, as set forth in its permit. (See 40 C.F.R. § 264.98(a).)

If a release of contaminants is detected through this initial-phase monitoring program, the facility must shift to a more detailed monitoring program designed to determine the precise nature and extent of contamination. (See 40 C.F.R. §§ 264.98(h)(4), 265.93(d).)

There are many potential monitoring problems you should be on the lookout for. Has the facility failed to do any monitoring at all? If it has monitored, has it drilled a sufficient number of wells? Are these wells appropriately located; specifically, is there an upgradient well, which can properly be used to establish background levels of contamination (that is, to determine what chemicals are in the groundwater *before* it enters the facility boundaries), and are there downgradient wells to detect any releases from the facility? Are samples taken from the wells with sufficient frequency? Are the samples tested for the appropriate range of contaminants? For further guidance on problems to check for, consult EPA's *RCRA Groundwater Monitoring Technical Enforcement Guidance Document*, EPA OSWER Directive 9950.1 (September 1986).

Groundwater monitoring deficiencies have plagued the RCRA program since its inception. A recent Government Accounting Office report reviewed 50 land disposal facilities and found that groundwater contamination data was inadequate at 78 percent of them.[6]

Design and Operating Requirements. Check to see that the facility is following the full range of design and operating requirements imposed by the facility's permit and the RCRA regulations. Such requirements are applicable to both permitted and interim-status facilities, although they are less strict for the latter. (See Figure 16.1.)

Limitation on Land Disposal of Certain Wastes. Check the facility's file

[6] "Hazardous Waste: Groundwater Conditions at Many Land Disposal Facilities Remain Uncertain," Document No. GAO/RCED-88-29 (February 1988), page 17. This report can be ordered at no charge from the U.S. General Accounting Office, P.O. Box 6015, Gaithersburg, Maryland 20877.

FIGURE 16.1
DESIGN AND OPERATING REQUIREMENTS

	Interim Status Facilities	Permitted Facilities
Containers	—Must be kept closed except when adding or removing waste —Must be inspected weekly —Must be located at least 50 feet from property line if contents are ignitable or reactive —Must be compatible with waste	—Same as Interim Status plus the following requirements: —Containers with free liquids must have containment system with impervious base —Must be designed to drain accumulated liquids —Run-on must be prevented —Spills or leaks must be removed promptly
Tank Systems	—Must have written assessment certified by independent engineer attesting to tank's integrity —Must have secondary containment system designed to prevent leaks, including leak detection system and one or more of the following: liner, vault, double-walled tank or equivalent device —Ancillary equipment must have secondary containment system except for aboveground pipes and pumps, which must be inspected on a daily basis —Must have spill prevention controls (e.g., check valves, dry discount couplings), overfill prevention controls (e.g., level sensing devices, high level alarms, automatic feed cutoff, or bypass to a standby tank), and sufficient freeboard (space above materials in tank) to prevent spills from waves or precipitation —Must do daily inspections of overfill/spill control equipment, aboveground tanks and surroundings, and data gathered from monitoring equipment; annual inspections of cathodic protection system; and bimonthly inspections of sources of impressed current	—Must have written assessment certified by independent engineer and by corrosion engineer (if metal component will contact soil or water) attesting to tank's integrity and installation —Must be tested for tightness Same as Interim Status —Must develop a schedule for overfill control inspection; conduct daily inspections of aboveground tanks and surroundings and the data gathered from monitoring equipment; annual inspections of cathodic protection system; and bimonthly inspections of sources of impressed current

	—If leak or spill occurs, tank or secondary system must be removed from service and either repaired or closed, Regional Administrator must be notified, and certificate from engineer must be obtained if major repairs were done	—Same as Interim Status
	—Special requirements for generators of between 100 and 1,000 kilograms of waste per month stored in tanks: two feet of freeboard unless there is a containment structure; a means to stop inflow if the waste is continuously fed; daily inspections of discharge control equipment, the data from monitoring equipment, and the level of waste; and weekly inspections of construction materials and surroundings	—No special requirements
Surface Impoundments	—May use any method approved by EPA which prevents migration of hazardous waste into groundwater or surface water at least as well as liners and leachate collection systems —Must maintain 60 centimeters (2 feet) of freeboard (space above materials) to prevent overtopping of dike by overfilling, waves, or storm; if less freeboard is maintained, engineer certificate is required —Must have cover on dike —Must conduct daily inspections of freeboard level and weekly inspections of surface impoundment	—Must use double liners, leachate collection system (EPA may waive this requirement for mono-fills) —Permit will specify design features to prevent overflow from operations, overfilling, wind and waves, rainfall, run-on, mechanical malfunction, and human error —Dikes are required —Must be operated to prevent overflow from operations, overfilling, wind and waves, rainfall, run-on, mechanical malfunction, and human error —Must inspect during installation, weekly thereafter and after storms —If impoundment is out of service for more than 6 months, engineer certificate regarding dike is required —Must be removed from service if level of liquids drops suddenly or dike leaks
Waste Piles	—Facility may elect to treat waste pile as a landfill —Must cover or otherwise protect pile against wind dispersal	—Must have run-on control system and run-off management system

Continued on next page

FIGURE 16.1 (continued)

	Interim Status Facilities	Permitted Facilities
	—Must either put pile on impermeable base with run-on and run-off systems and prompt emptying of collection systems, or protect pile from run-on by other means and add no liquids to pile —Collection systems must be emptied promptly	—Collection systems must be emptied quickly after storms —Must cover or otherwise protect pile to prevent wind dispersal —Must inspect weekly and after storms
Land Treatment	—Must have run-on control system, run-off management system, wind dispersal management, an unsaturated zone monitoring program, weekly inspections, and inspections after storms —Collection systems must be emptied promptly	—Same as Interim Status plus the following requirements: —Permit will be granted only after demonstration and will specify design features including: treatment zone to minimize run-off, run-on control system, and run-off management system
Landfills	—Liner or leachate collection system is required unless alternative design is shown to work as well —Must have run-on control system, run-off management system, and wind dispersal management —Collection systems must be emptied promptly —Must maintain maps showing location and contents of each cell —Free liquids are restricted —Containers must be reduced in volume before burial	—Must have double liners and leachate collection system; this requirement can be waived by EPA for mono-fill —Same as Interim Status plus the following requirement: —Must inspect weekly and after storms
Incinerators	—Results of waste analysis must be placed in operating record —May feed hazardous waste only if incinerator is in steady state condition of operation —Must monitor instruments every 15 minutes —Incinerator and alarms must be inspected daily —Must obtain certificate to burn certain wastes	—Must be designed to achieve a destruction and removal efficiency of 99.99% for each principal organic hazardous constituent —May burn only wastes allowed by permit except in approved trial burns —Must be operated in accordance with permit, including automatic cut of waste feed

Thermal Treatment	—Before adding waste, must bring process to steady state condition —Results of waste analysis must be placed in operating record —Must monitor instruments every 15 minutes —Emissions must be observed at least hourly for normal appearance —Must inspect thermal treatment process daily —Open burning or detonation is prohibited except for waste explosives —Certificate required to burn certain wastes	—Must have continuous monitoring of combustion temperature, waste feed rate, combustion gas velocity, and carbon dioxide —Must inspect incinerator and equipment daily —Must test emergency waste feed cut-off system and alarms weekly, and must conduct operational testing monthly —Monitoring and inspection data must be placed in operating record	—Considered to be miscellaneous treatment units; are subject to those permit conditions which regulatory authorities deem necessary to protect human health and the environment
Chemical, Physical, & Biological Treatment	—Must have waste feed cut-off or by-pass system —Whenever a new waste is to be treated, waste analysis plan must be placed in operating record —Must inspect discharge control and safety equipment and data from monitoring equipment daily, and construction materials and surroundings weekly		—Considered to be miscellaneous treatment units; are subject to those permit conditions which regulatory authorities deem necessary to protect human health and the environment

Additional requirements for ignitable, reactive, and incompatible wastes apply to all facilities. Ignitable and reactive wastes must not be placed in the facility unless (1) they are pretreated so that the resulting mixture is no longer ignitable and reactive and the operator takes precautions (documented in the record) to prevent reactions, (2) the wastes are handled separately, or (3) the component of the facility which handles these wastes is used only for emergencies. Incompatible wastes must not be placed in any facility unless the operator takes precautions (documented in the record) to prevent reactions.

Note also that expansions to interim status facilities are in some instances subject to stricter regulations that more closely correspond to those applicable to permitted facilities.

(especially its manifests) to make sure that it is not accepting wastes that are banned from land disposal. (See boxed text on page 217.)

Closure and Post-Closure Care. If the facility has stopped or is about to stop accepting wastes, make sure that all regulatory requirements are being obeyed. Closure of a TSDF must be carried out in accordance with a closure plan, which specifies how wastes will be removed from the facility or (if wastes are to be left onsite) how they will be contained so as to minimize the possibility of future migration. (See 40 C.F.R. §§ 264.111 to 264.112, 265.111 to 265.112.) If wastes are to remain onsite after closure, the facility must then begin obeying the terms of its post-closure permit, which specifies requirements for maintenance, groundwater monitoring, and corrective action in the event of a leak. (See 40 C.F.R. §§ 270.1 (c), 264.117 to 264.118, 265.117 to 265.118.) The postclosure requirements must be followed for at least 30 years after closure. (See 40 C.F.R. §§ 264.117 (a)(1), 265.117(a)(1).)

Unfortunately, EPA has been very slow in processing and issuing post-closure permits. Prod EPA to expedite its action on the facility that concerns you; if necessary, you may want to consider bringing a citizen suit against EPA to compel the agency to rule on a pending permit application. (See page 223.) In the meantime, until a permit is issued, make sure the facility is following its closure plan.

Financial Responsibility. Make sure the facility has obeyed RCRA's financial responsibility requirements. Each TSDF must show that it has (or has access to) sufficient funds both to compensate for accidents that cause bodily injury or property damage to third parties and to pay for closure and for postclosure care. (See 40 C.F.R. §§ 264.147(a), 264.147(b), 265.147(a), 265.147(b), 264.143, 265.143, 264.145, and 265.145.)

Loss of Interim Status. Check to ensure that the facility is not operating illegally. The 1984 amendments to RCRA required that each land disposal facility do two things by November 8, 1985: first, submit a complete permit application to the agency, and second, certify that the facility is in compliance with all groundwater monitoring and financial responsibility requirements. (See 42 U.S.C. § 6925(e)(2).) Of 1538 land disposal facilities across the nation, 995 failed to take the above actions and therefore lost their right to continue operations.[7]

Enforcement

If your investigations turn up an important violation of legal requirements, ask the agency to take enforcement action. (See Chapter 7.) Note

---

[7] General Accounting Office, "Hazardous Waste: Enforcement of Certification Requirements for Land Disposal Facilities," Document No. GAO/RCED 87-60BR, page 19.

The 1984 Amendments and Land Disposal

The 1984 Amendments to RCRA cracked down sharply on land disposal of hazardous wastes. Land disposal encompasses any placement of wastes in or on the land, including landfills, surface impoundments, waste piles, injection wells, land treatment facilities, salt dome or salt bed formations, and underground mines or caves. (See 42 U.S.C. § 6924 (k).)

None of the following wastes may be placed in a land disposal facility unless they have been pretreated in accordance with specific EPA regulations (pretreatment regulations are at 40 C.F.R. Part 268, Subpart D):

The so-called "California List" wastes, consisting of liquid hazardous wastes which are highly acidic (pH 2.0 or lower), or which contain cyanides, arsenic, cadmium, chromium, lead, mercury, nickel, selenium, thallium, polychlorinated biphenyls, or halogenated organic compounds, (See 42 U.S.C. § 6924 (d));

Certain spent solvent and dioxin-containing wastes (specifically, the ones numbered F001, F002, F003, F004, F005, F020, F021, F022, and F023 on EPA's RCRA regulatory list in 40 C.F.R. § 261.31), (see 42 U.S.C. § 6924 (e)); and

All other hazardous wastes included on EPA's RCRA regulatory lists, (See 42 U.S.C. § 6924 (g).)

Note that the land disposal ban does not apply to those wastes which EPA determines can be safely placed in land disposal facilities without any danger of migration for as long as the wastes remain hazardous. (See 42 U.S.C. §§ 6924 (d) (1), 6924 (e) (1), 6924 (g) (5).) EPA's determinations as to safety are to be made on a staggered schedule ending in May 1990. Check 40 C.F.R. Part 268, Subpart C for the current status of the chemicals that interest you.

The 1984 amendments also banned the placement in *landfills* of certain kinds of liquid waste. (See 42 U.S.C. § 6924 (c).) EPA has issued regulations providing additional detail on this subject. (See 40 C.F.R. §§ 264.314, 265.314.)

---

that even in states that have received authority to administer the RCRA program, EPA retains its enforcement authority. (See 42 U.S.C. § 6928(a).) In such states your information should be presented to the state agency *as well as* to EPA.

The enforcing agency has four options open to it.

Administrative orders. The agency can issue an administrative order requiring the violator to comply, assessing civil penalties up to $25,000 per day per violation, and suspending or revoking any permit that has been issued to the facility. (See 42 U.S.C. § 6928(a).) Indeed, RCRA provides that the agency *must* revoke a facility's permit if the agency determines that the facility has violated the TSDF regulatory program

RCRA and Federal Facilities

Federal facilities, such as military compounds, Department of Energy facilities and the like are subject to the full range of RCRA restrictions, including the permit requirement. (See 42 U.S.C. § 6961.) They are also among the worst offenders of RCRA requirements. The U.S. Department of Justice claims, however, that EPA does not have authority to issue administrative orders or go to court to correct regulatory violations at federal facilities. (EPA can, however, issue remedial orders against interim-status facilities.) Instead EPA, and sometimes the Department of Justice, will negotiate with the offending agency to remedy the problem. In states where a state agency is authorized to implement the RCRA program, the state agency can issue administrative orders or go to court to enforce RCRA requirements.

You can pressure EPA to take aggressive action at federal facilities but citizens are not likely to be given a direct role in this process. Where your state agency has the authority, you should press it to act. In many cases, given the limited ability of EPA to enforce RCRA against federal facilities, the best remedy may be to file a citizen suit against the facility yourself.

To identify federal facilities that may have hazardous waste TSDFs, ask EPA for a copy of the federal facilities hazardous waste inventory data for your area. The inventory should identify every facility that treats, stores or disposes of hazardous waste; it is updated every two years. (See 42 U.S.C. § 6937.)

(see 42 U.S.C. § 6925(d)), but the agency also has authority to reissue permits.

The facility to whom an enforcement order is addressed has the right to request a public hearing. (See 42 U.S.C. § 6928(b).) If such a hearing is held, you can request the agency to allow you to intervene or ask to participate as an *amicus*. (See Chapter 7.) Your rights to intervene or contribute as an amicus are not automatic, however, and the agency can keep you out if it wants. Once a decision is issued by the agency officer who presided at the hearing, any party (including citizen intervenors) may appeal to the head of the agency. (See 40 C.F.R. § 22.30(a).)

If the order survives the hearing and appeal process, it becomes final and can be challenged in court. If the facility does not challenge a final order, but does not comply, the agency can assess additional civil penalties and can suspend or revoke any permit issued to the violator (if it has not done so already). (See 42 U.S.C. § 6928(c).)

Civil lawsuits. Instead of issuing an administrative order, the agency can go directly to court to sue for an order requiring the facility to come into compliance and to pay civil penalties. (See 42 U.S.C. §§ 6928(a), 6928(g).)

Criminal prosecutions. If the facility violated regulatory requirements *knowingly*, the agency can bring criminal charges. RCRA's criminal penalties range up to $1,000,000 in fines and 15 years' imprisonment. (See 42 U.S.C. §§ 6928(d), 6928(e).) RCRA's civil and criminal penalties are not mutually exclusive, and the agency could seek both against a single violating facility.

Additional data collection. If the agency feels that there is a problem at the facility but that the evidence against the facility is not yet sufficiently developed to warrant immediate enforcement action, the agency can gather additional information on the facility through inspections or data requests.

If the agency refuses to take enforcement action, you have the right to bring a citizen suit in federal district court. The court can order the violator to comply, and can assess the same civil penalties that would be available in a government civil enforcement action. (See 42 U.S.C. § 6972(a).) Citizen suits can be expensive, largely because of the stiff resistance lodged by violating facilities. You should contemplate such a suit only if the violation is serious and your evidence is strong enough to make a favorable outcome likely.

Forcing Action Through Community Awareness

In addition to your efforts to communicate directly with governmental enforcement authorities or to bring your own enforcement suits, focusing widespread public attention on your problem can be a very effective means of spurring action. This is illustrated by the experience of Citizens for Control of Toxic Waste (CCTW), who are fighting to close a hazardous waste dump located near their agricultural community of Grand View, Idaho. According to CCTW, more than 78,000 tons of toxic chemicals, including acid, heavy metals, PCBs and DDT have been buried at the site, contaminating a shallow aquifer.

CCTW began its fight to close the dump by holding public meetings to raise community awareness. Although CCTW has only 15 members (out of a total Grand View population of 354), a meeting in a nearby town drew 300 people. Eventually, media attention attracted the interest of the state legislature and citizens from around the state.

CCTW, joined by the Idaho Conservation League and Idaho Fair Share, mounted a tenacious campaign to oppose the dump's permit renewal. Local and state legislators also joined the fight, prompted to action by CCTW's efforts. Eventually the county commissioners saw the light and made closing the dump a priority issue. The fight for Grand View's right to clean, safe groundwater continues, much strengthened by the dedicated efforts of a small number of citizen activists.

SETTING REGULATORY REQUIREMENTS

Citizen participation is not limited to ensuring that the regulatory rules are being obeyed. Rather, you can also participate in establishing what those rules will be, thus playing an important role in determining what level of protection our vital groundwater resources will receive. Your participation can take three main forms: involvement in the decision to issue a permit, overseeing closure of a facility, and policing exemptions that may be granted by the agency.

Operating Permits

As indicated previously, each TSDF must obtain a permit in order to conduct operations. The permit is a detailed document that specifies the rules the facility must follow. In writing these rules, governmental permit authorities need not restrict themselves to parroting the requirements set forth in the agency's regulations, but may also add additional or different rules tailored to the specific situation of the permit applicant. (See 40 C.F.R. § 270.32(b).)

Your participation in the permit process is crucial. In many cases, you may want to advocate that no facility should be allowed in your area because of groundwater concerns. But even when a permit is going to be issued, you need to be involved to make sure its restrictions are tough. In the absence of special circumstances, the rules enshrined in the facility's permit will last for several years.[8]

You have the right to review a proposed permit in draft form, to submit comments on it, and to speak at the public hearing if one is held. (See Chapter 7.) The following discussion gives you several ideas on how to structure your participation.

Reporting Requirements. The most stringent rules may be of little value if citizens (and indeed regulators) remain in the dark about whether these rules are being violated. Ask that the facility be required to report to regulatory authorities at frequent intervals on every key aspect of its operations. Of particular importance, you should make sure that the facility provides regulators with groundwater monitoring results, self-inspection reports, assurances of financial responsibility, and waste manifests.

Groundwater Protection Standards. A key facet of the permitting

---

[8] RCRA allows the agency to grant TSDF permits lasting for as long as ten years; land disposal permits must be reviewed and modified as appropriate once every five years. See 42 U.S.C. § 6925(c)(3).

process is the setting of "groundwater protection standards" (GPS), which are maximum levels of specified groundwater contaminants. (See 40 C.F.R. §§ 264.92, 264.94(a).) If these maximums are exceeded due to releases from the facility, then the facility must undertake corrective action. (See 40 C.F.R. § 264.99(i).)

First, you should make certain that GPS are established for *all* contaminants that could potentially be released from the facility. To determine which contaminants to worry about, check the permit application for data on the applicant's planned and past waste-disposal patterns. Additional information on past disposal patterns can be gleaned from the facility's manifests, manifest-discrepancy reports, and unmanifested waste reports.

Second, encourage the agency to follow the RCRA regulations by setting the GPS at (1) the specific maximum concentrations set forth in the regulations (generally, these will be the MCLGs set under the Drinking Water Act, see Chapter 11), or (2) the background levels present in nearby groundwater (for chemicals lacking a specific regulatory maximum). (See 40 C.F.R. § 264.94(a).)

Oppose attempts by the facility to advocate weaker "alternate concentration limits" (ACLs). To establish an ACL, a facility must prove to the agency that the contaminant concentrations permitted by the proposed ACL will not pose a substantial present or potential hazard to human health or the environment. In determining whether a proposed ACL would pose such a hazard, the agency must consider the nature of the facility's wastes; the local geology and groundwater flow characteristics; the current quality of the groundwater; the current and potential uses of the groundwater and the location and rates of any groundwater withdrawals; the risks to human health; the potential damage to wildlife, crops, vegetation, and physical structures; and the persistence and permanence of any adverse effects. (See 40 C.F.R. § 264.94(b).)

Cleanup of Past Releases. RCRA requires corrective action not only for those releases occurring after the date of permit issuance, but also for those that had occurred before. If corrective action for prepermit releases cannot feasibly be completed before the facility's permit comes up for issuance, then the permit must contain a compliance schedule as well as assurances that the applicant has sufficient financial resources to undertake the necessary corrective measures. (See 42 U.S.C. § 6924(u).)

Under EPA's national corrective action strategy,[9] the corrective action

---

[9] "EPA Draft Strategy for Carrying Out RCRA Provisions Requiring Corrective Action at Hazardous Waste Facilities," BNA Environment Reporter *Current Developments* (Oct. 31, 1986), at page 1068. EPA plans to replace this draft strategy with formal regulations in the near future.

process is divided into four stages. First, the agency completes a "RCRA Facility Assessment" (RFA), which is designed to identify whether any releases have occurred. The agency may rely on the applicant's data, but should perform its own investigation if that data is insufficient or unreliable.

Second, if the RFA indicates that a release has occurred, the agency may order the facility to conduct a "RCRA Facility Investigation" (RFI), which is designed to determine the nature, extent and rate of contaminant migration. The agency may also order that, while the RFI is being conducted, the facility undertake certain *interim* corrective measures to prevent the problem from becoming unmanageable before a final response can be devised.

Third, after the RFI has been submitted to and evaluated by the agency, the facility must then proceed to devise a set of corrective measures.

Fourth, if the agency determines the corrective measures to be satisfactory, the facility must carry them out.

EPA's national strategy suggests (although the matter is far from clear) that the agency plans to wait until the third step, the devising of corrective measures, before seeking comments from the public. You should certainly provide comments at that crucial stage, but you should also be involved much earlier. If you know that the RFA or RFI for the facility that concerns you have not yet been issued, ask the agency to send you these documents as soon as they are ready so that you can comment on them.

In structuring your participation in the corrective action process, you should discourage attempts by the agency or facility to sweep troublesome contamination problems under the rug, to settle for less than full cleanup, or to delay completion of corrective measures beyond the date of permit issuance.

Groundwater Monitoring. Make sure that the facility has designed a groundwater monitoring program that will be adequate to detect any release into groundwater (detection monitoring), and, once a release is found, to determine whether the contaminants exceed the GPS specified in the permit (compliance monitoring). Key points to check include the number and location of wells, frequency of sampling, analysis of samples (Is the laboratory testing for all contaminants for which GPS have been established in the facility's permit?), and frequency of reporting to the agency.

Design and Operating Requirements. Verify that the design and operating requirements imposed by the RCRA regulations have been incorporated into the facility's permit. (See Figure 16.1, on pages 212–215.) In particular, make sure that all facilities have employee-training programs adequate to ensure competence in waste-handling procedures; that they conduct inspections at frequent intervals, and file reports with the regulatory agency detailing the results; and that they have on hand a contingency plan describing how any emergency situations will be handled (if the plan designates a specific fire and police department for emergency use, check with the department to make sure it is willing and able to respond as described). Check that land disposal facilities have double liners and a leachate protection system and that waste storage containers are properly labeled.

Financial Responsibility. As indicated previously, facilities must demonstrate that they will be able to pay for both accidental releases and the cost of closure and post-closure care. Make sure that the amount of money involved is sufficient to cover all reasonably forseeable costs. Also, ask that the facility's financial assurances be reevaluated each year. Annual reevaluation will provide an early warning system to spot downturns in the net worth of "self-insuring" facilities (that is, facilities who rely on their own resources, rather than on a commercial insurance policy, to meet RCRA's financial responsibility requirement).

Regulatory Inertia. All the valuable protections offered by the permit system are of little value if the agency drags its feet in processing permits. The 1984 RCRA amendments require that the agency issue permits or permit denials for all pre-1984 interim-status land disposal facilities no later than November 8, 1988. (See 42 U.S.C. § 6925(c)(2)(A)(i).) If EPA has missed this deadline for the facility you are concerned about, you have the right to bring a citizen suit to force the agency to act. (See 42 U.S.C. § 6972(a)(2).) The availability of a federal citizen suit against an authorized *state* agency is less clear, but state law may give you the right to bring such a suit in state court.

Closure and Post-Closure

Your opportunities for participation in closure and post-closure plan development differ depending on whether the facility closing is permitted or in interim status. All facilities must have a proper closure plan. Those facilities where hazardous materials will remain after closure (e.g., a landfill) must have a post-closure plan to ensure future problems do not occur at the site. For permitted facilities, both closure and post-closure plans are developed as part of the operating-permit process, and become conditions of the operating permit when issued. (See 40 C.F.R. §§ 264.112(a)(1), 264.118(a).) You may comment on the drafts of these plans just as you would comment on any other provision of a draft permit.

For interim-status facilities, the owner is required to prepare the closure and post-closure plans while the facility is still in operation and to keep them on-site. (See 40 C.F.R. §§ 265.112(a), 265.118(a).) The closure plan need not be submitted to the agency until six months before closure is scheduled to begin. (40 C.F.R. §§ 265.112(d)(1).) At that point the agency releases the plan to the public for comment and in some cases holds a public hearing. Based on its review of the plan and the public comments, the agency can either approve or disapprove the plan. If the plan is disapproved, the facility is given a chance to amend it; if the facility fails to do so, the agency may make any necessary amendments itself. (See 40 C.F.R. §§ 265.112(d)(4).)

Post-closure at interim-status facilities involves not just preparation of a plan, but also initiation of the permit process. Specifically, any interim-status facility whose closure will be conducted so as to leave hazardous wastes onsite must apply for and obtain a post-closure permit. (See 40 C.F.R. § 270.1(c).) You have the same right to participate in the post-closure permit process as you would to participate in a RCRA operating-permit proceeding.

Timing of Public Participation. Permitted and interim-status facilities suffer from opposite problems of timing. Because closure and post-closure plans for permitted facilities are prepared at the time operating permits are issued, they may be badly out of date by the time the facility is actually ready to close many years later. In such circumstances you should request that the agency reevaluate the plans and allow additional public participation.

For interim-status facilities, the public is not involved until shortly before the plans are actually implemented. You should ask the agency to request that the facility submit a copy of the plans at a much earlier date, so that they will be accessible to public scrutiny well before implementation time arrives.

Areas of Concern. In commenting on closure and post-closure plans, focus on several key points. First, encourage the agency to consider closure and post-closure plans at the same time, rather than putting the post-closure plan aside to be dealt with later. This simultaneous approach will help ensure that valuable opportunities to assess and correct contamination will not be lost. For instance, if a facility is allowed to close and cap a landfill without investigating the current state of ground-water contamination or taking any steps to remedy such contamination, it will be much more difficult and expensive to conduct investigative and remedial action later on.

Second, make sure that those facilities that are capable of removing *all* contaminants during closure (waste piles, surface impoundments, containers and tanks) include adequate provision in their plan to do so. In the case of surface impoundments and waste piles, make sure that the facility has prepared a *contingent* post-closure plan detailing what will be done

Violations of Closure and Post-Closure Plans

The need for close citizen scrutiny of the closure process is illustrated by the experience of People Against Hazardous Landfill Sites (PAHLS), an Indiana citizens' group. Through its investigations PAHLS learned of various violations of regulatory requirements at a landfill in Wheeler, Indiana. After PAHLS brought these violations to the attention of the Indiana Board of Health, the Board issued a citation against the company. Nevertheless, the landfill's closure and post-closure plans failed to remedy the violations, which included:

- failure to develop a plan to collect and treat runoff of hazardous liquids;
- failure to describe a method to collect, remove, treat, and monitor leachate; and
- failure to provide a sufficient number of monitoring wells, or place them at the proper depth or location to yield background samples, and ensure immediate detection of hazardous wastes in the uppermost aquifer.

PAHLS followed up by sending a 60-day notice to the company, threatening to bring a citizen suit against the facility if these problems were not corrected. This notice prompted EPA officials to hold public hearings, at which PAHLS was able to voice its concerns. Ultimately, EPA incorporated nearly all of the group's suggestions into the closure and post-closure plans.

---

if efforts to remove all wastes prove unsuccessful. (See 40 C.F.R. §§ 264.228(c)(1)(ii), 264.258(c)(1)(ii).)

Third, for facilities that leave waste in the ground after closure, make sure that post-closure permits provide for adequate containment of wastes, a scientifically sound groundwater monitoring program to be conducted at regular intervals, and prompt preparation of a corrective-action plan in the event contamination is discovered. Verify that the site will be sequestered so that unauthorized persons or livestock cannot enter the waste area, and that no use will be made of the site (such as construction of buildings) that might interfere with the facility's ability to contain and manage the wastes.

Exemptions

Oppose efforts by the agency or the facility to weaken the RCRA regulatory program. EPA's regulations offer numerous opportunities for exemptions from key regulatory requirements: A facility may request that EPA delist its wastes (thus extricating it from the RCRA program entirely) (see 40 C.F.R. § 260.22); that the time period for completing closure be extended (See 40 C.F.R. §§ 264.113, 265.113); that the time period for post-closure care be reduced (See 40 C.F.R. §§ 264.117(a)(2)(i), 265.117(a)(2)(i)); that no groundwater monitoring be required (See 40 C.F.R. §§ 264.90(b), 265.90(c)); and so on.

Judicial Review of Permit Decisions

If you are dissatisfied with EPA's decision on a permit, you can petition for judicial review within 90 days in the U.S. Court of Appeals for the circuit where the permittee transacts business. (See 42 U.S.C. § 6976(b).) The court will almost certainly be unwilling to accept any new evidence from you, but instead will base its review solely on the administrative record compiled by EPA during the permit proceeding. Accordingly, make sure that your comments to EPA on the draft permit include all the evidence you will want the court to consider later on.

Permits issued by authorized *state* agencies may not be reviewed in federal court, but must be challenged in state court under state law.

NONHAZARDOUS SOLID WASTE FACILITIES

If the facility that concerns you accepts solid wastes that are not considered "hazardous" under EPA's regulations, check to see if the facility is obeying the solid waste regulatory program established under Subtitle D of RCRA. (See 42 U.S.C. §§ 6941 to 6949a.) The Subtitle D program is administered at the state level, with federal involvement limited primarily to providing funds to state regulatory authorities. (See Chapter 28 for more about state waste control programs.)

To locate facilities subject to Subtitle D, ask your state or EPA for a copy of the inventory of open dumps, which was prepared in 1985. (See 42 U.S.C. § 6945(b).) Unfortunately, this inventory cannot be considered comprehensive because (1) lack of funds limited the number of facilities that could be counted, and (2) with the passage of time, new dumps may have opened that did not exist in 1985. You should supplement the data in the inventory by making your own inquiries in your area. In particular, try to identify where household and business refuse is taken by both governmental and private trash disposal authorities.

Once you have focused on a facility, check to see whether it is an "open dump" (a facility operated in violation of EPA's regulations, see 40 C.F.R. § 275) or a "sanitary landfill" (a facility operated in compliance with EPA's regulations, see 40 C.F.R. § 275). Subtitle D requires that no new open dumps be permitted to begin operations, and that all existing open dumps be closed or upgraded in accordance with EPA regulations. (See 42 U.S.C. §§ 6943(a)(2), 6943(a)(3), 6944(b), 6945(a).)

Also check to see whether the facility accepts *hazardous* waste either from individual households, or from "small quantity generators" who are exempt from the duty to send their waste only to RCRA-permitted TSDFs. (See Chapter 17.) If the facility does accept such hazardous waste, it must obtain a special permit from the state. (See 42 U.S.C. § 6945(c).) EPA will soon be issuing revised regulations governing those facilities that are subject to this special permit requirement. These new regulations will require groundwater monitoring, appropriate siting of new and existing facilities, and corrective action. (See 42 U.S.C. § 6949a(c).) Unfortunately, however, they will apply only to municipal facilities; issuance of regulations on industrial facilities has been indefinitely deferred.

If your investigations reveal violations of either the open dumping or special permitting requirements, ask the state to take appropriate enforcement action. Should the state fail to do so, you can bring a citizen suit to enforce the ban on open dumping (the permitting requirement, by contrast, is not subject to federal citizen suit enforcement). (See 42 U.S.C. §§ 6945(a), 6972(a)(1)(A).) Your enforcement powers are actually greater than EPA's, because that agency may not enforce even the open dumping ban, except in those states that fail to adopt an adequate solid waste regulatory program—see 42 U.S.C. § 6945(c)(2).

In 1988, EPA proposed new regulations governing the disposal of nonhazardous waste. Though the proposed rules are an improvement over the existing regulations, which essentially fail to protect groundwater, EPA's proposal has been severely criticized for not going far enough. Check with EPA about the status of the proposed solid waste rules. The rules are likely to be debated for much of 1989 and may, in the end, be preempted by new legislation from Congress on the subject. Strong regulations controlling solid waste disposal are a critical part of an effective program to protect groundwater quality. You should participate in the rulemaking process to convince EPA to strengthen its proposed rules and work with Congress to get tougher laws passed.[10]

[10] EPA's proposed solid waste regulations can be found at 53 *Fed. Reg.* 33314 (August 30, 1988).

REFERENCES

WASTE DISPOSAL FACILITIES

EPA Draft Strategy for Carrying Out RCRA Provisions Requiring Corrective Action at Hazardous Waste Facilities. 17 Environmental Reporter (BNA) 1068. October 31, 1986.

EPA, Office of Groundwater Protection. *EPA Activities Related to Sources of Groundwater Contamination.* Washington, D.C. February 1987.

EPA, Office of Policy, Planning and Information, Office of Solid Waste. *1986 National Screening Survey of Hazardous Waste Treatment, Storage, Disposal, and Recycling Facilities.* Washington, D.C. (Prepared by the Center for Economics Research.) December 1986.

EPA, Office of Solid Waste and Emergency Response. *National Survey of Hazardous Waste Generators and Treatment, Storage, and Disposal Facilities Regulated Under RCRA in 1981.* Washington, D.C. April 1984.

EPA, Office of Solid Waste and Emergency Response. *RCRA Groundwater Monitoring Compliance Order Guidance.* Washington, D.C. August 1985.

EPA, Office of Solid Waste and Emergency Response. *RCRA Orientation Manual.* Washington, D.C. January 1986.

EPA, Office of Solid Waste and Emergency Response. *Report to Congress: Solid Waste Disposal in the United States.* Washington, D.C. 1988. For a copy of the Executive Summary (EPA/530-SW-88-011A), call the RCRA Hotline (see Appendix D).

EPA, Office of Solid Waste and Emergency Response. *The Hazardous Waste System.* Washington, D.C. June 1987.

Fortuna, Richard C., and David J. Lennett. *Hazardous Waste Regulation, the New Era: An Analysis and Guide to RCRA and the 1984 Amendments.* McGraw-Hill. New York, N.Y. 1987.

Goldman, Benjamin A., James A. Hulme, and Cameron Johnson, Council on Economic Priorities. *Hazardous Waste Management: Reducing the Risk.* Island Press, Washington, D.C. 1986.

Harris, Christopher, William L. Want, and Morris A. Ward. *Hazardous Waste: Confronting the Challenge.* Quorum Books. New York, N.Y. 1987.

O'Leary, Philip R., Patrick W. Walsh, and Robert K. Ham. "Managing Solid Waste," *Scientific American*, December 1988, page 36.

Peterson, Jonathan M., *RCRA Enforcement Provisions After the 1984 Amendments*, 5 Virginia Journal of National Resources Law 323 (Spring 1986).

Quarles, John. *Federal Regulation of Hazardous Wastes: A Guide to RCRA.* Environmental Law Institute. Washington, D.C. 1982.

Stever, Donald W. *Law of Chemical Regulation and Hazardous Waste.* C. Boardman Co. New York, N.Y. 1986.

U.S. Congress, Office of Technology Assessment. *Serious Reduction of Hazardous Waste: For Pollution Prevention and Industrial Efficiency.* Washington, D.C. 1986.

U.S. General Accounting Office. *Hazardous Waste: Enforcement of Certification Requirements for Land Disposal Facilities.* Washington, D.C. January 1987.

U.S. General Accounting Office. *Hazardous Waste: Groundwater Conditions at Many Land Disposal Facilities Remain Uncertain.* Washington, D.C. February 1988.

LAWS AND REGULATIONS

U.S. Code Citation

Resource Conservation and Recovery Act, 42 U.S.C.A. §§ 6901-6991(i) (West 1983 & Supp. 1988).

Code of Federal Regulations Citations

40 C.F.R. Parts 256–257.
40 C.F.R. Parts 260–271.
40 C.F.R. Parts 22–24.

CHAPTER 17

Industrial and Commercial Sources

INDUSTRIAL AND COMMERCIAL facilities contain many potential sources of groundwater contamination. Many factories and businesses use chemicals in manufacturing processes and in performing commercial services. These businesses may store hazardous chemicals and waste materials on-site in drums, storage tanks, waste piles, and surface impoundments; they may dispose of wastes in landfills, treatment facilities, or injection wells on site or contract for their treatment and disposal at facilities elsewhere. Wastewater discharged from factories and businesses into surface streams or sewers may contain toxic or hazardous pollutants. Each of these activities poses risks to groundwater.

Just as no single comprehensive federal law addresses groundwater contamination, no single regulatory program controls the wide array of activities at industrial and commercial sources that may cause such contamination. Some of these activities, such as the disposal or treatment of hazardous waste or the storage of chemicals in tanks on site, are regulated by specific programs that are discussed separately in other chapters of this book. (See Chapter 16 on treatment, storage, and disposal facilities; Chapter 18 on injection wells; Chapter 19 on storage tanks; and Chapter 22 on transportation of hazardous materials.)

This chapter addresses general issues related to groundwater contamination from commercial and industrial sources: how you can identify businesses in your community that use hazardous chemicals that may cause groundwater contamination; how you can obtain remedial action for spills or leaks at such facilities; and long-term measures you can take to prevent contamination of your groundwater from such sources. Be sure to consider the possibility of obtaining information or remedial action related to industrial and commercial sources under your state and local laws, discussed in Chapter 28.

IDENTIFYING POSSIBLE INDUSTRIAL SOURCES

Groundwater contamination may arise from spills, leaks, or routine discharges of toxic or hazardous chemicals and wastes from a wide array of industries or businesses in your area. Businesses that use or create such materials include manufacturers of chemicals, pesticides, paper, and paints, oil and gas refineries and distributors, metal plate companies, manufacturers of "high-tech" electronic components, body shops, public utilities, dry cleaners, and dye works, among others. Knowing which businesses handle such materials, and the types of chemicals and wastes that are present in local factories, is essential in planning for accidental spills or leaks and monitoring compliance by these companies with regulatory requirements that will reduce risks of contamination to your groundwater. Such knowledge may also permit you to trace the sources of any existing contamination in your groundwater.

You may already be familiar with many businesses or industries in your area that use chemicals in their work. For further information, various public sources may be helpful:

- Local agencies concerned with public safety, building codes, or worker safety, such as the fire department, may maintain a list of businesses handling dangerous chemicals.
- Chemical manuals and reference works, such as the *Industrial Directory*, identify chemicals manufactured in the United States or in individual states, and provide a partial listing of facilities that handle such chemicals.
- The Standard Industrial Classification (SIC) Code, a system that groups industries by code number based on their type of activity, contains an explanation of the types of products and wastes generated by each category, including descriptions of their toxicity, corrosivity, reactivity, and ignitability.[1]

Another major source of information will be the EPA regional office for your area or, in states where responsibility for regulation of hazardous waste has been delegated from EPA to the state, your state environmental agency. Pursuant to the Resource Conservation and Recovery Act (RCRA), any business that handles hazardous wastes (see pages 232–233) must notify EPA (or the responsible state agency) of its location and activity, and must identify the wastes it handles, whether the business generates the wastes itself, transports them for other companies, or operates a treatment, storage, or disposal unit. This information may identify many

[1] The *1982 Census of Manufacturers and Census of Mineral Industries*, published by the Department of Commerce, Bureau of Census, contains in-depth descriptions of code categories and wastes generated by each category.

Materials Regulated by Federal Statutes

The federal statutes regulating chemicals and chemical wastes—the Resource Conservation and Recovery Act (RCRA), the Toxic Substances Control Act (TSCA), the Emergency Planning and Community Right-to-Know Act (EPCRA), the Occupational Safety and Health Act (OSH Act), and the Superfund law (CERCLA)—employ varying definitions for the materials they regulate. Those terms can be quite confusing. Brief summaries of the regulatory definitions used in each statute follow:

Hazardous wastes (RCRA)	Discarded or accidentally released materials that can pose a substantial threat to public health or the environment if improperly managed. There are two categories of hazardous wastes: (1) wastes listed by EPA, either as general types of waste streams (spent solvents, wastewater treatment sludges, etc.) or as specific chemicals, and (2) wastes that do not appear on EPA's list but that, when tested by EPA or the company creating them, are found to be ignitable, reactive with other chemicals, corrosive, or toxic. See Chapter 16 for a more detailed explanation.
Extremely hazardous substances (EPCRA)	366 toxic, reactive, volatile, dispersable, and flammable chemicals that can cause serious irreversible health effects from accidental releases. A list of such substances is published in EPA's regulations (see 40 C.F.R. Part 355, Appendix A). This list may be expanded or modified by the EPA.
Hazardous chemicals (OSH Act, EPCRA)	More than 50,000 chemicals that cause acute health effects, cancer, or that are flammable, explosive, or reactive with other chemicals. OSH Act's definition is set forth in regulations at 29 C.F.R. § 1910.1200(c).
Toxic chemicals (EPCRA)	Chemicals listed in the Senate committee report for EPCRA, reprinted in 52 Fed. Reg. 21168 (June 4, 1987). EPA can add chemicals to this list if they cause acute or chronic human health effects or if their toxicity and persistence can

	cause serious adverse impacts on the environment. EPA can also delete chemicals from this list if they do not meet these criteria.
Hazardous substances (CERCLA)	A broad array of chemicals and wastes defined as hazardous or toxic under the Clean Water Act, Clean Air Act, RCRA, or TSCA, or designated by EPA under CERCLA itself as presenting substantial danger to public health, welfare, or the environment when released into the environment.
Chemical substances (TSCA)	Any organic or inorganic substance of a particular molecular identity (excluding mixtures, pesticides, tobacco, nuclear materials, and foods, drugs, or cosmetics).
Pollutant (Clean Water Act)	Dredged spoil, solid waste, sewage, garbage, chemical wastes, and industrial, municipal, or agricultural waste discharged into water.

possible sources within your community, as EPA estimates up to 96 percent of all hazardous waste is treated, stored, or disposed of at the point of its generation.[2]

If a business does not dispose of its own wastes on-site, it must prepare a manifest for wastes it ships to another facility for treatment or disposal that identifies the chemicals in the wastes and their destination. Review of those manifests, which the business must file with your EPA regional office or your state environmental agency, will disclose many businesses that generate hazardous wastes in your community, and provide information regarding the chemicals such businesses are using. The RCRA program, and ways in which you can monitor its requirements, are discussed below.

EPA's regional office or your state environmental agency will also have information identifying businesses that discharge pollutants into surface waters, indirectly threatening groundwater. Such discharges require a permit, known as a National Pollutant Discharge Elimination System (NPDES) permit, under the federal Clean Water Act, and the permit files maintained by EPA or your state environmental agency should prescribe

[2] U.S. EPA, *National Survey of Hazardous Waste Generators and Treatment, Storage, and Disposal Facilities, Regulated Under RCRA in 1981* (1984).

maximum levels of specific pollutants that a particular business may discharge and contain actual discharge-monitoring reports filed by the company. That information can identify companies that use hazardous chemicals, document violations of such companies' permits, and reveal potential threats to groundwater from the discharge of such chemicals into surface waters that feed aquifers. The Clean Water Act is also discussed further below.

The Emergency Planning and Community Right-to-Know Act (EPCRA)

The best source of information concerning the presence and characteristics of hazardous chemicals in your community, however, are new programs established by a recent federal law, the Emergency Planning and Community Right-to-Know Act (EPCRA), passed as Title III to the Superfund Amendments and Reauthorization Act of 1986. Enacted in response to the Bhopal disaster, EPCRA establishes an emergency-planning process at the state and local level to prepare for accidental spills and leaks of dangerous chemicals. The Act also requires businesses to disclose to governmental officials and the public extensive information concerning the types and amounts of chemicals present at their plants. As the Act's procedures are implemented, local officials and the public will gain a detailed picture of the industrial and commercial activities within each community that may contaminate groundwater.

EPCRA's emergency-planning procedures, discussed below, focus upon accidents involving "extremely hazardous substances," chemicals that may cause immediate injury to public health (see pages 232–233). Local planners may obtain from businesses and factories that handle such dangerous chemicals any information needed to develop emergency response plans. That information is likely to include locations and amounts of particular dangerous chemicals, health risks and proper medical treatment for exposure, and methods for cleaning up spills. Local emergency plans incorporating such information are available to the public, and will be helpful in assessing risks to groundwater from chemical spills and accidents.

In addition to emergency planning, EPCRA establishes broad "community right-to-know" procedures to increase public knowledge and access to information on the presence of chemicals in the community and releases of these chemicals into the environment. First, the Act extends previous employee right-to-know procedures developed by the Occupational Safety and Health Administration (OSHA) to the community as a whole. OSHA requires businesses that handle hazardous chemicals (see pages 232–233) to make available to employees Material Safety Data

Sheets (MSDSs) that identify dangerous chemicals in the workplace, describe their potential health risks and physical hazards, and specify precautions for their handling. EPCRA now requires such businesses to provide MSDSs (or a list of chemicals for which an MSDS is available) to state emergency response commissions, local emergency-planning committees, and local fire departments. If the company provides only a list, the committee may require it to file the actual MSDSs. MSDSs received by the local committees are available to the public, and you may require the committee to obtain the actual MSDSs if it has not done so.

Second, businesses that handle hazardous chemicals must annually submit chemical inventories to the state committee, local committee, and the local fire department. Such inventories must contain estimates of the amounts of chemicals in each of five OSHA health hazard categories (acute health hazards, chronic health hazards, fire hazards, explosion hazards, or hazards from reaction with other chemicals) held within the plant, and the general location of such chemicals within the facility. At the request of the state or local committees or the fire department, businesses must provide more detailed information, known as Tier II information, identifying specific chemicals, amounts of each chemical, its manner of storage, and its location within the facility. Tier II information held by state or local committees is available to the public, and the public can request that those committees seek such information from businesses if they have not already done so.[3]

Finally, EPCRA requires businesses handling substantial amounts of "toxic chemicals" (see definition in box) to submit annual reports of their releases, both accidental and routine, of such chemicals into the environment.[4] Such reports must identify the use of each chemical within a factory, the amounts present at the plant, any waste treatment or disposal methods used, and the annual quantity of the chemical entering each environmental medium (i.e., emissions into the air, water, or on land). This information is available to you from your local planning committee, which is required to create a public file containing these reports, and is also available on a national computerized toxic chemical data base managed by EPA.

---

[3] If the business in question stores more than 10,000 pounds of the chemical in question, the state or local committee must ask it for Tier II information in response to a public request. If the business stores less than that amount, a committee may, but need not, ask for such information in response to a public request.

[4] Annual reports must be submitted by businesses in Standard Industrial Classification Codes 20–39 (principally manufacturing companies) that manufacture more than specified amounts of a particular chemical. The initial threshold, for 1988, included companies that manufactured 75,000 pounds of a chemical, or that used 10,000 pounds. The manufacturing threshold declines to 25,000 pounds by 1990.

For more information concerning EPCRA's requirements, you can call EPA's Emergency Planning and Community Right-to-Know Information Hotline (ouside 202 area code, dial toll free (800) 535-0202; within 202 area code, call 479-2449). the hotline provides assistance on compliance with EPCRA and can help citizens determine what information businesses must report and how they can obtain access to such data.

RESPONDING TO SPILLS OF CHEMICALS THAT MAY THREATEN GROUNDWATER

Spills or leaks of chemicals within your community should be addressed promptly to prevent contamination of groundwater supplies. The primary response mechanism in your community is the local emergency response plan required for each local planning district under EPCRA. The Act requires each state to appoint a statewide emergency response commission, which in turn designates local emergency-planning districts (usually divided along county or other political lines). The state commission must appoint a local emergency-planning committee for each district, made up of public officials, environmental and public safety personnel, and representatives of the media, community groups, and local businesses that use dangerous chemicals. The local planning committee is responsible for developing and implementing a comprehensive emergency response plan for spills or leaks of "extremely hazardous substances" (see pages 232–233) from local businesses or from transportation through the district.

The local emergency response plan must identify all business sites within the area that contain extremely hazardous substances and any transportation routes likely to be used to transport such substances, designate an emergency response coordinator for the district, and describe how businesses and local emergency and medical personnel should respond to any release of such chemicals. The local plan must provide for prompt public notice of any release and establish evacuation plans. Businesses that use extremely hazardous substances must cooperate with the local committee in the development of the plan, by providing any requested information and by appointing a facility emergency coordinator to work with the committee and to direct emergency action by the company itself.[5]

The local emergency response plan is triggered when a business (or a transporter) suffers a spill or leak of more than a specified amount of any

[5] Businesses are subject to these requirements if they contain on-site extremely hazardous substances in an amount that equals or exceeds a threshold-reporting quantity established by EPA for that chemical. Transportation facilities are generally excluded from the Act's planning and public disclosure requirements, although they must report accidental spills of extremely hazardous substances.

extremely hazardous substance or of any other substance that is required to be reported to the National Response Center under the Superfund law.[6] The business must provide immediate notice of the spill, by telephone or in person, to the state emergency response commission and to the emergency coordinator designated by the response plan for the local district.[7] The initial notice must identify the chemical or chemicals released, describe the circumstances of the spill or leak, identify health risks it may pose (together with appropriate medical treatment), and specify proper precautions, including evacuation if necessary. The business must submit a written follow-up notice describing actions taken to respond to or contain the release and containing additional information regarding health risks and medical treatment.

Spills of chemicals listed under § 103(a) of CERCLA, the Superfund law, must also be reported to the National Response Center. (See Chapter 13.)

If you discover a spill or leak of any chemicals, promptly notify your local emergency coordinator. Your local emergency coordinator, working with your state environmental and public safety agencies and the EPA, can take immediate action to identify the chemicals involved and to initiate an appropriate response to protect public safety. Follow up to confirm that appropriate action has been taken in accordance with your local emergency response plan.

If you do not believe that the spill has been cleaned up properly, or if you believe that chemicals have been spilled or dumped in your community in the past without remedial action, you should contact the National Response Center as well as state and local environmental and public safety officials. Chapter 13 describes the Superfund program. You may also try to persuade EPA to take action under its imminent hazard authority under RCRA, TSCA, and the Clean Water Act. Chapter 14 describes EPA's imminent hazard authority and how you can invoke it.

Preventing Groundwater Contamination from Industrial Sources

There is much that you and other citizens can do to reduce the risk of contamination to your community's groundwater from commercial and industrial sources.

[6] Because the purposes of the two Acts differ, the list of chemicals defined as "extremely hazardous substances" under EPCRA differs from the list of hazardous chemicals that companies must report to the National Response Center under the Superfund law (CERCLA). There is substantial overlap, however: more than 100 chemicals listed under EPCRA are also listed under CERCLA.

[7] Spills or leaks that do not release chemicals beyond the boundaries of a plant do not have to be reported. Spills that occur during transportation of chemicals may be reported to the emergency telephone operator, rather than the state and local committees.

Emergency Planning

First, you can become involved in the emergency planning for your community under EPCRA to ensure that chemical spills and leaks are cleaned up and aquifers protected. The Act mandates an open planning process: local emergency-planning committees must include representatives of community organizations and the media, and information they obtain from local businesses regarding their use of chemicals must be available to the public. Make sure that the community is in fact fairly represented on your committee; some states have named persons closely connected with local industry as community representatives. You can petition the state emergency response committee to modify the membership of your local committee, but there is no explicit mechanism under the Act to challenge the local committee's makeup in court. (A state committee's decision to reject your petition may be reviewable under state law, however.)[8]

You should carefully review the emergency response plan the committee prepares. A good emergency response plan should contain a thorough analysis of the community's risk from chemical accidents: hazard identification (the identity, location, and nature of chemical hazards within the community), vulnerability analysis (the areas and populations that may be affected by an accidental spill or leak), and risk analysis (the potential for accidental releases from local industrial practices and likely severity of their effects). Plans should identify hazardous materials transportation routes as well as businesses using such chemicals.

The plan should contain explicit direction for containing and cleaning up spills of each chemical or category of chemicals found in the community, including precautions for handling particular chemicals and guidance for medical treatment of persons exposed to particular chemicals. Contingency plans should address "worst case" accidents involving large releases, releases of multiple chemicals, and accidents that involve simultaneous fires, power outages, etc. Plans must establish clear quick procedures for giving immediate notice of spills to the affected public and detailed evacuation plans.

Of direct relevance to your groundwater, plans should identify vulnerable aquifers, community wells, and surface waters that may carry chemi-

---

[8] A different problem concerning community representation has arisen in states that have divided the state into very large planning districts, effectively restricting community involvement; several states have created one "local" committee for the entire state. You can petition the state committee to establish smaller planning districts, although there is no explicit judicial review for these decisions.

cals into aquifer recharge zones. Cleanup methods should not call for indiscriminate flushing of chemicals into ditches or down storm drains or sewers. The movements of potential accidental releases in water or air should be projected in "plume maps" showing the neighborhoods and water supplies likely to be affected.

If you believe that your local committee has failed to obtain adequate information regarding the use of chemicals in local businesses, ask it to acquire all relevant information. EPCRA authorizes the local committee to obtain *any* information it needs from local businesses to develop its plan. Many companies have prepared "hazard assessments" for safety audits, internal management of accidents, and insurance purposes; such assessments contain detailed information on chemicals, handling practices, safety measures, and emergency responses. If the committee has not requested internal hazard assessments, safety audits, and emergency plans from local businesses, ask that it do so.

Your committee is also authorized to obtain detailed "Tier II" chemical inventory information, specifically identifying all hazardous chemicals present at a plant, instead of the more general categorical information available under "Tier I." Similarly, the committee is entitled to ask for Material Safety Data Sheets (MSDSs) for every toxic chemical handled in local businesses, rather than a mere list of chemicals for which such sheets are available. If your committee has not obtained such information, request that it do so; it must ask for such information and make it available to you.[9]

Review carefully any claims by local companies that the identities of chemicals they use are protected "trade secrets." EPCRA allows companies to withhold the identities of particular chemicals from local and state committees (and the public) if:

- The chemical's identity has been kept confidential by the company;
- It is not required to be disclosed under other laws;
- It is not readily discoverable through "reverse engineering" (analysis of the company's discharges or products); and
- Its disclosure would cause substantial competitive harm.

You can challenge a trade secret claim by petitioning EPA to disclose the identity of the specific chemical; you should include any information that disproves any of the factors above.[10]

[9] Local committees may, but are not required to, obtain Tier II information upon public request from companies holding less than 10,000 pounds of a particular chemical.

[10] Specific chemical identities must be provided to health care professionals regardless of claims of trade secret, where necessary for treatment of a person exposed to a chemical. The company can require the doctor to sign a confidentiality agreement.

You should also review carefully reports of accidental releases of extremely hazardous substances filed by local businesses. Major chemical
accidents will receive substantial public attention, but minor releases
must also be reported (if over the reporting threshold for a particular
chemical). Review of these reports may reveal improper handling or
safety procedures at a local company or ineffective emergency response
by public officials.

Finally, EPCRA authorizes citizen suits to enforce its terms. If a local
company fails to submit accident notices or required chemical information, you may sue the owner or operator of the plant. You may sue EPA if it
fails to respond to petitions to add or delete chemicals, or to disclose the
identity of a chemical claimed as a trade secret. You may sue EPA, the
state, or the state emergency response commission for failing to make
public information available to you. Citizen suits can be a valuable tool
for securing business compliance with the Act's provisions, particularly
if your local committee is ineffective.

Other Regulatory Programs

There are also opportunities to reduce risks to your groundwater through
the general federal programs that regulate businesses that use or manufacture chemicals—the Resource Conservation and Recovery Act (RCRA),
the Clean Water Act, and the Toxic Substances Control Act (TSCA).

THE RESOURCE CONSERVATION AND
RECOVERY ACT (RCRA)

RCRA is the comprehensive federal program that regulates solid wastes
and hazardous wastes. Businesses that treat, store, or dispose of hazardous wastes on-site, in surface impoundments, underground tanks,
injection wells, or other means are subject to close regulation under
RCRA. That program, and opportunities for your involvement, are described in Chapter 16. Businesses that only generate hazardous wastes,
shipping them off-site to be treated, stored, or disposed of, are required to
track and report the movement of their wastes with manifests.[11] When

---

[11] RCRA's manifest requirements apply to businesses that generate more than 100 kg of
hazardous waste, equivalent to 15 gallons of acid or water contaminated by such waste,
per month. Businesses may store an unlimited amount of wastes on-site for 90 days; if
they exceed this period, they must obtain a permit as a treatment, storage, and disposal
facility, and are subject to close regulation as described in Chapter 17. Businesses that
generate less than 1,000 kg of waste per month (small-quantity generators) may store
wastes for a longer period to permit accumulation of amounts that are efficient to transport.

wastes arrive at their final destination at a treatment, storage, or disposal facility, that facility must return a copy of the manifest to the originating business. Any breakdown in this "cradle-to-grave" tracking system must be reported by the originating business to your state or regional Office of Solid and Hazardous Waste.

Agency oversight of compliance with manifest requirements is poor; EPA's enforcement goal, for example, is to inspect only four percent of generators and transporters regulated annually, leaving many violations undetected. Your review of local businesses' compliance with RCRA's manifest requirements may thus reveal substantial problems with waste handling in your community that would otherwise escape attention. Repeated exception reports filed by a company indicating that it has not received a return manifest from a treatment facility may reveal a major break in the tracking system, or even an illegal disposal operation. You may uncover companies that appear from other sources of information to be using chemicals and generating hazardous wastes, yet neither file manifests nor hold a permit to treat, store, or dispose of their wastes on-site, thus violating RCRA. In such cases, notify EPA's regional office and your state environmental agency, and demand an inspection.

Make sure that local businesses do not improperly "delist" wastes as hazardous, thereby allowing them to treat the wastes with few controls. Businesses can petition to exclude their wastes from regulation under RCRA, even though those wastes are listed as hazardous by EPA, by demonstrating that variations in their raw materials, processes, or other factors make the waste nonhazardous as handled at their particular facility. A waste cannot be delisted unless it does not exhibit any of the characteristics (toxicity, corrosivity, ignitability, or reactivity) for which it was listed, nor can it contain any other constituent that could make it hazardous. Notice of a petition to delist a waste is provided in the Federal Register, though you can receive direct notice of a petition by asking to be put on a mailing list by your EPA regional office or your state environmental agency. Information on how to file comments and who to contact for additional information will also be provided in this notice.

If a local business does seek to delist its wastes, ask EPA or your state agency to conduct a facility inspection to sample and analyze wastes and check the accuracy of the petitions. A report to Congress by the General Accounting Office found that conditions were significantly misrepresented at over 70 percent of delisting sites where waste analyses were completed.[12]

---

[12] U.S. General Accounting Office, Hazardous Waste: EPA Has Made Limited Progress in Determining Wastes to Be Regulated (December 1986), 43.

THE CLEAN WATER ACT (CWA)

Other opportunities to guard your groundwater exist under the Clean Water Act, the federal program established to clean up the nation's surface waters. The primary regulatory mechanism of the Clean Water Act is the National Pollutant Discharge Elimination System (NPDES) permit program, which requires businesses to obtain a permit to discharge any pollutant into streams, lakes, or other waters. The Act has been largely ineffective in dealing directly with groundwater contamination,[13] but efforts to reduce pollution of surface waters will indirectly benefit your groundwater, since streams and lakes often recharge aquifers.

Check the NPDES permits for local businesses and the compliance files maintained by your regional EPA office of water quality or, in states where authority to administer the permit program has been delegated to the state, your state environmental agency. The file will contain Discharge Monitoring Reports (DMRs), which dischargers must file monthly with EPA or the responsible state agency, listing pollutants that must be controlled along with the level at which they were actually discharged. Violations of discharge levels must be reported; if you find repeated violations, ask EPA or your state to take action to enforce the permit conditions. The compliance file should also have reports of inspections conducted by EPA or your state, which will further document discharge violations and explain enforcement action taken or the lack of such action. Your regional EPA office will have records of your state's enforcement action progress as well; this may include reports of violations found where no action has been taken, indicating a need for citizen enforcement.

To bring the facility into compliance with the permit, ask EPA or your state to conduct an inspection and issue an enforcement order. This request must be in writing and must outline the basis for the request. Use the information from the compliance file and explain any physical signs of uncontrolled discharge that you discovered at the site. Include the name of the generator, and the location, dates, and times of any discovered problem.

If your state agency has primary enforcement authority, but refuses to conduct an inspection, or you are dissatisfied with its findings after an inspection, ask your regional EPA office to inspect the site. Your request should include the information described above, along with an explanation of why the state's actions were unsatisfactory.

---

[13] Some states, however, require permits for discharges into groundwater. See chapters 23 and 28.

If EPA does not respond, consider a citizen suit. Citizen suits brought to enforce violations of an NPDES permit can be very successful because proof of a violation can be demonstrated on the face of the discharger's own monitoring reports. You must give 60 days' notice to EPA, the state, and to the violator itself before bringing suit, and you cannot bring suit if EPA or your state is itself diligently prosecuting an enforcement action in court to require the company's compliance.

Make sure the NPDES permits you review address all discharges to surface waters of pollutants from the site. Discharges from any pipe, ditch, surface impoundment, or trough are subject to regulation, including any runoff from a waste unit that might occur because of heavy rainfall (the discharge does not have to be deliberate to require an NPDES permit). Your research should help you determine whether or not the permit has included all pollutants likely to be found in discharges from the site, based upon the chemicals found on-site. The pollutants that dischargers are required to test for are grouped by industry category. The NPDES permit application testing requirements are contained in 40 CFR Part 122, Appendix D. The *Quality Criteria for Water 1986,* EPA's guide to the potential environmental effects of pollutants in surface water, will also suggest chemicals of considerable concern to your health that should be controlled by NPDES permits. Failure to include all discharges is a ground for you to seek a permit modification through administrative channels. (See Chapter 7.) If this is unsuccessful, a citizen suit may be appropriate.

You should also monitor and participate in proceedings to issue new NPDES permits or to renew existing permits. Ask EPA or your state agency, if it has primary responsibility for the NPDES permit program, to notify you of any new permit applications or renewal proceedings in your community. Review those applications in the same manner as you do existing permits, to ensure that they address all pollutants likely to be discharged by the business involved and set appropriate limits for such pollutants. Where a business seeks to enlarge its existing discharge limits, look to see that its request is based on legitimate grounds, such as new processes or equipment or increased production capacity. Chapter 7 describes how to participate in the administrative process by which permits are granted.

The NPDES permit system applies only to facilities that discharge pollutants directly into surface waters. Many companies, both small and large, discharge waste water into sewers; those wastes are then treated and discharged by publicly owned treatment works (POTWs), which must obtain their own NPDES permit. "Indirect dischargers" that deliver their wastewater to a POTW are not subject to an NPDES permit, but are required to pretreat their waste if it is not susceptible to treatment by the

POTW, or if the wastes would interfere with the POTW treatment system. EPA's general pretreatment standards are set out in 40 C.F.R. Part 403; categorical standards, listed at subchapter N of 40 C.F.R., apply to particular industries.

EPA's pretreatment standards prohibit discharges into POTWs that will interfere with or pass through the facility untreated, and require industrial users of municipal sewage plants to monitor and report their discharges. In addition, large POTWs (with a design flow of 5 million gallons daily), and smaller facilities with recurrent problems, are required to develop their own local pretreatment programs to ensure that indirect dischargers are complying with pretreatment requirements. Under such local programs, POTWs are required to identify industrial users that are subject to pretreatment requirements, monitor their discharges to the facility, and enforce compliance. POTWs can require local industrial dischargers to obtain a permit.[14]

Contamination problems may arise when industrial dischargers fail to pretreat their waste, pouring chemicals, heavy metals, and other toxic wastes down sewer drains. Enforcement against indirect dischargers is difficult, as it is very hard to trace untreated waste to a particular source. Violations are more likely to be found by looking to the industrial site itself. Usually, it is the government owner of the POTW that goes after indirect dischargers, as the POTW will ultimately be held responsible for violations of its NPDES permit caused by indirect dischargers.

If you discover a POTW that is seriously or continuously violating its NPDES permit conditions, complaints to EPA or your state agency could lead to enforcement action, or even to forfeiture by the POTW of federal grant money, the major funding received by these systems under the Clean Water Act. If EPA or your state will not take enforcement action, you may sue the POTW directly to enforce compliance with its NPDES permit. If the POTW is designed to treat 5 million gallons daily, or if it is having substantial problems complying with its NPDES permit because of indirect discharges, it should establish a pretreatment program, as described above, to ensure compliance by industrial dischargers. Review that program to ensure that it identifies all likely dischargers, monitors the nature and amounts of their discharges, contains adequate enforcement authority, and is supported by adequate funding and personnel. If the POTW's records identify industrial users that fail to pretreat their wastes, seek enforcement action from EPA or your state agency. The Clean

---

[14] If your state has primary enforcement authority for the NPDES program, it can assume responsibility for implementing pretreatment standards instead of making the POTW take responsibility for doing so.

Water Act provides civil and criminal penalties for violations of pretreatment requirements.

THE TOXIC SUBSTANCES CONTROL ACT (TSCA)

TSCA authorizes EPA to regulate the manufacture of chemical substances (excluding pesticides, foods, drugs, cosmetics, and tobacco); EPA has identified more than 60,000 such chemicals. Under TSCA, EPA can require manufacturers to test chemical substances to determine their toxicity and chemical properties, and can take regulatory action to control the manufacture, distribution, use, or disposal of chemicals where such activities present "an unreasonable risk of injury to health or the environment."

Regulatory action under TSCA must be based on comprehensive risk/benefit analysis, and can include prohibiting manufacture or distribution of substances, restricting the uses for which they can be sold, requiring warning labels, and regulating the manner of their disposal. Businesses are required to file premanufacture notices (PMNs) notifying EPA of their intention to manufacture new chemicals; EPA may take action to regulate or to require additional testing of such chemicals. Manufacturers are required to notify EPA immediately of any information indicating that a chemical presents a substantial risk to health or the environment, and to maintain other records, which EPA may inspect. Finally, Congress has specifically regulated two substances: polychlorinated biphenyls (PCBs) and asbestos, under TSCA.[15]

EPA has not to date exercised the full scope of its authority under TSCA to regulate the manufacture and use of dangerous chemicals. If your investigation of hazardous chemicals present in your community reveals that little is known regarding the toxicity or chemical properties of a chemical manufactured by local companies, or that manufacture, use, or disposal of the chemical poses significant risks to human health or the environment, you can petition EPA to require testing of the chemical's properties or to regulate its manufacture, distribution, or disposal. Your petition must include substantial information regarding the chemical's properties, health effects, and public exposure and should be prepared

[15] Because of widespread public concern about environmental contamination by PCBs, a highly toxic and carcinogenic group of chemicals widely used in electrical transformers and equipment, Congress prohibited their manufacture and distribution, and restricted their use in existing applications. EPA responds to spills of PCBs using its authority under TSCA. If you discover that PCBs have been spilled in your community, contact your regional EPA office immediately.

Similar public concerns led Congress to regulate asbestos in public schools under TSCA in 1986.

Citizen Action Against Chemical Releases in Their Community

Citizens in Toms River, New Jersey, became aware of dangerous chemicals stored at the Ciba-Geigy plant in their community only after a pipeline carrying hazardous waste for ocean dumping offshore broke underground, contaminating groundwater. They not only learned that the pipeline was owned by Ciba-Geigy, but that ten miles of the line ran through their township. Concern led to formation of the Ocean County Citizens for Clean Water and the Save Our Ocean Committee, both of which decided to investigate the facility and take action to prevent further hazards.

First, they found out when the facility's NPDES permit was up for renewal and opposed its reissuance in the administrative hearing. Though they were unsuccessful, greater limits on pollutant discharges were written into the permit. Publicity surrounding this effort led the environmental law clinic of Rutgers University Law School to file suit to overturn the issuance of the permit, as it did not address the effects of ocean discharges.

Ocean County Citizens for Clean Water also decided to investigate the plant's waste activity and discovered that a landfill in their area owned by Ciba-Geigy was a designated Superfund site. From records analyzing wastes from the 100,000 drums illegally dumped at the landfill, they discovered the company was manufacturing phosgene gas, an extremely toxic substance. Because the landfill was located near a senior citizens' home, the group was very concerned about the potential health risks. They pressured Ciba-Geigy to stop manufacturing the gas, arguing it should only be used as needed and not stored at the facility. Public outcry was so great from this discovery and from citizen inspections of the plant (conducted at Ciba-Geigy's request) that the company decided to stop storing phosgene gas. No administrative or legal action was needed; intense public pressure led to this decision.

Upon the realization that their awareness of groundwater contamination was triggered only after a chemical disaster, the groups decided preventive measures were needed. Almost 100 percent of Ocean County relies on groundwater for domestic use, a resource threatened by the 12 Superfund sites already designated in the county, with dozens more under investigation. Ocean County Citizens for Clean Water and the Save Our Ocean Committee worked with the Ocean County Board of Health to pass a local ordinance requiring home owners with private wells to test their groundwater for the 129 priority pollutants listed by the Clean Water Act. If a problem exists, it must be corrected or the home must be hooked to the public water system before it can be sold. As a result of this law, groundwater contamination was discovered in hundreds of wells throughout the county.

For more information on these actions, contact Ocean County Citizens For Clean Water, P.O. Box 4724, Toms River, New Jersey, 08754-4724.

with expert scientific and legal assistance. (Such assistance may be obtained from local universities or from state or national environmental organizations.) EPA does not generally respond positively to such petitions, but denial of your petition may be reviewed in the federal courts through a citizen suit.

Should your investigation of chemicals present in your community reveal that a chemical manufacturer is violating restrictions or requirements imposed by EPA under TSCA, notify EPA and ask for enforcement action. If EPA fails to take action, you can bring a citizen suit.

REFERENCES

COMMERCIAL AND INDUSTRIAL SOURCES

Burcat, Joel, and Arthur Hoffman. "The Emergency Planning and Community Right-to-Know Act of 1986: An Explanation of Title III of SARA." Environmental Law Reporter (Environmental Law Institute). Washington, D.C. January, 1988.

Environmental Law Reporter. *Clean Water Deskbook.* Environmental Law Institute. Washington, D.C. 1988.

Environmental Law Reporter. *Community Right-to-Know Deskbook.* Environmental Law Institute. Washington, D.C. 1988.

EPA, Office of Solid Waste. *RCRA Orientation Manual.* Washington, D.C. 1986.

LAWS AND REGULATIONS

U.S. Code Citations

Clean Water Act (Federal Water Pollution Control Act), 33 U.S.C.A. §§ 1251–1387 (West 1986 & Supp. 1988).

Emergency Planning and Community Right-to-Know Act (Title III of Superfund Amendments and Reauthorization Act), 42 U.S.C.A. §§ 11001–11050 (West Supp. 1988).

Toxic Substances Control Act, 15 U.S.C.A. §§ 2601–2654 (West 1982 & Supp. 1988).

Resource Conservation and Recovery Act (Subchapter III of Solid Waste Disposal Act), 42 U.S.C.A. §§ 6921–6939b (West 1983 & Supp. 1988).

Code of Federal Regulations Citations

40 C.F.R. Part 122 (National Pollutant Discharge Elimination System.)

40 C.F.R. Subchapter N (effluent guidelines and standards).

40 C.F.R. Part 403 (pretreatment standards).

40 C.F.R. Part 355 (emergency planning and notification).

40 C.F.R. Part 370 (hazardous chemical reporting: community right-to-know).

40 C.F.R. Part 372 (toxic chemical release reporting: community right-to-know).

CHAPTER 18

Waste Disposal Wells

RESIDENTS OF LAKE CHARLES, Louisiana discovered that hazardous wastes were being injected into disposal wells in their neighborhood. Their discovery eventually led to the formation of a citizens' group called CLEAN (Calcasieu League for Environmental Action Now). The group captured the attention of the media to spread word of the threat and organized grass-roots protests in the area. The Governor of Louisiana gave CLEAN money to hire experts to study the site's effects on the Chicot Aquifer, the source of drinking water for 1 million residents of Louisiana and Texas. Contaminants were found below the surface and had entered the aquifer. After dozens of hearings and the intervention of the local district attorney on CLEAN's behalf, action has been taken to clean up some problems at the site, but members of CLEAN and other citizens are still fighting to stop these wells from destroying their water quality.

The use of wells to dispose of liquid and other wastes in the ground is growing in this country—particularly in the oil-producing states of the South and Midwest. Federal regulatory programs are just beginning to stop some of the most abusive practices, but in many ways the injection of wastes, including hazardous wastes, into wells remains one of the least carefully regulated disposal practices. And because these methods leave wastes in the ground essentially uncontained, groundwater contamination from these wells is justifiably a major concern for citizens. This chapter provides suggestions for using the federal regulatory program to halt or prevent groundwater contamination from these sources.

HOW TO FIND DISPOSAL WELLS

The first step in taking action to protect against groundwater contamination from disposal wells is to locate the wells in your area and learn about the kinds of wastes the wells handle. The best place to begin is the regulatory agency responsible for implementing the underground injection control program of the federal Safe Drinking Water Act. This is either a state agency (in states with an EPA-approved program) or the regional EPA office (in those states without an approved program). Request from the agency information about active wells in your area. The agency should be able to provide you with information about the location, ownership, and type of many of the active disposal wells in the state. The inventory data should also disclose what kinds of wastes may be going into the wells.

The agency depends in large part on self-reporting for certain types of wells, however, and its lists may be incomplete as a result, so you should also conduct your own investigation. That search should be broad. Waste disposal wells come in many forms, ranging from high-pressure deep injection systems that may pump wastes into deposits thousands of feet below the surface to cesspools that lie just a few feet underground. Indeed, federal regulations defining disposal wells include any hole dug deeper than it is wide.

To lay the foundation for your search for information, you need to become familiar with EPA's well classification system. EPA recognizes five classes of wells:

Class I: These wells are used to inject hazardous waste and nonhazardous municipal and industrial wastes. These are usually deep wells and, when within ¼ mile of an underground source of drinking water (USDW), must inject wastes below the drinking-water aquifer. (USDW is defined to mean an aquifer, or part of an aquifer, that serves now or may serve in the future as a drinking-water source.) Wells that fail to meet this ¼ mile restriction to protect USDWs are now banned. (See Class IV below.) Approximately 200 wells nationwide are used to inject hazardous waste and most are operating under permit. The present estimate of the number of such wells that inject waste defined as nonhazardous (but still a potential threat to groundwater quality and your health) is about 400.

Class II: These wells are used for a variety of purposes during oil and gas operations, primarily to reinject unwanted brine brought to the surface during drilling and production. There are approximately 160,000 of these wells concentrated in about 170 wellfields, and new ones are put into use on a regular basis.

Class III: These wells are used in the process of extracting minerals, including sulfur, uranium, salts, and potash. There are approximately 10,000 of these wells across the country.

Class IV: These wells were once used to dispose of hazardous and radioactive waste. They differ from Class I wells in that the Class IV category covers wells where wastes are disposed into or above, rather than below, a formation within ¼ mile of a USDW. Class IV wells are now banned, with few exceptions.

Class V: This catchall category includes any well that does not fit into Classes I–IV, including stormwater runoff wells, cesspools, and wells accepting discharges from agricultural practices. Though these wells are not to be used for disposal of hazardous wastes, many kinds of municipal, industrial, and agricultural wastes that could present health threats if present in groundwater are injected or collected in these wells. Many are shallow and often not much more than a dug hole. These wells, which number at least in the hundreds of thousands and perhaps as high as 1 million, are subject to almost no federal regulation.

The regulatory agency is likely to have a nearly complete inventory of information about Class I, II, and III wells. Many of these will be operating under a permit or by rule issued by the agencies; most of the others should have applied for a permit. The agency may not be aware of all Class IV wells, in particular those that may have been improperly closed or abandoned. Class V wells are likely to present the biggest problem, in part because many operators of such wells may not even know that they must report to the regulatory agency.[1] To identify all the Class V wells in your area, it is particularly important that you conduct your own search. You may want to focus your investigation in areas particularly sensitive to contamination, such as aquifer recharge zones and wellhead areas.

TAKING ACTION AGAINST RELEASES AND CONTAMINATION

Are Wells Leaking?

Once you have identified the disposal wells in the area, the first thing you will want to do is to discover which wells may already be contaminating groundwater. The basic question is whether fluids are leaking from the well (or migrating from the injection zone where they were intended to stay) into an aquifer that may provide a water supply.

---

[1] For recent information on Class V wells, request a copy of EPA's 1987 "Report to Congress on Class V Injection Wells—Current Inventory, Effects on Ground Water, and Technical Recommendations" from the EPA Regional Office (EPA Document No. EPA 570/9-87-006). Summaries of the report are available. If they do not have copies available for your review, you can order the report from the National Technical Information Service, 5285 Port Royal Road, Springfield, VA, 22161, (703) 487-4650. NTIS Order No. PB 88 111 596 (price $56.95 in 1988).

In addition to this EPA report, some states with EPA-approved programs have prepared their own detailed Class V well reports. Ask the state regulatory agency if such a report was done for your state.

Begin by requesting from the regulatory agency the monitoring reports filed by well operators. All Class I, II, and III wells are generally required to conduct pressure tests, which may reveal a leak in the well structure or potential migration of the injected fluid from its intended geologic formation. Some wells may even be required to use additional test wells to monitor groundwater quality directly.

You may also be able to identify problem wells by tracing contaminants discovered in water quality tests to wastes disposed of in nearby wells. Water quality testing may be conducted by federal, state, or local agencies monitoring ambient groundwater quality (see Chapter 4), public water suppliers performing tests required by the SDWA (see Chapter 11), or private citizen testing of drinking-well water (see Chapter 4). If the same chemicals injected into disposal wells are showing up in groundwater quality tests, you may, using information about the hydrogeology of the area, be able to make a strong case that a particular well is causing a contamination problem.

You may find helpful information in quarterly compliance reports that should be submitted by the regulatory agency to EPA. These reports should describe, for major facilities, regulatory violations and other problems, including potential contamination.

A final important step, especially if the evidence you have does not conclusively demonstrate a problem at a particular well, is to request the regulatory agency to conduct its own investigation of the facility. In any state where EPA is the regulatory agency, EPA has authority to require the well operator to conduct groundwater monitoring and submit reports to EPA. In addition to groundwater monitoring, you can request that EPA conduct an inspection of the facility and review its records. State agencies in those states with authorized programs are likely to have similar authority.

Effective Response

Once you have established that a disposal well is leaking, you can seek an enforcement action to correct the problem and, if necessary, clean up existing contamination. The regulatory agency has authority to take a variety of actions depending on the problem. As a general rule, the agency can impose additional construction, operation, and monitoring requirements, require the operator to take corrective action, and even close down the well. But for Class V wells, the EPA is authorized to take such enforcement action only if the contaminants have migrated into a drinking-water aquifer and may cause a violation of primary drinking-water standards (see Chapter 11) or may adversely affect human health. (See 40 C.F.R. § 144.12.) Some states may have special enforcement programs designed to attack less severe Class V well problems.

Agency action may take a variety of forms, ranging from administrative orders and permit modifications to court actions. You generally have a right to participate in and influence the outcome of these processes once they are initiated. For specific advice about how to take full advantage of these processes, turn to the enforcement discussion later in this chapter.

You may want to consider bringing a citizen suit directly against the well operator for groundwater contamination if the regulatory agencies fail to act or do not respond adequately, but such a suit is likely to be dependent on expert testimony and evidence and will be difficult. (See the discussion of citizen suits at Chapter 8.)

If the contamination caused by a particular well creates a substantial and imminent health threat and/or involves a hazardous substance, pursue the remedies available under the SDWA and Superfund emergency response authorities (see Chapter 14). For abandoned wells that have created a contamination problem, you may be able to initiate long-term cleanup action under the Superfund program (see Chapter 13). Do not forget to look for options under state law (see Chapters 8 and 29).

ENSURING COMPLIANCE WITH OPERATING REGULATIONS

In many cases, there may not be strong evidence of contamination from a particular well, but you can still take action to ensure that wells are being operated properly and according to legal requirements, and thereby minimize the risks of future contamination. This section focuses on requirements set out in federal regulations. In those states with an authorized injection well control program, there may be additional requirements. Check with the regulatory agency.

Check Permit Status

A relatively easy and important step is to ask the regulatory agency for information about the permit status of disposal wells. Most Class I, II, and III wells are required either to have a permit from the regulatory agency or have applied for one by a specific date in order to be operating legally.[2] If

---

[2] Those wells operating without a permit are often referred to as wells operating "by rule" because they are authorized by regulation. Such wells are required to comply with certain general operating, monitoring, and closure requirements. See next section.

you discover that a Class I, II, or III well is operating without a permit and without a timely application on file, notify the regulatory agency and request that it begin an enforcement action to close the well. If the state regulatory agency or EPA does not take action, consider bringing a citizen suit against the well operator. For this type of violation, a citizen suit is likely to be relatively straightforward.

Another important step is to ensure that no wells are still operating that inject hazardous waste into or above a drinking-water aquifer (Class IV wells). Any wells discovered in your investigations that may fit the Class IV definition, and are not properly closed, are probably operating illegally and should be brought to the attention of the regulatory agency for enforcement actions and closure. Again, if the agency does not take action, this type of violation can be attacked with a citizen suit.

Review Basic Regulatory Requirements

Once you have checked the well's permit status, the next step is to ensure that it is meeting minimum operating regulations. Only Class I, II, and III wells are now subject to such restrictions under the federal program. For Class V wells, look to state laws. (See Chapter 29.)[3]

Whether operating under a permit or authorized by regulation pending a permit decision, disposal wells must meet certain minimum standards for operation, monitoring, reporting, and closure. Permitted wells are also subject to construction standards. Permits may have special conditions that are more specific and tailored to the particular circumstances of a given well than are the general standards of the EPA regulations.

Citizens can gather information about compliance with these general regulatory requirements and specific permit conditions in a variety of ways. You can find some information in the regulatory agency files, ask the agency to conduct an inspection of a particular facility, or request information from the facility itself. Even though in most cases the facility is not required to give you information, some operators will cooperate with informed citizen groups to reduce potential conflicts. To monitor compliance with special permit conditions, you must get a copy of the permit from the regulatory agency. First we will summarize the major points you should be looking at and then discuss how to take action if you discover problems.

---

[3] All wells are subject to the fundamental requirement that they not cause a violation of a primary drinking-water standard or adversely affect health by contaminating an aquifer. If any well—even a Class V well—violates this requirement, the regulatory agency can take action against it.

OPERATING, MONITORING, AND REPORTING REQUIREMENTS

All Class I, II, and III wells must comply with regulations that govern injection pressure and establish monitoring and reporting schedules. (See 40 C.F.R. § § 144.28, 146.13, 146.23, 146.33.) Pressure limitations should be set to prevent failure of the injection system and fracturing of the injection formations. Injection pressure must be monitored on a regular basis and reviewed for unusual pressure changes that may disclose leaks in the system. (Class I wells are subject to special requirements to permit closer monitoring of the injection zone.) Some wells, particularly Class I wells, may be required by permit or special order to conduct direct monitoring of the injection zone or nearby aquifers with separate test wells to detect migration of injected fluids.

Pressure reports and information about the kinds of wastes injected into the well must be regularly submitted to the regulatory agency. Request copies of these reports from the agency and make sure they are completed by the operator fully and on time. Look for special monitoring requirements, and information in the reports to verify that any such requirements are being followed.

CONSTRUCTION REQUIREMENTS

For wells operating under a permit, it is important to ensure that minimum construction requirements set forth in the regulations and any more detailed requirements in the permit are being followed. Minimum standards for the three classes of wells are contained at 40 C.F.R. § § 146.12, 146.22, 146.32. Focus on the requirements that address these problems:

- Proper well depth and injection into formations adequately separated from usable aquifers.
- Casing and cementing around the well to prevent leakage to other formations (and for Class I wells, special tubing and packer requirements to minimize the risks of migration).
- Corrosion prevention to stop the injected fluids from destroying the well linings.

These violations will be difficult for citizens to detect without help. If you discover other problems at the facility that may be related to construction inadequacies, ask the agency to conduct an inspection of the well.

PLUGGING AND CLOSURE REQUIREMENTS

You can also ensure that Class I, II, and III wells are closed and plugged in accordance with minimum requirements. Well operators are required to maintain sufficient financial resources to close wells properly and must develop adequate closure plans. (See 40 C.F.R. § § 144.28(d), 144.60-.70.) Make sure that any operating wells are in compliance with these requirements by reviewing the financial information and closure plans submitted to the agency by the operator. Request to be notified of well closures and review the reports filed by the operator to ensure that the plugging and abandonment complies with approved plans. Wells that have not filed closure reports and are not completing regular operating reports should be investigated for possible failure to meet plugging requirements.

PROBLEMS WITH ADJACENT WELLS

Contaminants from a well may escape the injection zone and move into usable aquifers by moving along pathways created by other wells, particularly old or abandoned wells, in the area. This is most common where

FIGURE 18.1
REGULATORY PROGRAMS FOR STATES WITH ACTIVE HAZARDOUS WASTE INJECTION WELLS

State	Program Implemented by		Number of active wells
	State	EPA	
Texas	X		68
Louisiana	X		54
Arkansas	X		2
Oklahoma	X		4
Florida	X		1
Alabama	X		2
Kansas	X		5
Illinois	X		6
Alaska		X	1
California		X	1
Indiana		X	8
Kentucky		X	2
Michigan		X	13
Ohio	X		14
Total			181

Source: *Government Accounting Office*

nearby wells are improperly sealed or closed. For example, Louisiana has 38,700 oil and gas wells, often old or abandoned, many of which could create conduits to drinking-water aquifers for the 4 billion gallons of hazardous waste that are injected annually in the state (see Figure 3.4, on page 31). To prevent such problems, you can take steps to enforce a corrective-action plan.

Applicants for permits for most Class I, II, and III wells must identify other wells within the area of review (the area surrounding the well likely to be affected by the injection of fluids and into which migration is likely). For those wells that may be improperly sealed or closed, the applicant must develop a plan for corrective action to prevent contamination. (See 40 C.F.R. § 144.55.) Operators of existing wells seeking permits will be required to correct the problem on a specific schedule. New wells are not allowed to begin operations until the corrective action is complete. Whenever you know or suspect that a disposal well is located near other wells (or that a proposed well will be), ask the agency whether corrective-action plans were developed and carried out for adjacent wells. If necessary, ask the agency to conduct an inspection to make sure.[4]

SPECIAL HAZARDOUS WASTE REQUIREMENTS

You should pay special attention to the dangers presented by wells injecting hazardous waste. In 1981, EPA estimated that 60 percent of land-disposed hazardous wastes ended up in injection wells. That figure is even higher today. (See Figure 18.1, on preceding page.)

In addition to complying with injection well rules, Class I wells used to dispose of hazardous waste are required to meet some but not all of the requirements for hazardous waste facilities described in Chapter 16. For instance, hazardous waste wells that do not yet have full permits must comply with some of the restrictions for interim-status hazardous waste facilities. These include preparedness plans for accidental releases, contingency and emergency procedures, and manifest, reporting, and record-keeping requirements. (See 40 C.F.R. § 265.430.) Such wells must also, like other interim-status hazardous waste facilities, have certified their compliance with applicable groundwater monitoring[5] and financial re-

---

[4] One significant exception exempts existing Class II wells (those related to oil and gas production) from these requirements. Since most Class II wells are found in large wellfields, this is likely to pose a serious problem. If there are special circumstances surrounding a particular existing Class II well that make groundwater contamination likely, press the agency to include this adjacent well-plugging requirement in its permit.

[5] In most cases, groundwater monitoring is not required by federal regulations. In those states with EPA-approved programs, however, there may be groundwater monitoring requirements imposed on Class I wells. If the well operator fails to comply with these state standards, the state can revoke the well's interim status and close it down.

sponsibility requirements (see 40 C.F.R. § § 144.28(g), (d)) by November 1985 or have stopped operations and closed. Wells that have permits must follow the requirements of the hazardous waste tracking system. (See 40 C.F.R. § 144.14.)

As a general rule, however, injection wells that handle hazardous wastes are not required to conduct groundwater monitoring or follow the rigorous closure and post-closure requirements established for other hazardous waste facilities. This is a major weakness of federal Class I well regulation. You can, however, press the regulatory agency to use its authority to impose such requirements on a case-by-case basis in the permits for these wells.

Finally, injection wells are subject to the rules, referred to in Chapter 16, that automatically prevent land disposal of certain particularly toxic hazardous wastes unless the EPA specifically authorizes further disposal by regulation. The first of these prohibitions for a specific list of wastes was due to go into effect in August 1988, but EPA has since granted a two-year variance. The prohibition will apply to additional wastes on a staggered schedule. Check with the regulatory agency about the status of these restrictions and police Class I wells for injection of prohibited wastes by reviewing reports filed with the agency.[6]

HOW TO USE AGENCY ENFORCEMENT PROCEDURES

This section describes how to initiate and take part in the various actions the regulatory agency may take to address groundwater contamination problems or violations of the regulatory requirements. You should first refer to the general suggestions for initiating and participating in agency enforcement and remedial actions in Chapter 7. Then follow the specific information about the injection well program below. Remember that if a state agency is implementing the program, there may be some variations in the process. Remember also to check for any state agency-EPA enforce-

---

[6] Even if EPA decides to ban disposal of certain wastes by underground injection, operators can petition for an exemption based on a showing that there will be no migration of the hazardous waste from the injection zone. Make sure you are on the agency's mailing list to receive notice of such petitions and that you participate in the comment process to prevent such exemptions where, as is likely to be true in most cases, the no-migration showing is based on speculative models. The agency should follow the process for permit modifications in deciding such petitions. Once granted, these exemptions can be challenged in court but, unless you can identify procedural errors or significant information the agency failed to consider, a lawsuit is likely to become a difficult battle of experts where the odds are against you.

ment agreements, which may identify enforcement priorities in your state that you can use to put pressure on the agency to take action at the facility that concerns you.

Administrative Orders

To correct permit violations, to require compliance with special conditions, or to address contamination problems, the agency may issue administrative orders to well operators. For most violations, your first step toward spurring action should be to ask the agency to issue an order. The agency will circulate proposed orders for review and comment by the public (make sure that you are on the agency's mailing list to receive notice of proposed orders). If the violator requests one, an informal hearing must be held and, if you comment on proposed orders, you have a right to participate in the hearing and present evidence. Challenges in court to final agency enforcement orders must be filed within 30 days of the issuance of the order.

Permit Actions

You can also request the agency to use its permitting authority to take action against problems and violations. Wells operating without permits (Class V wells and wells operating "by rule") may be required to obtain a permit in the event of groundwater contamination or failure to comply with regulatory requirements. For wells that already have permits, the agency can modify or "revoke and reissue" the permit to address new circumstances or problems not considered in the original permitting process, or to respond to leaks and contamination. Permits can be terminated entirely for noncompliance with regulations or permit conditions or because of contamination problems. (See 40 C.F.R. § § 144.39, 144.40, 144.41.) The agency will follow basically the same procedure for all these actions. Chapter 7 describes how to become involved in and use the permit process.

Court Actions

In the appropriate case, the agency can bring a court action to compel compliance with the program. Request the agency to initiate such actions where necessary, particularly to enforce the terms of an administrative order being ignored by the violator. You have a right to intervene in civil lawsuits brought by the agency to the extent that your interests are not represented adequately by the agency or other parties.

State and EPA Enforcement Roles

The underground injection control program is designed to be operated by the states. In states that still have EPA-run programs, it is the regional EPA office that must enforce the program requirements and from which you should seek action. In those states with EPA-approved programs (programs that meet the minimum federal requirements discussed here), pursue enforcement options with the state agency first. Even if the state has an approved program, however, the regional EPA office still retains enforcement authority. In fact, the EPA is required to take enforcement action if, after 30 days notice from EPA of a program violation, the state agency does not take adequate enforcement action. So, even in states with an approved program run by a state agency, it is good idea to notify the regional EPA office of violations at the same time you seek action from the state agency.

CITIZEN SUITS

If neither the state nor the regional EPA office takes action, either at the administrative level or in court, or if the action the agency takes is inadequate, you can bring a citizen suit for violation of any requirement or condition of the SDWA, including a failure to comply with outstanding enforcement orders. In fact, citizen suits are likely to be important enforcement tools. Violation of injection well requirements is widespread and agency inspection and enforcement efforts are generally inadequate. EPA estimates nationwide noncompliance with underground injection regulations to be as high as 40 percent. In Louisiana, for example, only half of the 2,932 oilfield waste wells in the state were inspected by the Office of Conservation in 1984 (perhaps because they only had five inspectors on staff) and 30 percent of those violated the program requirements.

PREVENTING FUTURE PROBLEMS BY USING THE PERMIT PROCESS

By participating in the permit process and advocating permit denial or stringent, protective permit conditions, you can often stop problems before they start. Remember, however, that under the federal SDWA pro-

The Law Has Teeth

In February 1987, in the first criminal action under the Safe Drinking Water Act, a federal grand jury indicted Jay Woods Oil, Inc., located in Harrison, Michigan, and its Vice President, Edmund Woods, for blocking four underground injection wells to prevent EPA from conducting leak detection tests. Conviction carries a maximum five-year jail sentence and a $1.25 million fine.

---

gram, as a general rule only Class I, II, and III wells must apply for a permit. The more common Class V wells can be required to apply for and comply with a permit, but only if they are already the cause of a contamination problem that may violate water quality standards or create a health hazard. State laws, on the other hand, may require that certain types of Class V wells have a permit. (See Chapter 29.)

To issue injection well permits, EPA follows the same procedures used for several other pollution control permits. State agency procedures are likely to be similar. Review the discussion of this process and our suggestions for taking full advantage of opportunities for public involvement in Chapter 7.

Review the permit application to ensure that the applicant has submitted all the detailed information required by the regulations. (See 40 C.F.R. § § 144.31 and 146.14, 146.24, 146.34.) Pay particular attention to information about the area surrounding the well, other wells in the area, the geologic setting (including drinking-water aquifers and testing results), complete design and construction plans, and the characteristics of the wastes. Press the regulatory agency to require more data from the applicant if what you review is incomplete or inadequate.

Make sure that any permit issued by the regulatory agency contains at least those conditions and terms required by the regulations. (See 40 C.F.R. § § 144.51, 144.52, 144.54 and Part 146.) As you become familiar with the more technical aspects of disposal wells (seek the help of experts at local universities, for example), attempt to get the regulatory agency to use its authority to add special restrictions which, though not generally required by the regulations, may be necessary to protect groundwater quality in the particular case (see 40 C.F.R. § 144.52(b)(1)).

We suggest that you begin by focusing your attention on the issues listed below and make sure they are addressed to your satisfaction in the permit process. This list is not complete, but does suggest major issues that are likely to be important in most cases.

If, despite your best efforts in the permit process, a bad permit is issued, you can generally appeal the decision (or particular conditions of the permit) to a higher level within the agency. (See Chapter 7 for general

Questioning Waste Management

Waste Characterization

What wastes are to be deposited in the well?

Are the wastes hazardous?

Have tests been done to ensure that the wastes will not react with the injection zone formation to cause plugs in the formation, dissolve the formation and create migration pathways, or even cause blowouts? Does the permit require continued testing of the waste to ensure compatibility with the formation and other injected wastes? (Existing regulations do not require specific compatibility tests, but you should urge that such a requirement be included in the permit.)

Location and Geology

How close to a usable aquifer or USDW is the well? (Remember, Class I wells, if within ¼ mile of an USDW, must inject below the lowermost USDW.)

Is the injection zone deep enough?

How thick, and of what type, is the rock formation separating the injection zone from the usable aquifer?

Have all the wells within the area of review been identified and properly plugged? If not, has an adequate corrective-action plan been developed? (Existing Class II wells are exempt from this general requirement. Argue for its inclusion as a special permit condition.)

Is the area's geology adequately understood? Will unknown faults or fissures create migration routes?

Construction and Operation

Will the well have adequate casing, cementing, and special injection tubing to insulate groundwater from the waste?

Have measures been developed to prevent corrosion of the well structure by the injected waste?

Will the integrity of the well be checked often? (Minimum requirements in the regulations—five years in some cases—are too lenient. Advocate more regular testing.)

For hazardous waste wells, will there be adequate cleanup of other waste management units at the facility that have caused releases in the past? (See Chapter 16)

Will drinking-water aquifers, confining formations, and the injection zone be monitored directly with test wells to disclose migration? (Regulations, even for Class I hazardous waste wells, do not require such monitoring in the general case. This is a major loophole in the regulations. The pressure monitoring now required for Class I, II, and III wells is unlikely to detect slow leaks of wastes from the well or injection zone. Press for an adequate groundwater monitoring and reporting plan.)

Are monitoring wells located properly to detect contamination?

Closure

Has the operator demonstrated an ability to pay for closure and plugging? Advocate posting of a bond or insurance rather than reliance on financial assets.

Does the operator have an adequate plan to plug and close the well?

Is there provision for monitoring after closure to ensure wastes do not migrate after the well closes? (This is not required by the regulations, but probably necessary, especially for Class I wells.)

suggestions on this process.) If appeals are not successful, you can challenge the permit in court.

PREVENTING AQUIFER EXEMPTIONS

Be on the lookout for petitions to exempt aquifers from protection under the SDWA. Petitions can be filed with the regulatory agency to designate certain aquifers as exempted aquifers if they meet specific criteria. (See 40 C.F.R. § 146.4.) Basically, the petition must show that the aquifer is not now and will not in the future serve as a drinking-water source. If the agency agrees, wells could be allowed to inject wastes, even hazardous wastes, directly into the aquifer. Exemptions can apply even to portions of aquifers, from which wastes could migrate to usable portions of aquifers. Even if aquifers are isolated, it is short-sighted to assume that they will never be usable and to allow their further degradation.

Ask to be notified of any exemption petitions to the regulatory agency, and take part in the public comment process. The EPA Administrator must approve all aquifer exemptions, even in those states with delegated programs. (See 40 C.F.R. § 144.7(b).) If the regulatory agency grants an exemption when it should not, present your arguments to the EPA Administrator to prevent approval.

REFERENCES

WASTE DISPOSAL WELLS

Arbuckle, J. Gordon, et al. *Environmental Law Handbook.* Government Institutes, Inc. Rockville, Md. 1987.

Bloom, Jane, and Wendy Gordon. *Deeper Problems: Limits to Underground Injection as a Hazardous Waste Disposal Method.* Natural Resources Defense Council. New York, N.Y. 1985.

Citizen's Clearinghouse for Hazardous Wastes. *Deep Well Injection: An Explosive Issue.* Arlington, Va. August 1985.

EPA, Office of Drinking Water. *Report to Congress on Injection of Hazardous Wastes.* Washington, D.C. May 1985.

Fortuna, Richard C., and David J. Lennett. *Hazardous Waste Regulation, the New Era: An Analysis and Guide to RCRA and the 1984 Amendments.* McGraw-Hill. New York, N.Y. 1987.

Gordon, Wendy. *A Citizen's Handbook on Groundwater Protection.* Natural Resources Defense Council. New York, N.Y. 1984.

Kelly, William J. *The Safe Drinking Water Act Amendments of 1986, A BNA Special Report.* Bureau of National Affairs. Rockville, Md. 1986.

U.S. General Accounting Office. *Hazardous Waste: Controls Over Injection Well Disposal Operations.* Washington, D.C. August 1987.

LAWS AND REGULATIONS

U.S. Code Citation

Safe Drinking Water Act, 42 U.S.C.A. §§ 300f-300j (West 1982 & Supp. 1988).

Code of Federal Regulations Citation

40 C.F.R. Parts 144–147.

CHAPTER 19

Underground Storage Tanks

ONLY WHEN THERE was an explosion in a home in Belleville, Indiana, did residents of that town realize that leaking underground storage tanks had severely contaminated their drinking wells. Six inches of gasoline was found floating on top of one well. In Provincetown, Massachusetts, a leak discovered at a Standard Oil service station released 3,000 gallons of gasoline into surrounding groundwater. As a result the town had to reduce pumping at its well by 80 percent. In New Mexico, investigators found groundwater samples with benzene concentrations of 23 mg/liter, 1,000 times the standard set by the New Mexico Water Quality Control Commission. Benzene is a particularly troublesome component of gasoline; because it mixes with water rather than floating on top, complete groundwater decontamination is nearly impossible.[1]

You might discover the effects of leaking underground storage tanks near your own home by less drastic means than the unfortunate homeowner in Indiana, but even very low levels of contamination of drinking water can be dangerous to human health and safety. Gasoline, for example, can make water unsafe to drink at only one part per million, that is one gallon of gasoline in one million gallons of groundwater.[2] But gasoline tanks are not the only problem. There are at least 1.5 million underground tanks in the United States that are used to store petroleum products and hazardous substances, and some estimates range as high as ten million.[3] Many businesses, small and large, use underground tanks to store solvents and other hazardous chemicals.

This chapter is all about using the federal LUST (Leaking Underground Storage Tank) program, adopted by Congress as part of the 1984 amendments to RCRA, to fight groundwater contamination from underground

[1] Environmental Task Force, *Cities Nationwide Affected by LUST*, Washington, D.C.

[2] Gopnik, Morgan. "Subterranean Saboteurs: Leaking Underground Storage Tanks." *Re: Sources*, Vol. 7, no. 1. Environmental Task Force. Washington, D.C. 1987.

[3] "Plugging the Leak in Underground Storage Tanks: The 1984 RCRA Amendments," Vermont L. Rev., Vol. 11, p. 267 (1986).

tanks. This program applies to most tanks containing petroleum or other hazardous substances, except hazardous wastes.[4] Tanks containing hazardous wastes are regulated by the hazardous waste program under RCRA (See Chapter 16.) Like other programs, the LUST program is designed to be implemented by the states with EPA approval and based on minimum requirements established by EPA.

DISCOVERING LOCATIONS

To prevent the potential health, safety, and economic disaster of leaking underground storage tanks in your community, you must first identify tank locations, both active and abandoned. The first place to look is state or local offices designated by the governor of each state to receive tank information. (See Appendix E.) All tank owners (with certain exceptions; see below) must notify these designated offices of the age, size, type (including any release detection devices), location, and content of their tanks.

Owners of underground storage tanks taken out of operation but still in the ground must also notify the designated agency of each tank's age, date taken out of operation, size, type, location, and contents (the type and quantity of substances left in the tank). This information is vital because even tanks that have been pumped "dry" may still contain hundreds of gallons of liquid. Even the slightest evidence of corrosion may be cause for alarm. Residents of East Setauket, Long Island, were horrified to learn in February 1988 that a pin-sized hole in a storage tank system owned by Northville Industries had caused 800,000 gallons of leaded gasoline to

---

[4] Tanks sited above ground are not regulated under the under-ground storage tank program described in this chapter. Above-ground tanks may pose a serious threat to ground and surface waters, as recently demonstrated by the 1988 collapse of an Ashland Oil storage tank near Pittsburgh, Pennsylvania, which created a massive oil spill on the Monongahela and Ohio Rivers. Above-ground tanks holding oil or gasoline are loosely regulated, together with large below-ground tanks, under regulations promulgated in 1973 by EPA under the authority of § 311(j) of the Clean Water Act, 33 U.S.C. § 1321(j). Those regulations, set out at 40 C.F.R. Part 112, are applicable to perhaps 600,000 facilities whose location creates a risk of discharge of harmful quantities of oil into surface waters. EPA's regulations require such facilities to prepare spill prevention, control, and countermeasure plans, including plans for dikes or other containment measures; plans must be certified by a professional engineer as having been prepared "in accordance with good engineering practices." Spill prevention and control plans must be kept on-site, and made available to EPA for on-site review, but are not otherwise required to be submitted to EPA for approval. EPA inspects approximately 1,000 facilities under this program annually.

In response to the Ashland Oil spill, legislation was introduced in Congress to more rigorously regulate above-ground tanks, but was not enacted.

leak into groundwater near their homes over a ten-year period. The company estimated that the plume, which was 100 feet underground and seven feet thick in places, had spread under 30 acres.[5]

Unfortunately, this notification requirement does not extend to tanks left in the ground but taken out of service before 1974. To obtain information on these tanks, contact the local or state fire marshall's office, which may have information on these tanks as part of its fire prevention program. Your state may also have its own underground storage tank program with reporting or inventory requirements. (See Chapter 30).

Some sleuthing may be necessary to find out where abandoned tanks might be. If you suspect gasoline or other hazardous substances are contaminating your groundwater and can identify no current likely source, check zoning records of areas near your wells to find out what sort of businesses or industry might have once been located there. Building officials and current business owners may also have information on these tanks. Gasoline service stations are clearly the most common sites of underground storage tanks, but underground tanks are used by a wide variety of concerns, from department stores to schools, from construction companies to farms. Any enterprise that once maintained a number of vehicles or used heavy equipment is likely to have had an underground storage tank for gasoline on the premises. Talk to long-time residents who may remember where businesses used to be located. Public libraries and museums may have copies of old business directories or even maps. Any of these may lead to the discovery of an abandoned leaking tank.

In your investigation of tank locations, you should also be aware of those underground storage tanks—even those that are still in active use—that are not subject to the notification requirement or any other regulation of the federal LUST program. Check with local or state agencies to find out whether they regulate tanks not covered by the federal program. Among the tanks not included in the federal program are:

- Farm and residential tanks holding less than 1,100 gallons of motor fuel used for noncommercial purposes
- Tanks storing heating oil for burning on the premises
- Systems for collecting stormwater and wastewater
- Septic tanks

Once you have identified the tanks in your area, the next important step is to try to discover which tanks may be leaking. You can request from the agency or the tank owner information that may identify problem tanks. Under EPA's new LUST regulations, tank owners are required to monitor

[5] Schmitt, Eric. "Gasoline Leak Threatens Quiet L.I. Area." *New York Times*, B1 (February 10, 1988).

for releases and report possible releases to EPA (or the state agency implementing an EPA-approved state program). Existing tank owners must have a monitoring system in place no later than December 1993. New tanks must have a monitoring system when they are installed.

You can also try to find leaking tanks by comparing groundwater quality test results from wells near suspected tanks to reports available at the agency describing the content of the tanks. If you have enough evidence to substantiate concern but not enough to prove a leak, request the agency to inspect the tank or conduct its own monitoring.[6]

TAKING ACTION AGAINST IMMEDIATE THREATS

If your investigation leads you to believe that an underground storage tank is leaking and may contaminate groundwater, there are several steps that you can take to stop the release of contaminants and, if necessary, initiate cleanup action.

Corrective-Action Plans

The first option is to get the tank owner to take corrective action. Corrective action happens in two phases.

The first phase involves the tank owner and the regulatory agency. LUST regulations require that all underground storage tank owners report any spills or leaks to the agency within 24 hours, or within a reasonable period of time as determined by the implementing agency. (Chemical tanks must also report above-ground spills to the National Response Center, see Chapter 13.) The owner must take immediate steps to stop the leak, clean up visible contamination, and assess the site for damage to soil and groundwater. Within 20 days of a leak, the owner must notify the agency of immediate actions taken. The owner must also report to the agency the results of the investigation into possible soil or groundwater contamination.

Because these steps are to be conducted within days of the leak, you will have little opportunity for involvement, unless, of course, you find that the owner failed to take these measures. In that case, you should inform the tank owner of the existing legal obligations and, if necessary, ask the agency to step in to order the owner to take action.

The second phase is critically important and you should take an active role. Whenever there is any evidence of remaining soil or groundwater

---

[6] In 1986, EPA completed a survey of more than 1,000 leaking USTs around the country to develop information about the causes of releases. This survey, called the Release Incident Survey, may contain information about tanks in your area.

contamination at the site, you should pressure the agency to request the owner to conduct an additional investigation and prepare a corrective-action plan to remove the contamination. These plans could require any of the following:

- Containing the contamination plume with trenches and barriers
- Removing and/or treating contaminated soil
- Providing alternative water supplies to your community
- Restoring the groundwater resource
- Restricting use of contaminated groundwater
- Conducting periodic assessments of the plan to ensure that it is effective

Ask the agency for copies of the plan and submit your comments on its adequacy. In formulating your comments you should consider such site-specific factors as the hydrogeology surrounding the tank, the homes and businesses in the area affected by the spill, and the size of the release. Also, make sure you are given full access to all tank information, including age, size, type, history of past violations, leak detection and monitoring records, insurance coverage, and the like. This information will help you formulate comments on the adequacy of any cleanup plan, including the ability of the owner or operator to carry it out. You should also ask the agency to hold public meetings before a proposed corrective-action plan is approved.

If the agency approves a plan, follow up on its implementation by the owner to make sure it is properly completed. You can request that the agency modify the plan if it proves to be inadequate to address the problem. You may also be able to appeal decisions to approve plans or challenge them in court.

Corrective-Action Orders

If the owner of a tank fails to take proper corrective action, request the agency to issue an administrative order requiring the owner to initiate responsive action. The agency has the authority to require the owner or operator to take a variety of actions, such as making exposure assessments to determine potential effects and cleaning up releases. See page 269 for more suggestions about using this process.

Agency Action Under the LUST Trust Fund for Petroleum Releases

For releases of gasoline and other petroleum products, request the agency to take corrective action itself with money from the LUST Trust Fund. (Remember that for nonpetroleum releases, you should seek action under the Superfund program or other imminent hazard authorities. See Chapters 13 and 14.) The Fund can be used for investigation and cleanup, including provision of alternate water supplies and temporary or permanent relocation of residents. The agency can use the Fund only when a solvent owner cannot be found, prompt action is necessary, or the owner refuses to comply with corrective-action requirements or orders. Priority is given to the leaks that pose the greatest threats. To convince the agency of the need for action, try to develop and present as much evidence as possible, including:

- Samples taken from your well and other drinking-water wells in the area that show the presence of gasoline components, including benzene, xylene, toluene, and lead.
- The presence of fumes in your home or public buildings.
- Physical evidence of a leak near a business known to own underground storage tanks of petroleum, including puddles containing an iridescent film, fumes, or discolored soil.
- Records of explosions or fires that may be traced to the leak.
- Inventory records from your designated state agency to show whether abandoned tanks are causing the problem.
- Information on the tank owner showing financial ability to pay for cleanups from leaks or to compensate you for damages to your health, home, or property (If the owner does not have insurance or another guarantee to pay for cleanups, provide alternative water supplies, drill new public or private wells, or relocate your home, insist that the agency take action.)

Once you convince the agency to begin an investigation or corrective action with Trust Fund money, the next step is to make sure that what the agency does is sufficient. Ask for information about the plans for investigations or cleanup and give the agency your comments. Follow up on the action and request further work if the agency stops short of a full cleanup.

Citizen Suits and Common-Law Actions

If the owner fails to respond to a release and the agency (the state or EPA) does not take action, you may consider filing a citizen suit under RCRA against the owner or operator of the tank for failure to follow the corrective-action regulations. A court can impose stiff financial penalties on the violator and order it to take action. The most difficult aspect of such suits is likely to be proving that a particular tank is leaking (unless the owner has reported a leak and not responded). If the release involves a waste, including hazardous wastes, you may be able to bring a court action against the owner under the RCRA imminent hazard provision to force cleanup action. (See Chapter 14.)

If a leak from an underground tank actually results in the contamination of your wells, you may also consider filing an action for nuisance, negligence, or trespass under the common law. (See Chapter 8.)

State LUST Programs

Always check whether there is a state program that may apply, independent of the federal LUST program. Many states now have their own underground storage tank regulations and require spill or leak reporting and cleanup action. (See Chapter 30.)

ENFORCING LUST REGULATIONS

Even when underground tanks appear to be free from leaks, you can take action to make sure they stay that way by enforcing the LUST regulations. By pressuring the agency to take enforcement action or, in some cases, by doing it yourself, you can stop violations before they develop into groundwater problems. (Remember that not all tanks are covered by the federal program. See page 266.)

You can find information about compliance with these requirements at the agency, in some cases, or at the facility itself. Designated state or local agencies (see page 265 and Appendix E), usually the state UST office, will have notification forms with basic tank information for all regulated tanks, including newly installed tanks. Certain operating and maintenance data is kept by the owner. When the tank owner will not give you his records, ask the agency to request them from the owner. When there is evidence of a violation, ask the agency to inspect the tank or conduct its own monitoring to make sure.

Interim New Tank Requirements

If local tanks were installed between May 7, 1985, and the effective date of EPA's LUST regulations (December 22, 1988), you can take steps to ensure that they meet certain interim standards. Tanks installed in this interim period must comply with the following general requirements:

- Tanks must prevent releases of the stored substances due to corrosion or structural failure.
- The stored material must be compatible with the tank material so that the tank will not corrode.

LUST Regulations

EPA's regulations establishing controls for new and some existing underground storage tanks became effective in December 1988. Following is a list of important requirements you should monitor for tanks in your area.

DESIGN, CONSTRUCTION, AND INSTALLATION

- Tanks and piping may be constructed of fiberglass, coated steel, or metal without additional corrosion protection. If constructed of coated steel, the tanks and piping also must contain corrosion protection devices, such as a cathodic protection system. Such corrosion protection devices must be regularly inspected. If constructed of metal without corrosion protection, records must be maintained showing that a corrosion expert has determined that the site is not so corrosive as to cause leaks during the tank's operational life. Tanks and piping must be installed properly and precautions must be taken to prevent damage.
- Information on design (including corrosion protection), construction, and installation of tanks and piping is available on the notification forms that are on file in the state office designated to receive these forms, usually the state UST office. Records of inspections, monitoring, and testing of corrosion protection devices are on file at the UST site or must be made available to the implementing agency upon request.
- These standards apply to new UST systems. Existing UST systems have until December 22, 1998, to meet these standards.

SPILL AND OVERFILL CONTROL

- With the exception of systems filled by transfers of no more than 25 gallons at one time, UST systems must use one or more overfill prevention devices. These devices include sensors to detect tank capacity level, automatic flow shutoff valves, and spill catchment basins.
- Owners and operators must report spills and overfills to the implementing agency within a reasonable time period.
- The standards apply to new UST systems and some existing tanks. Existing UST systems must have complied with these standards by December 22, 1998.

REPAIRS

- Tank repair must be conducted in accordance with a code of practice developed by a nationally recognized association or an independent testing laboratory. Following completion of the repair, tightness tests may be required.
- Owners must maintain repair records at the site or must make them available to the implementing agency upon request.
- These repair standards apply to new and existing UST systems.

FIGURE 19.1
LEAK DETECTION ALTERNATIVES

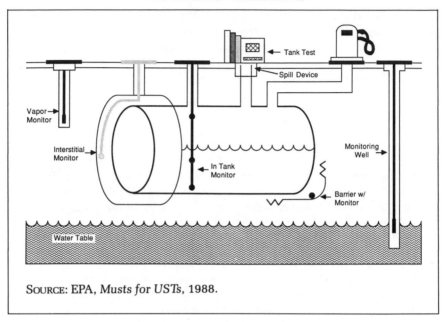

SOURCE: EPA, *Musts for USTs*, 1988.

LEAK DETECTION, REPORTING, AND INVESTIGATION

- Release detection requirements differ between petroleum USTs and hazardous substance USTs. Petroleum UST systems may choose from among five primary release detection methods, such as: 1) automatic tank gauging that tests for loss of product and conducts inventory control; 2) testing or monitoring for vapors within the soil gas of the tank area; and 3) testing or monitoring for liquids in the groundwater. Hazardous substance tanks must use a secondary containment system (a double walled tank or an external liner), unless the owner has obtained a variance.

  You should watchdog carefully any chemical tank that has received a variance and oppose variance applications that rely on less effective systems.

- Owners must maintain records pertaining to the system's leak detection methods, the results of recent tests to detect possible leaks in the system, and maintenance records of release detection equipment, and must make them available to the agency on request. Information on the type(s) of leak detection method(s) used by a UST system is also on the notification form on file in each state's UST office.

- These leak detection standards apply to new UST systems. Existing systems must comply between December 22, 1989, and December 22, 1993, according to a phase-in schedule based on the age of the UST system.

OUT-OF-SERVICE SYSTEMS AND CLOSURE

- During a temporary closure the owner must continue all usual system operation, maintenance, and leak detection procedures, and must comply with release reporting and cleanup regulations if a leak is suspected or confirmed. Release detection, however, is not required if the UST system is empty.

  If a UST system is taken out of service for longer than 12 months, it must be permanently closed unless it meets certain performance standards and upgrading requirements. Before a tank may be permanently closed, the owner or operator must test for system leaks; if a leak is found, the owner or operator must comply with the corrective-action regulations. Once the tank is permanently out of service, it must be emptied, cleaned, and either removed from the ground or filled with an inert solid material.
- Information on closure procedures is available at each tank site or must be made available to the implementing agency upon request.
- These closure standards apply to all new and existing UST systems.

FINANCIAL RESPONSIBILITY

- These regulations specify the amount and scope of coverage required for taking corrective action and for compensating third parties for bodily injury and property damage from leaking tanks. Minimum amounts of coverage required vary depending on the number of tanks owned, from $1,000,000 for up to 100 USTs to $2,000,000 for more than 100 USTs.

  Various mechanisms may be used to fulfill the coverage requirements, such as self-insurance, indemnity contracts, insurance, standby trust funds, or state funds. Quick action on your part may help ensure that available funds are directed toward coverage. Pay particular attention to tanks with self-insurance. Even large companies may go bankrupt and leave you unprotected.
- Information on financial responsibility for new tanks must be filed with the EPA regional office. Owners also must maintain evidence of financial responsibility at tank sites or places of business, or make such evidence available upon request of the implementing agency.
- These financial responsibility standards apply to most new and existing UST systems.

Getting Enforcement Action

ADMINISTRATIVE ORDERS

If you discover violations of the LUST regulations (or are trying to initiate corrective action as discussed above), the first step toward achieving a response is to request the agency to issue an administrative order to the owner. As always, the key to success is presenting as much evidence as possible of the problem or violation to the agency. If necessary, you can also ask that the agency conduct an investigation or an inspection of the tank to verify the existence of a problem. If your state has an EPA-approved program, request action first from the state agency, but inform the regional EPA office as well. EPA retains parallel authority to enforce the LUST requirements even after it approves a state program. The agency can order the violator to take corrective action for releases and comply with regulatory requirements, and can impose fines (up to $10,000 a day for violating regulations and up to $25,000 a day for failing to comply with orders).

There is no formal procedure established to give you an opportunity to comment on orders before they are issued, but you should always make your concerns and recommendations known in written comments (see Chapter 7). For EPA-issued orders, EPA must hold a public hearing if the violator requests one. Make sure you are given notice of any such hearing and exercise your right to participate. Remember, state agencies may follow slightly different procedures in issuing LUST orders. Check with your state agency if you live in a state with an EPA-approved program.

COURT ACTIONS AND CITIZEN SUITS

EPA can file a civil action in court to force a tank owner to comply with regulatory requirements or take corrective action. EPA is most likely to use this option where the owner refuses to comply with an administrative order. You may be able to intervene in such an action if EPA is not protecting your interests.

You can file your own lawsuit if EPA and the state agency fail to take action (or take inadequate action) against a tank owner who is in violation of LUST requirements. In some cases you may be able to substantiate a violation with reports that the tank owner must file with the agency. (See Chapter 8.)

DEVELOPING AN EFFECTIVE STATE PROGRAM

EPA encourages states to implement the LUST regulations and assume primary enforcement authority for the LUST program. Contact the responsible state agency to determine whether, and when, it plans to apply for primary enforcement authority for this program. Federal law requires that the public be given notice of such an application and an opportunity to participate in a comment period, public hearing, and informal meetings with state regulatory personnel. Your object should be to achieve a state program that is better than the minimum federal requirements.

Use the sources of information described in the first section of this chapter to develop information about underground storage tanks in your state and the potential for groundwater contamination, including:

- The vulnerability of areas to groundwater contamination due to local hydrogeology.
- The severity of the problems experienced in your state.
- The time, cause, and nature of releases from tanks and the corrective measures taken.

Especially where groundwater is concerned, prevention is far better than an attempted cure. Cleaning up leaks is expensive and nearly impossible to do thoroughly. Therefore you should focus on the provisions of your state's proposed regulations most directly related to preventing leaks, as opposed to merely detecting them, crucial though the latter is. The Environmental Action Foundation (formerly the Environmental Task Force), a nonprofit organization in Washington, D.C., made the following recommendations for regulation of underground storage tanks in its comments to the EPA. These recommendations can also serve as a useful guide for your own comments on proposed state regulations.

TECHNICAL REQUIREMENTS

- Require secondary containment for all underground storage tanks to prevent leaks into groundwater.
- Require automatic overfill shutoff mechanisms.
- When a leak is confirmed from an underground storage tank, require a tank tightness test for all underground storage tanks at the same site, which are of the same type and age.
- If a tank is removed from the ground, it should be cleaned, not simply emptied, with the hazardous waste disposed of according to RCRA standards.
- Underground storage tanks abandoned before regulations went into effect should be inspected and closed by removing the tank, decontaminating all soil and groundwater and cleaning the tank of any residue.

NOTIFICATION AND RECORDKEEPING REQUIREMENTS

- Notices of leaks or any related problems should be kept on record for the life of the tank.
- Site plans, testing results, and monitoring records should be regularly reported to your state and kept on file until tank closure.

PUBLIC INFORMATION AND PARTICIPATION

- The public should have access to all tank information, including location, age, and type of tank, and an explanation of its contents; inventory infor-

mation; leak detection and groundwater monitoring records; and insurance and other financial responsibility requirements.
- The public should have the right to participate in formulating corrective-action plans. Once a leak is discovered, tank owners must take immediate measures to contain the leak, but permanent cleanup plans should guarantee the public the opportunity to participate.

GROUNDWATER CLASSIFICATION

- Aquifers should not be declared unpotable and therefore exempt from cleanup.

FINANCIAL RESPONSIBILITY

- Federal regulations authorize self-insurance for large corporations. Very large companies have in the last few years filed for bankruptcy, however, and self-insurance provisions could leave the burden of thousands of leaking tanks on taxpayers and public funds. It is better to have some form of secured coverage.
- An annual check on the financial responsibility of the tank owner should be required.

REFERENCES

UNDERGROUND STORAGE TANKS

Commons, Geoffrey. *Plugging the Leak in Underground Storage Tanks: The 1984 RCRA Amendments*, 11 Vermont Law Review 267 (Spring 1986).
Conservation Council. *Preventing Petroleum Storage Tank Leaks: A Citizen's Guide.* New Brunswick, Canada. April 1987. Order from Conservation Council, 180 St. John Street, Fredericton, New Brunswick, Canada E3B 4A9.
East Michigan Environmental Action Council. *Community Involvement in Underground Fuel Storage Tank Mapping.* Birmingham, Mich. 1985. To order, contact: East Michigan Environmental Action Council, 21220 West Fourteen Mile Road, Birmingham, Mich. 48010. (313) 258-5188.
Environmental Defense Fund. *Secondary Containment: A Second Line of Defense.* Washington, D.C. 1988.
EPA, Office of Public Affairs. "EPA Proposes Leak Detection and Cleanup Rules for Underground Storage Tanks." *Environmental News*, April 2, 1987.
EPA, Office of Underground Storage Tanks. *Designing and Installing Underground Storage Tanks Under the New Federal Law.* Washington, D.C. January 1987.
EPA, Office of Underground Storage Tanks. *More About Leaking Underground Storage Tanks.* Washington, D.C. 1987.

EPA, Office of Underground Storage Tanks. *Musts For USTs—A Summary of the New Regulations for Underground Storage Tank Systems.* Washington, D.C. September 1988.

EPA, Office of Underground Storage Tanks. *Proposed Regulations for Underground Storage Tanks: What's in the Pipeline?* Washington, D.C. April 1987.

EPA, Office of Underground Storage Tanks. *Proposed Regulations for Underground Storage Tanks: Your Financial Responsibilities.* Washington, D.C. April 1987.

Gopnik, Morgan. "Subterranean Saboteurs: Leaking Underground Storage Tanks." *Re: Sources.* Vol. 7, No. 1. 1987. To order, contact the Environmental Task Force, 1012 14th Street, N.W., 15th Floor, Washington, D.C. 20005. (202) 842-2222.

New England Interstate Water Pollution Control Commission. *LUSTline.* Bulletins Nos. 2–6. Boston, Mass. December 1985–July 1987. To order, call the RCRA/Superfund EPA Hotline at (800) 424-9346 or (202) 382-3000 and ask for copies of the LUSTline bulletins.

"New Underground Storage Tank Rules Are Unlikely to Protect Nation's Drinking Water Supplies." *Hazardous Waste News,* #24. Environmental Research Foundation. Princeton, N. J. May 11, 1987.

Schmitt, Eric. "Gasoline Leak Threatens Quiet L.I. Area." *New York Times,* B1. February 10, 1988.

LAWS AND REGULATIONS

U.S. Code Citation

Resource Conservation and Recovery Act, LUST Program, 42 U.S.C.A. §§ 6991-6991i (West Supp. 1988).

Code of Federal Regulations Citation

40 C.F.R. Part 280 (Underground Storage Tanks).

CHAPTER 20

Pesticides and Other Agricultural Problems

PESTICIDES HAVE BECOME an increasingly prominent source of ground-water contamination. According to a special report published by the Environmental and Energy Study Institute, EPA has documented ground-water contamination in 23 states by 17 pesticides from normal agricultural use, with an Iowa study showing low pesticide concentrations in 70 to 80 percent of wells sampled.

Pesticide use is extremely decentralized, both in terms of the large number of individuals who apply pesticides and the huge expanse of land on which application occurs. This feature of pesticide use poses special obstacles to both identification and redress of groundwater problems.

IDENTIFYING THE PROBLEM

Narrowing the Field of Pesticides

Because currently available pesticide products number in the thousands, a necessary first step in problem identification is determining which specific pesticides are being used in your area. Local county Cooperative Extension Service offices or agents, or your state agricultural agency should have information on which products are appropriate for use on local crops. Further information may be obtained by going directly to local pesticide marketers (or, in the case of suburban use, to lawn care companies, which not only provide pesticides but also apply them). Bear in mind that marketers will be most generous with information when talking with individuals whom they view as potential customers. You should also consider making inquiries directly of farmers. Finally, some states have programs for monitoring pesticide levels in drinking wells. If the area tested by such a program is analogous to yours in terms of agricultural practices, the test results may help you to zero in on suspect pesticides.

Determining the Leaching Potential of Pesticides

Once you have determined which pesticides are used in your area, try to find out whether they have the potential to leach into groundwater. If the pesticide has recently undergone registration or reregistration by the EPA Office of Pesticides and Toxic Substances, that office will have data on leachability. (See Figure 20.1.) Unfortunately, such data will be lacking if the pesticide was registered long ago. Another agency to consult is the Agricultural Research Service of the U.S. Agriculture Department.

Even if specific data on chemicals of interest to you are unavailable from either of these sources, you may be able to build up a general idea of leaching potential by applying the criteria listed at page 36 of EPA's May 1986 *Pesticides in Ground Water: Background Document*. To apply these criteria, you will first need to obtain information on the chemical composition of the pesticide. You may also require the assistance of an expert to help you apply EPA's criteria correctly.

Determining the Vulnerability of Soils

The threat pesticides pose to groundwater is a direct function of the consistency and composition of the soil. For information on this subject, contact firstly the nearest district office of the Water Resources Division of the United States Geological Survey, and secondly the county office of the Soil Conservation Service, U.S. Department of Agriculture. If these two agencies do not have specific information on the soils in your area, you can obtain at least a general idea of vulnerability by applying the criteria on pages 38–39 of EPA's *Background Document* to your knowledge of local soils. For more detailed information, gather soil samples and take them to a hydrologist.

Ascertaining Whether Your Groundwater Is Contaminated by Pesticides

If you are served by a public water system, your initial step should be to check the monitoring records kept by the water authority. Currently these records are required to include data on only six pesticides, but EPA should add at least 20 additional pesticides to this list by 1989.

If you live in a rural community that relies on private wells, check to see if your state has a program for monitoring pesticide levels in such

FIGURE 20.1

EPA's List of Pesticides Most Likely to Leach into Groundwater

Acifluorfen	Disulfoton
Alachlor	Disulfoton sulfone
Aldicarb	Diuron
Aldicarb sulfone	Endrin
Aldicarb sulfoxide	Ethylene Dibromide
Ametryn	ETU
Atrazine	Fenamiphos sulfone
Atrazine, dealkylated	Fenamiphos sulfoxide
Baygon	Fluormeturon
Bromacil	Heptachlor
Butylate	Heptachlor epoxide
Carbaryl	Hexachlorobenzene
Carbofuran	Hexazinone
Carbofuran-30H	Methomyl
Carboxin	Methoxychlor
Carboxin sulfoxide	Methyl paraoxon
Chloramben	Metolachlor
alpha-Chlordane	Metribuzin
gamma-Chlordane	Metribuzin DA
Chlorothalonil	Metribuzin DADK
Cyanazine	Metribuzin DK
Cycloate	Nitrates
2,4-D	Oxamyl
Dalapon	Pentachlorophenol
Dibromochloropropane	Pichloram
DCPA	Pronamide metabolite, RH24,580
DCPA acid metabolites	Propachlor
Diazinon	Propazine
Dicamba	Propham
5-Hydroxy Dicamba	Simazine
3,5-Dichlorobenzoic acid	2,4,5-T
1,2 Dichloropropane	2,4,5-TP
Dieldrin	Tebuthiuron
Diphenamid	Terbacil
Dinoseb	Trifluralin

SOURCE: EPA, *Office of Pesticide Program*

wells. If so, and especially if you can make a convincing case that your local risk of contamination is high, request that monitoring be done in your area.

If you are unable to interest governmental authorities in doing monitoring, you may want to consider taking your own water samples and sending them to a laboratory for analysis. This could be prohibitively expensive, however, unless you have narrowed down the list of potential pesticide contaminants to a fairly small number.

Also relevant to your data collection activities is EPA's ongoing National Pesticide Survey, which will sample hundreds of public and private drinking wells across the country and analyze the data obtained. The study will be completed in 1990; in the meantime check with EPA to see if the agency has any preliminary results that can be released to you.

REMEDYING CURRENT CONTAMINATION

If your investigations disclose the presence of pesticides in your groundwater, your remedies under federal law are, unfortunately, limited. If you are served by a public water system, the water utility is required to clean up your water to remedy any excess above maximum pesticide levels specified in the EPA National Primary Drinking Water regulations. (See Chapter 11.) If you are served by a private well, however, remedial action may be impracticable. Cleanup of groundwater is difficult and expensive, even when the contamination emanates from a geographically small source like a hazardous waste dump; in the case of agricultural pesticides, which may be leaching into groundwater over a land area covering hundreds and thousands of acres, cleanup may be effectively impossible.

If your contamination problem is especially egregious, however, and results from a geographically confined source—such as a dump where empty pesticide containers have been placed—then you may want to consider requesting action from EPA. This action could take the form either of an "imminent hazard" order under the SDWA or Superfund directing the responsible party to undertake cleanup or of a response action by EPA itself under Superfund. (See Chapters 13 and 14.) (While Superfund prevents EPA from imposing financial liability on farmers for

Fighting Current Contamination

Citizens living near Trenton, New Jersey, have been fighting to remedy pesticide contamination of groundwater caused by a mixing plant for fertilizers and pesticides. Citizens around the plant first noticed a problem because of strange odors in their air. Testing revealed chloroform, lindane, chlordane, and trichloroethylene (TCE). After a two-year study, the New Jersey Department of Environmental Protection finally ordered the chemical company that then owned the facility to lay a water line. Even this did not end the township's woes; contractors refused to install a line because the soil was too contaminated. Finally, a citizen decided to file suit to compel the company to provide them with potable water and prevent future contamination.

the lawful use of pesticides, it does not bar EPA from undertaking its own cleanup actions.) Finally, if your contamination can be traced to the storage of pesticides in underground tanks, EPA may be able to order remedial action under RCRA. (See Chapter 19.)

PREVENTING FUTURE CONTAMINATION

Given the essential impossibility of cleaning up groundwater contamination that encompasses a large geographic area, EPA has emphasized that "prevention is the key."

Spill Response

If you become aware of a pesticide spill, report it immediately to the National Response Center. If the pesticide in question is one of those that are listed as hazardous substances under Superfund, and if the spill involves more than the established "reportable quantity" of the pesticide, then EPA has authority to undertake cleanup. (See Chapter 13.)

Regulatory Action

You can also take action to prevent groundwater contamination from pesticides by using the federal regulatory process. First, you can police compliance with restrictions on the use of pesticides. Second, you can try to get changes in the use restrictions for particularly threatening pesticides in your area. And finally, you can make sure pesticide residues are disposed of properly. (See Figure 20.2.)

LABEL VIOLATIONS

Restrictions on use of a pesticide, whether devised by the manufacturer or by EPA, are reflected in the "label," which must be provided to any purchaser of that pesticide. The label contains detailed instructions on how to apply the pesticide, how much to apply, when to apply, and what precautions to take. If these instructions are phrased in mandatory terms (for example, if they are prefaced with phrases like "do not . . ." or "apply only to . . .", rather than more equivocal language like "should not be applied to . . ." or "avoid applying to . . ."), they are federally binding requirements, which must be obeyed by the user. (See 7 U.S.C. § 136j(9)(2)(G).)

Some label violations may pose a direct or indirect threat to groundwater. Examples of such violations could include application of too much

The Clean Water Act and Groundwater Protection

The 1987 amendments to the Clean Water Act authorized EPA to fund state efforts to protect groundwater from "nonpoint" pollution sources such as agriculture. (See 33 U.S.C. § 1329(i).) Money is available for "research, planning, groundwater assessments, demonstration programs, enforcement, technical assistance, education, and training." Encourage your state to devise a nonpoint source program covering pesticides in groundwater, and to submit it to EPA for funding.

pesticide per acre; application to wetlands or marshy areas; disposal of empty pesticide containers that have not been properly rinsed out; or application of a "restricted use pesticide"—that is, one that may only be used by a certified pesticide applicator—by an individual who lacks such certification. To find out the requirements applicable to the pesticides that

FIGURE 20.2
POTENTIAL SOURCES OF PESTICIDE CONTAMINATION OF GROUNDWATER

	Manufacturers/ Formulators	Dealer	Industrial User	Land Application
Spills and Leaks				
Storage Areas	X	X	X	X
Storate Tanks/				
Pipelines	X	X	X	
Loading/Unloading	X	X	X	X
Transport Accidents	X	X	X	X
Disposal				
Process Waste	X		X	
Off-specification				
material	X			
Cancelled products	X	X	X	X
Containers	X	X	X	X
Rinsate				X
Land Application				
Leaching*				X
Backflow to				
irrigation well				X
Run-in to wells,				
sinkholes				X
Mixing/loading				
areas				X

*Leaching potential affected by chemical-physical properties of pesticide, hydrogeologic setting, and application and cultivation practices.
SOURCE: EPA, Pesticides in Groundwater: Background and Document

interest you, contact EPA or your state pesticide agency and ask for a copy of the approved label for that pesticide.

If you become aware of any label violations, promptly bring them to the attention of EPA. In some states, under cooperative agreement with EPA, state agencies may also have authority to enforce label restrictions. If you live in such a state, notify both agencies of the violation. If EPA decides to take enforcement action, it will first issue a proposed order (you should submit your comments on the order to EPA) and, if the violator requests one, EPA will hold a hearing before the order becomes final. The hearing is usually a formal, trial-type process in which citizens can request to intervene or participate as amicus. (See Chapter 7.) EPA's own authority is limited to issuing orders that impose financial penalties. By filing a lawsuit in a federal district court, however, EPA can obtain an order directing the violator to comply with the requirements at issue. For state enforcement procedures, contact your state agency.

CHANGES IN OR CANCELLATIONS OF REGISTRATION

Many pesticides can pose a threat to groundwater even if pesticide label instructions are followed to the letter. In such cases the best regulatory remedy is to seek to limit use of the pesticide or, in extreme cases, to cancel its registration altogether. The most practical focus would be to request EPA to impose finely tuned restrictions on use of the offending pesticide in your county or state.

Such restrictions are imposed by EPA pursuant to the Federal Insecticide, Fungicide, and Rodenticide Act (FIFRA). Under FIFRA a pesticide may not be marketed unless it has successfully passed through EPA's registration process. (See 7 U.S.C. § 136a(a).) This process can result in a decision to either (1) grant the registration in the form requested by the manufacturer, (2) deny the registration outright, or (3) grant the registration but specify limitations on the pesticide's use. FIFRA specifies that EPA's decision between these three options must be guided by the goal of avoiding "unreasonable adverse effects on the environment." (See 7 U.S.C. § 136(a)(c)(5).) In determining whether effects are "unreasonable," EPA must consider not only the harm resulting from pesticide use, but also the benefits (for example, does the pesticide make a necessary contribution to production of an important agricultural commodity?) and the availability of alternative pesticides to achieve those benefits.

Where necessary to avoid such "unreasonable adverse effects," EPA may not only refuse to register a new pesticide, but also can restrict or ban the use of a previously registered pesticide. This action can be taken by the agency either on its own initiative or upon request by a concerned citizen.

Under EPA's traditional approach to pesticide registration, however, the practicality of such citizen requests was questionable. First, limiting use of a pesticide requires that EPA initiate complicated procedures, which give pesticide manufacturers all-too-ample opportunity to delay or even defeat needed action. Second, the FIFRA registration process has traditionally been used by EPA to address *nationwide* concerns about pesticide use. Accordingly, localized concerns about groundwater contamination in one geographic area might not be considered sufficiently important to warrant a change in EPA registration if they were not shared by a substantial number of other communities.

EPA has proposed a new groundwater strategy that may help to remove this second obstacle. In its December 1987 *Agricultural Chemicals in Ground Water: Proposed Pesticide Strategy,* the agency has suggested a new approach allowing different use restrictions to be imposed in various states and even counties. Moreover, the strategy would encourage states to adopt pesticide management plans which would regulate use down to the subcounty level (if a state refused to adopt such a plan, EPA would consider banning use of the pesticide in that state). The strategy would also place increased responsibility on the pesticide user by imposing restrictions whose applicability would depend on the user's interpretation of local field conditions. Finally, the strategy would involve pesticide manufacturers by requiring that they conduct groundwater monitoring where necessary to collect vital information.

The proposed strategy would seek to protect all groundwater that is a current or potential source of drinking water or that is vital to fragile ecosystems. Stronger protection would be given to "high priority groundwater," which serves as an irreplaceable source of drinking water for sizeable communities or is vital to the continued survival of endangered species or critical ecosystems.

The goal of EPA's strategy would be to ensure that groundwater contamination does not exceed maximum contaminant levels (MCLs) specified under the SDWA. (See Chapter 11.) Where no MCL has been established for a given pesticide, EPA would rely on interim drinking-water standards (known as "Health Advisory Levels").[1] (See Figure 20.3.)

You can begin to take advantage of this new approach to pesticide regulation by asking EPA to take action on specific pesticides that affect you. For instance, if your county suffers from present groundwater contamination or a high susceptibility to such contamination, and if a pesticide with demonstrated toxicity to humans is being used there in

---

[1] Health advisory levels (HALs) for pesticides and other contaminants are set by EPA and are designed to specify contamination levels that would pose an unacceptable risk to human health. HALs are supported by information on health effects that you can request from EPA for particular pesticides. You should also encourage EPA to establish HALs for pesticides that are used in your area, but not regulated under the SDWA.

FIGURE 20.3
PESTICIDES FOR WHICH HEALTH ADVISORIES HAVE BEEN PREPARED BY THE
EPA OFFICE OF DRINKING WATER*

Acifluorfen	Paraquat
Ethylene Thiourea	Dacthal
Amethryn	Picloram
Fenamiphos	Dalapon
Ammonium sulfamate	Prometon
Fluometuron	Diazinon
Atrazine	Pronamide
Fonofos	Dicamba
Baygon	Propachlor
Glyphosate	1,3-Dichloropropene
Bentazon	Propazine
Hexazinone	Dieldrin
Bromacil	Propham
MCPA	Dimethrin
Butylate	Simazine
Maleic hydrazide	Dinoseb
Carbaryl	Tebuthiuron
Methomyl	Diphenamid
Carboxin	Terbacil
Methyl parathion	Disulfoton
Chloramben	Terbufos
Metolachlor	Diuron
Chlorothalonil	2,4,5-Trichlorophenoxyacetic acid
Metribuzin	Endothall
Cyanazine	Trifluralin

*53 Fed. Reg. 565 (Jan. 8, 1988)

significant quantities, you may want to request EPA to impose a county-wide ban on that pesticide. Alternatively, you may want to ask that EPA require your state to implement a management plan to keep contamination within safe levels. The process EPA will be likely to use to make these decisions is described in the box on page 287.

An example of how EPA's new pesticide strategy will be applied in practice is furnished by the agency's recent proposal concerning aldicarb, a widely used pesticide that is toxic to humans in extremely low concentrations. EPA's proposal presents a three-pronged strategy to prevent groundwater contamination.

First, a nationwide label change would (1) ban application, mixing, or loading of aldicarb within 300 feet of any drinking-water well, and (2) convert aldicarb to a "restricted use" product, which may only be applied by certified pesticide applicators. Second, those states that EPA has identified as being at high risk of groundwater contamination from aldicarb (based on the hydrogeology of the area and the amount of aldicarb used)

EPA Pesticide Registration Process

The basic process for altering or revoking a registration begins with what is called a "special review." This procedure, described in EPA's regulations at 40 C.F.R. Part 154, involves (1) issuance of a preliminary notification of EPA's intent to initiate special review, (2) a public comment period on that notice, (3) a final notification of EPA's decision to initiate special review, (4) a public comment period on the advisability of denying or changing the pesticide's use, (5) meetings with interested parties, (6) an optional informal public hearing, (7) a notice of preliminary determination by EPA, (8) an additional public comment period, (9) referral to the Secretary of Agriculture and a scientific advisory panel, and (10) a notice of final determination by EPA.

A special review may result in a decision by EPA to require states to develop special management plans for that pesticide. If the state fails to act, however, this is not the end of the matter. Rather, in that situation EPA must also initiate a further proceeding to cancel the registration or alter the label restrictions. Under this process, EPA must issue a notice of intent to cancel the registration of the pesticide or to change its classification. The manufacturer has a right to challenge this notice by filing written objections with EPA. In addition, the agency must, upon request of the manufacturer, convene a formal hearing before an administrative law judge. If such a hearing is held, citizens can participate. Such a hearing is closely analogous to a full-blown court trial, with discovery, pretrial conferences and motions, presentation and cross-examination of witnesses, post-trial briefs, and an initial decision by the judge. This decision may be appealed by the manufacturer or other participants to the EPA Administrator (or the Administrator may choose to review it on his own initiative), resulting in a final decision by the agency. This in turn may be challenged in a federal court of appeals. (See 7 U.S.C. § 136n.)

There are ways in which this formidable procedural tangle can be trimmed somewhat. A citizen need not request special review at all, but can, with sufficient evidence, ask EPA to proceed directly to initiate cancellation proceedings. Moreover, if there is an imminent hazard necessitating prompt action, EPA may issue a notice of intent to suspend a pesticide registration. (See 7 U.S.C. § 136d(c).) This notice triggers a hearing which is more streamlined than the cancellation hearing described above. (See 40 C.F.R. §§ 164.102–164.122.) If EPA determines that there is an emergency, it can ban the use of the pesticide during the period when the suspension hearing is continuing. (See 40 C.F.R. § 164.223.) It is important to bear in mind, however, that a suspension proceeding results in a *temporary* EPA order whose sole purpose is to keep the pesticide from being used during the time when full cancellation hearings are ongoing. Suspension is thus a prelude to, not a substitute for, cancellation.

must prepare and implement management plans to prevent such contamination from occurring. If such plans are not forthcoming from a given state, EPA will initiate proceedings to cancel use of aldicarb in that state entirely. Third, areas which EPA has identified as having medium risk of aldicarb contamination will be subject to a groundwater-monitoring program designed to determine the need for regulatory restrictions.

Finally, do not forget that state governments retain important authority over pesticide use. States can (1) ban or restrict pesticide uses that EPA's registration process has permitted, or (2) allow pesticide uses that EPA has not yet registered, where such uses are necessary to meet "special local needs" within the state. Participation in state pesticide programs can thus be an essential part of your efforts to protect groundwater. (See Chapter 31.)

DISPOSAL

Check to see if the pesticide that interests you is considered hazardous within the meaning of RCRA. If so, disposal of pesticide residues and containers is governed by the detailed requirements of that law. (See Chapter 16.)

Nonregulatory Action

Given the large number of individuals who use pesticides and the huge geographic area over which this use occurs, regulatory approaches to pesticide use (short of total ban of the product) will be difficult to enforce. Perhaps the most practicable preventative measure of all, then, is to educate pesticide users concerning the threat facing groundwater and the pesticide application methods by which this threat can be avoided.

At the federal level, technical assistance programs for farmers are operated by three divisions of the Agriculture Department: the Cooperative Extension Service, the Soil Conservation Service, and the Agricultural Stabilization and Conservation Service. Your comments and pressure can help influence these programs to pay more attention to groundwater contamination. Make sure that farmers are being told both how to use pesticides safely, and how to minimize use of pesticides by application of alternative methods of pest control (these alternatives, which include such techniques as biological controls and altered cultivation practices, are known as "integrated pest management"). If you do not get satisfactory results from governmental education programs—many of which are unfortunately dominated by staunchly pro-chemical individuals—start your own in your community.

FERTILIZERS AND FEEDLOTS

While pesticides are the most serious groundwater contaminants produced by agriculture, they are not the only ones. Groundwater can also be tainted both by fertilizers from cultivated fields and by manure from "feedlots" (i.e., enclosures where large numbers of livestock are kept prior to slaughter). For example, a USGS study conducted in 1985 showed widespread drinking-water contamination by nitrates or nitrogen; in Nebraska, 70 percent of the wells tested exceeded the SDWA standards for nitrates. Both of these problems are addressed primarily under state law, but you should be aware that nitrates from fertilizer are regulated under the SDWA, so if you are served by a public water system, your water authority must ensure that your water is free from excessive nitrate concentrations. You should also know that feedlots that discharge manure into surface waters are required to have permits under the Clean Water Act, and you can file a lawsuit in federal district court against the feedlot owner if he has failed to obtain such a permit. (See Chapter 17.)

REFERENCES

PESTICIDES AND OTHER AGRICULTURAL PROBLEMS

Concern, Inc. *Pesticides: A Community Action Guide.* Washington, D.C. May 1985.

The Congress: Past Imperfect, Future Tense, 17 Envtl. L. Rep. 10010 (January 1987).

EPA, Office of Ground Water Protection. *Pesticides in Ground Water: Background Document.* May 1986.

EPA, Office of Ground Water Protection. *State Program Briefs: Pesticides in Ground Water.* Washington, D.C. May 1986.

Gottlieb, Bob, and Peter Wiley. "Pesticides Hit Home." *Sierra,* January/February 1986, page 48.

Holden, Patrick W., for the Board on Agriculture, National Research Council. *Pesticides and Groundwater Quality: Issues and Problems in Four States.* National Academy Press. Washington, D.C. 1986.

Journal of Pesticide Reform. Vol. 7, nos. 1 and 2, 1987.

Moore, Monica. *The "Dirty Dozen." Information Kit.* Pesticide Action Network (PAN) International. 1987.

U.S. Congress, Office of Technology Assessment. *Protecting the Nation's Groundwater from Contamination.* Washington, D.C. October 1984.

U.S. General Accounting Office. *Nonagricultural Pesticides: Risks and Regulation.* Washington, D.C. April 1986.

U.S. General Accounting Office. *Pesticides: EPA's Formidable Task to Assess and Regulate Their Risks.* Washington, D.C. April 1986.

U.S. Council on Environmental Quality. *Contamination of Ground Water by Toxic Organic Chemicals.* Washington, D.C. January 1981.

Using CERCLA to Clean Up Groundwater Contaminated Through the Normal Use of Pesticides, 15 Envtl. L. Rep. (Envtl. L. Inst.) 10100 (April 1985).

LAWS AND REGULATIONS

U.S. Code Citations

Federal Insecticide, Fungicide, and Rodenticide Act, 7 U.S.C.A. §§ 136–136y (West 1980 & Supp. 1988).

CWA Nonpoint source management programs, 33 U.S.C.A. § 1329 (West Supp. 1988).

Code of Federal Regulations Citation

40 C.F.R. Part 164.

CHAPTER 21

Mining

MINING, BOTH UNDERGROUND and on the surface, can have disastrous effects on groundwater. Mines can drastically change groundwater flow patterns and introduce toxic materials into groundwater. This chapter focuses on how to use the one major federal mining regulatory scheme, established by the Surface Mining Control and Reclamation Act (SMCRA), to protect groundwater. Unfortunately, this law applies only to coal mining. Other hard-rock and mineral mining on federal land must be approved by the federal agency that manages the lands. Regulation of hard-rock and mineral mining that is not on federal land is left largely to state law. (See Chapter 32.) This chapter begins with a brief reference to the few federal laws that govern coal and non-coal mining.

USING THE CLEAN WATER ACT

All mines that discharge pollutants into surface water must have a discharge permit (called an NPDES permit) issued under the Clean Water Act. The permit limits the types and quantities of pollutants that can be discharged into surface water at the mine site. Since in many areas, and particularly disturbed mining areas, pollution of surface water may lead quickly to groundwater contamination, you can take advantage of this regulatory program to protect groundwater from contamination by coal and other mines.

For active, ongoing, mining operations you can police the mine's compliance with the pollutant levels set in its NPDES permit and look for unpermitted discharges. You can initiate agency enforcement action (or bring your own lawsuit if necessary) to stop excessive pollution from permitted discharge areas at the mine and to stop altogether any unauthorized discharges. (Refer to Chapter 17 for detailed suggestions about how to take these steps.)

To prevent contamination from future mines, you can participate in the process leading to the issuance of an NPDES permit to ensure that stringent conditions are applied to the mine. The pollution limits that should apply to a mine will vary widely depending on its type and location, but some general issues to pursue are:

- Have all potential discharge areas at the mine been regulated?
- Does the permit cover all pollutants expected to be released by the mine (particularly any toxic metals resulting from acid mine drainage)?
- Are discharge limits set to reflect the cumulative impact of other mines and pollution sources in the area?

Make sure you find out which agency issues the NPDES permit for mines (in some states it is the mining regulatory agency; in others it may be a separate water quality agency) and then follow the suggestions in Chapter 17 for using the process.

MINING ON FEDERAL LANDS

Most mining operations on federal lands are subject to some kind of federal regulation to protect natural resources and there are opportunities within the process for citizens to advocate protection of groundwater. Mining on federal lands is particularly important in the western states where the large majority of land is owned and controlled by the federal government. Citizen rights and the regulatory scheme vary depending on the mineral or resource being mined, so the first step is to identify the type of mineral involved in a particular case.

One class of minerals, those that can be claimed by simply staking out an area and looking for them (gold, silver, lead, and others are in this category), is subject to a general requirement that the miner submit to either the Forest Service or Bureau of Land Management (depending on which agency manages the lands to be affected) a plan of operations for approval prior to mining. Ask the local offices of these agencies to put you on their mailing list for notice of such activities. Once submitted, plans of operation are available for public review and are usually subject to some environmental analysis under the National Environmental Policy Act (NEPA). Generally, a NEPA environmental review will include public notice and an opportunity for comment on the agency's analysis. Review the plan and the environmental analysis for discussion of groundwater impacts and advocate stricter control measures in the plan of operation if necessary. (See 36 C.F.R. § 228 (Forest Service lands), 36 C.F.R. Part 9 (National Parks), and 43 C.F.R. § 3809 (BLM lands).)

The other major class of minerals, those that are leased to the miner by the government (coal, phosphate, oil, and gas are in this group), is regulated by special conditions attached to the lease. There are differences in the process depending on the minerals involved, but the critical juncture for groundwater protection is when lease conditions are developed. Again, ask the agency (either the Forest Service or Bureau of Land Management) for notice that the leasing process is beginning, review the agency's environmental analysis (such an analysis, and an opportunity for public comment on it, is usually required by NEPA), and the proposed lease conditions to ensure that they address adequately any groundwater contamination problems, and ask for changes where they are needed. (See 43 C.F.R. § § 3400–3500.)

Decisions by the Forest Service or the Bureau of Land Management to either approve mining plans of operation or issue leases can be challenged by filing an administrative appeal (see 43 C.F.R. Part 4, 36 C.F.R. Part 211) and then by filing a lawsuit in federal court. (See Chapter 7.)

In both cases, you can also oversee the operation once it begins and alert the agency to the need for enforcement if lease conditions or plan of operation requirements for groundwater protection are violated by the miner.

COAL MINING

This section describes action you can take under the Surface Mining Control and Reclamation Act (SMCRA) against groundwater contamination from coal mines. (See Figure 21.1 for the location of coal fields in the United States.) The heart of the program is the mining permit. You can oversee the enforcement of permit requirements and influence the issuance of new permits to ensure that they include special conditions to minimize the potential for groundwater contamination.

Like most other federal pollution programs, this one is designed to be implemented by state agencies once equivalent state programs are enacted. Of the active coal-mining states, only Tennessee, Washington, North Carolina, and California do not administer the SMCRA program, as of the present writing. For these states, citizens should turn to the federal Office of Surface Mining (OSM), which implements the program, to carry out the suggestions here. In the other active coal-mining states, the state agency is the primary enforcer of the program and should be the first target for citizen action, but OSM retains some authority to act when the state fails.

Other Federal Mining Restrictions

For abandoned or other mine sites that contain hazardous substances and may threaten groundwater, you should follow the suggestions for getting cleanup action under Superfund. (See Chapter 13.) Also, do not forget possible immediate action under the imminent hazard authorities discussed in Chapter 14.

Uranium mill tailings piles at uranium processing facilities are subject to some federal control under the Uranium Mill Tailing Radiation Control Act. (See 42 U.S.C. §§ 7901–7942.) For facilities that have uranium mill tailings piles (all but one are in the West), groundwater threats can be severe. There are approximately 50 such facilities, half of them abandoned and half currently licensed by the Nuclear Regulatory Commission. Contact the regional office of the Nuclear Regulatory Commission for information about how to participate in license proceedings for the facilities and in cleanup decisions at sites causing groundwater problems.

Taking Action Against Coal Mines That Have Contaminated Groundwater

The most immediate problem is the coal mine that has already caused groundwater problems. To identify and take action against such mines, begin by reviewing groundwater-monitoring reports filed by the mine operator with the regulatory agency. For mines that may cause groundwater problems, the mining permit issued by the agency should require the operator to conduct tests of groundwater quality throughout the active mining and reclamation period. These monitoring results must be reported to the agency at least four times a year. The agency is required to let you review these reports for a particular mine. Look for increases in the tested parameters, which usually include at least acids, iron, and total dissolved solids, and look to see that the monitoring and reporting is being done on schedule.

You may also be able to identify mine-generated contamination by conducting your own sampling and testing of water wells near the mine site. Even surface water contamination in particularly sensitive areas may effectively indicate a high potential for the existence of groundwater contamination. In most cases, mine operators will have been required to test pre-mining samples of groundwater and even nearby drinking-well water to establish a background level, against which you can compare monitoring reports and your own sampling data.

If a review of the operator's groundwater-monitoring reports or your own testing discloses a potential contamination problem, ask the agency for information about the implementation of remedial measures at the

FIGURE 21.1
COAL FIELDS OF THE UNITED STATES

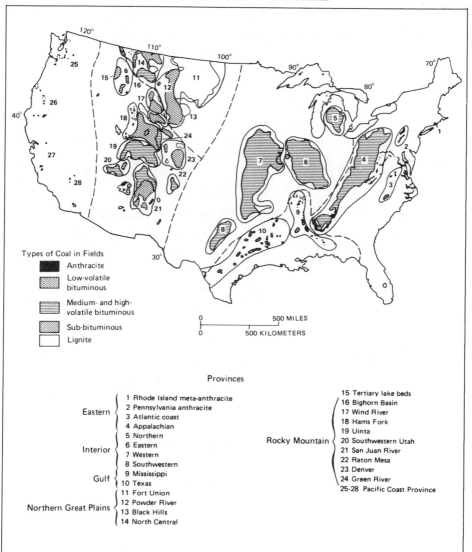

Types of Coal in Fields

- Anthracite
- Low-volatile bituminous
- Medium- and high-volatile bituminous
- Sub-bituminous
- Lignite

0   500 MILES
0   500 KILOMETERS

Provinces

Eastern
- 1 Rhode Island meta-anthracite
- 2 Pennsylvania anthracite
- 3 Atlantic coast
- 4 Appalachian

Interior
- 5 Northern
- 6 Eastern
- 7 Western
- 8 Southwestern

Gulf
- 9 Mississippi
- 10 Texas

Northern Great Plains
- 11 Fort Union
- 12 Powder River
- 13 Black Hills
- 14 North Central

Rocky Mountain
- 15 Tertiary lake beds
- 16 Bighorn Basin
- 17 Wind River
- 18 Hams Fork
- 19 Uinta
- 20 Southwestern Utah
- 21 San Juan River
- 22 Raton Mesa
- 23 Denver
- 24 Green River
- 25-28 Pacific Coast Province

SOURCE: National Research Council, Committee on Ground Water Resources in Relation to Coal Mining, Coal Mining and Ground-Water Resources in the United States (Washington, D.C.: National Academy Press, 1981). Reprinted courtesy of The Conservation Foundation and the National Academy Press.

mine. Any groundwater contamination should trigger specific remedial measures already included in the permit to eliminate the problem. Such measures may include:

- Increased monitoring to determine the extent of the contamination
- Public warnings if health is in imminent danger
- Installation of runoff control and treatment systems, modification of the mining operation, and, in some cases, the closing of the mine.

In addition, the mine operator is required to provide affected water users with alternate water supplies where necessary. If measures specified in the permit or by the agency are not being taken or alternate water supplies are not provided, you should initiate the inspection and enforcement process described below.

Policing the Mining Operation to Prevent Contamination

Even coal mines that cannot be linked definitively to groundwater problems may be operating in a way that is likely to cause groundwater contamination at some future time. By examining inspection records and observing the operation yourself, you may be able to discover such problems at the mine and get them corrected before they lead to groundwater contamination.[1]

WHAT TO LOOK FOR

Your inquiry into problems at the mine-site should focus on those violations most likely to affect groundwater quality. What those problems are will vary depending on the mine and location, but some of the general issues likely to be of concern at most mines are:

- Design and operation of runoff treatment and control systems
- Disposal of overburden (rock and soil overlying the coal formation) and other mine wastes away from sensitive recharge areas
- Stabilization of the surface area of the mine site to prevent erosion and uncontrolled runoff
- Design and operation of wells used for disposal of liquid wastes
- Limitations on blasting near sensitive recharge areas or other areas where groundwater is vulnerable
- Compliance with groundwater-monitoring plans

[1] One easy but important step is to make sure the coal mine has a mining permit. There are very few exceptions to the permit requirement and even some exploration activities require a permit. If the operation looks as if it is mine-related and it is unclear whether the operator has a permit, request the agency to inspect the site and determine whether a permit is required. Mining without a permit is a major offense.

To learn about the specific problems likely to be encountered at a particular mine, review, with the help of an expert if possible, the permit and permit application that are on file at the regulatory agency. You may be able to find help from local environmental organizations, university professors, or environmental attorneys. The application will contain background information about the hydrogeology of the area and the operator's assessment of the potential impact of the mine on water resources, including groundwater. The permit, and particularly the hydrologic reclamation plan, should describe the measures that must be followed by the operator to prevent groundwater contamination.

WHERE TO LOOK

To find out whether any of these important permit conditions are being violated at the mine or whether other significant problems exist, the place to start is the inspection report file at the agency. The regulatory agency is required to inspect the operation regularly—at least every three months. Reports of these inspections must be available for review at the agency and may disclose problems not yet corrected at the site. Ask the agency about any enforcement action taken for violations reported in inspection reports. Request to see enforcement orders and ask the agency whether the operator has complied fully with any such orders.

You can also conduct your own investigation of the mine by looking for obvious problems. Of course, your access to the mine may be limited— and you should never go without an operator's permission or an inspector—but even without direct access you may be able to observe problems such as:

- Discolored nearby streams or dying vegetation (you could also sample and test nearby stream water for acid drainage and other contaminants)
- Blasting in unauthorized areas
- Overflowing or leaking sediment ponds or ponds over filled with sediment
- Uncontrolled water runoff

Initiating Enforcement Actions

There are numerous opportunities to initiate and participate in the enforcement process if any of your investigations or reviews uncover potential problems. In fact, the SMCRA provides for more elaborate citizen participation mechanisms than most other federal regulatory systems.

If you believe the mine presents a threat to groundwater and/or is in violation of a permit requirement, begin by asking for an inspection of the mine. The request can be directed to the state agency or the federal Office of Surface Mining. The agencies must respond to an inspection request within a specified time and, if an inspection is conducted, must explain the results and any enforcement action taken. You can accompany the inspector to the mine or keep your request confidential.

If the agency refuses to inspect the mine or does not take adequate enforcement action, you can have the decision reviewed. The first step is an informal conference or review that will allow you to present your concerns to the agency without necessarily having legal or expert assistance. If your concerns are not cured after the informal review, you can request a formal, trial-type adjudicatory hearing. Here help from lawyers and experts will be important in most cases. The hearing process is also available to the operator. You can intervene in appeals brought by the operator to ensure that your interests are protected.

In some circumstances, unexpected problems or new information may require that the permit be modified, or even revoked and reissued with substantially different conditions. If problems disclosed by inspections are not addressed by the existing permit, press the agency to initiate the process for permit modification. This process, the same as that for issuing a new permit, is described in the next section.

Citizens may bring lawsuits directly against the operator for violations of the permit if the agency does not act and, unlike any other federal program, are permitted to sue for money damages for injury to property (including groundwater resources). The steps and key points of the enforcement process are summarized in Figure 21.2.

Making Sure Mines Are Reclaimed

To minimize the potential for groundwater contamination from mines that are no longer active, you can take steps to see that mines are properly reclaimed—that is, the land and water are restored to the pre-mine condition. Each mine operator is required to develop a reclamation plan and post a money bond before operations begin. When the mining operation is finished, the operator must complete the measures set out in the reclamation plan before he can recover the bond money. Look for notice in the local newspaper that the operator is going to request release of the bond, ask the agency for information about the thoroughness of the reclamation, ask to inspect the site yourself with the agency, and participate in the public hearings before the agency. Urge that the bond be retained until reclamation is complete. To challenge an agency decision to release prematurely a reclamation bond, follow the enforcement procedure by filing a request for informal review.

To have restoration done on surface mines abandoned before reclamation requirements were passed in 1977, contact the regulatory agency

FIGURE 21.2
THE SMCRA ENFORCEMENT PROCESS

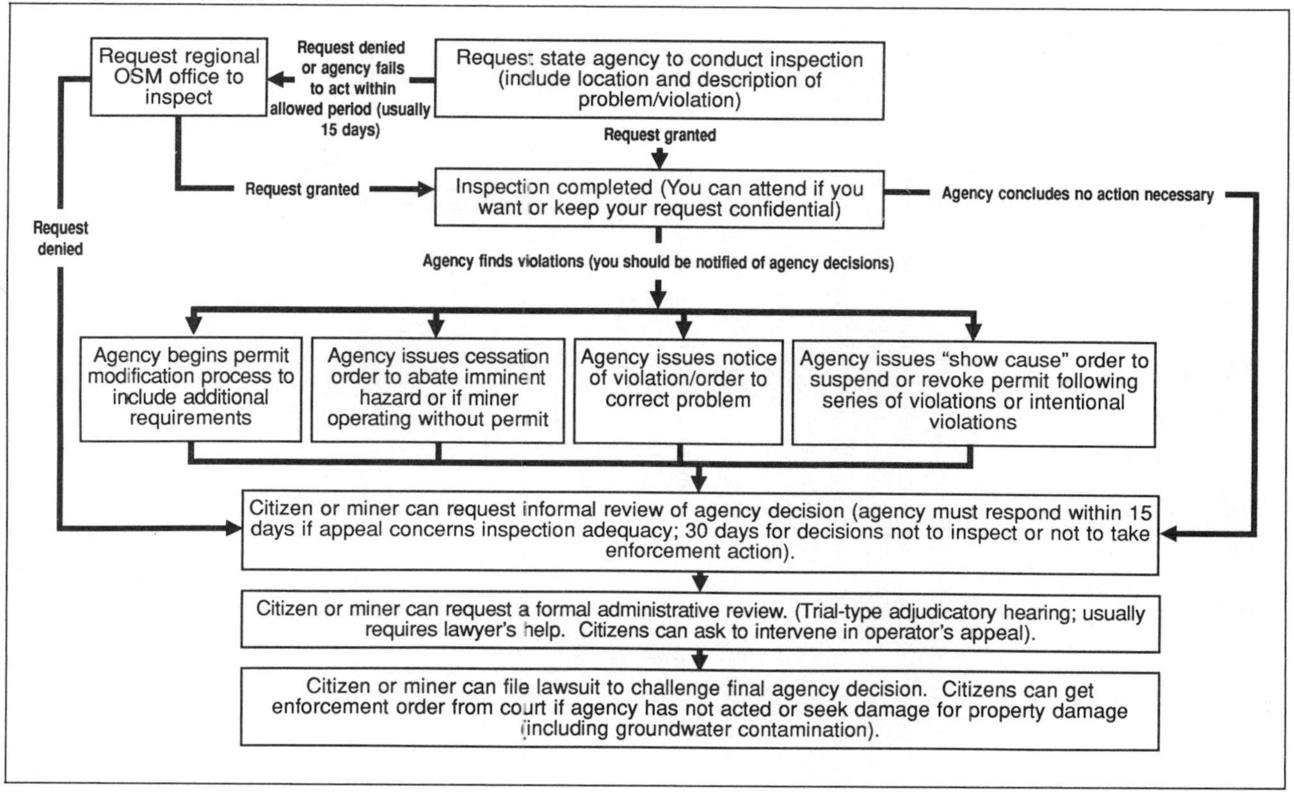

Request regional OSM office to inspect

Request denied or agency fails to act within allowed period (usually 15 days)

Request granted

Request denied

Reques: state agency to conduct inspection (include location and description of problem/violation)

Request granted

Inspection completed (You can attend if you want or keep your request confidential)

Agency concludes no action necessary

Agency finds violations (you should be notified of agency decisions)

Agency begins permit modification process to include additional requirements

Agency issues cessation order to abate imminent hazard or if miner operating without permit

Agency issues notice of violation/order to correct problem

Agency issues "show cause" order to suspend or revoke permit following series of violations or intentional violations

Citizen or miner can request informal review of agency decision (agency must respond within 15 days if appeal concerns inspection adequacy; 30 days for decisions not to inspect or not to take enforcement action).

Citizen or miner can request a formal administrative review. (Trial-type adjudicatory hearing; usually requires lawyer's help. Citizens can ask to intervene in operator's appeal).

Citizen or miner can file lawsuit to challenge final agency decision. Citizens can get enforcement order from court if agency has not acted or seek damage for property damage (including groundwater contamination).

about reclamation with funds from the Abandoned Mines Land Fund. It is unlikely that these funds will be provided to pay for public water systems to replace private supplies that have been contaminated, because it is extremely difficult to get the required proof that the mining operations caused the contamination. One thing you can do to facilitate reclamation that may prevent later contamination of groundwater, is to make sure your state is receiving these funds in proportion to the number of abandoned mines within its borders. Members of Virginia Citizens for Better Reclamation (VCBR) were plagued by acid mine drainage from abandoned mine lands, which contaminated their groundwater supplies, forcing them to haul water from other sources. Staffing and funding problems prevented the state agency from compiling a complete listing of abandoned mines, so volunteers documented unreported abandoned mines. They used this information to obtain from the federal Abandoned Mine Lands Program more reclamation money that was used, in part, to reclaim mine sites in their area.

Participating in the SMCRA Mine Permit Process

You can act to prevent groundwater contamination from future coal mines by participating in the SMCRA mine-permitting process. The permit should require the mine to be operated in a way that minimizes its impact on groundwater quality. You have several opportunities to ensure that the regulatory agency imposes the necessary conditions.

Begin by looking in the local newspaper for notice of an operator's intention to mine. It should contain the name of the person owning the land to be mined, the location of the land, and the boundaries of the area to be mined.

The next step is to review the permit application filed with the regulatory agency by the operator. The application describes in detail the area to be mined, the methods to be used and, among other things, proposals for preventing adverse impacts to groundwater. You have at least 30 days to evaluate the application and prepare comments to submit to the agency. (Look for the comment deadline in the newspaper notice or ask the regulatory agency.) You can also request that the agency hold an informal public hearing to air citizen concerns about the mine. To help you evaluate the mining proposal, you can ask for an opportunity to inspect the proposed mine site with agency personnel.

Evaluating the potential for groundwater impacts from a mine will be difficult as those impacts are largely dependent on local conditions. Many of the issues may require that you find expert assistance (hydrogeologists or other mining experts) to help assess the potential for contamination from a given mine and to suggest ways to minimize that threat. Some of the critical issues to focus on during a review of mining permit applications are:

ASSESSING THE IMPACT OF THE MINE ON WATER RESOURCES, INCLUDING GROUNDWATER (CALLED THE PROBABLE HYDROLOGIC CONSEQUENCES (PHC) DETERMINATION)

- Did the operator take sufficient groundwater samples to determine pre-mine groundwater flow, quantity, and contamination levels? Were samples taken over a long enough period to account for seasonal variations (experts say at least one year)? Were samples taken at varying locations (on and off the proposed mining site) and depths to locate all affected aquifers? Were samples taken at nearby drinking-water wells to establish background quality levels?
- Does the application contain accurate, detailed surface drainage maps and cross-sectional maps of the aquifers and geologic formations of the area? Are the locations of proposed surface water pollution discharges and the link to groundwater supplies established by the maps?
- Is there a complete survey of water users and uses in the area that may be affected by the mine?
- Have the underlying rock formations been chemically analyzed to determine whether acid- or toxic-forming drainage could result from exposure to air and water?
- Has the amount of waste to be generated by the mine been accurately projected to provide an adequate basis for developing permit controls?

EXAMINING THE MINING OPERATION

- Is the proposed mining method the most protective of groundwater resources? Are other less damaging methods feasible?
- Does the application specify where wastes and overburden are to be collected or disposed?
- Is there a blasting plan that provides for protection against groundwater disruption and pre-blasting surveys of local areas, including water resources?
- Does the application explain permit violations the operator may have incurred at other mines? (You can double-check this information with the agency—permits should not be issued to persistent or wilful permit violators.)

CRITIQUING THE METHODS PROPOSED TO PREVENT WATER RESOURCES DAMAGE (CALLED THE HYDROLOGIC RECLAMATION PLAN)

- Is there a complete groundwater-monitoring plan that includes on- and off-site monitoring wells (including nearby water wells)? Does the plan require testing for all expected contaminants and frequent (at least quarterly) reporting to the agency?

- Are provisions made for sealing boreholes, shafts, or wells to prevent drainage into groundwater? Will sediment controls (such as sediment ponds, drainage systems, and embankments) prevent runoff that could leach into groundwater? Will liquid wastes be adequately treated before discharge?
- Are specific remedial measures outlined to address groundwater contamination caused by the mine, such as provision of additional containment systems or buffering of acid mine drainage? Does the plan outline how the operator will provide alternative water supplies?

EVALUATING THE RECLAMATION PLAN PROPOSED TO RESTORE THE MINE SITE

- Will the land be planted with grasses and trees to prevent rainfall from washing exposed rock and minerals into streams and groundwater?
- Are drainageways designed to catch and treat runoff from mining operations that cannot otherwise be prevented? (These can include sediment ponds, ditches, and the like.) Will buffer zones be maintained to decontaminate runoff from drainage areas?
- Will all underground mines, shafts, and other land disruptions be filled or sealed?
- Will solid wastes at the site be taken to an approved disposal facility?

Based on the application, the agency should assess the cumulative impact to water resources from the proposed mine together with all other existing and potential mines in the area. This assessment, called the cumulative hydrologic impact assessment, is an important factor in the permitting process and deserves careful review to ensure it is complete.

After the close of the public comment period and any hearings held on the application, the agency will decide whether to grant a permit and what conditions the operation must meet. In some cases the agency may require the operator to resubmit the application with new or different information. Such a resubmission triggers a new public comment period.

If the agency approves a permit and concerns you raised in comments are not addressed adequately, you can challenge the decision by asking, usually within 30 days of the decision, for a formal hearing before the agency. Like the hearing described for the enforcement process, these are adjudicatory-type proceedings that will probably require the assistance of a lawyer and expert witnesses.

If the formal hearing does not result in changes in the permit, you can bring a lawsuit to challenge the permit decision. This too will require the assistance of a lawyer and, probably, expert witnesses to explain the inadequacies of the permit.

Using Appeal Process to Challenge Permits

Sometimes, using appeal procedures can give the citizen group leverage to strike a compromise with the operator and agency. Citizens concerned about the Left Beaver Coal Company in Kentucky sought a formal hearing to have the company's permit denied when they discovered it had poured sludge from its operations into an underground mine, causing contamination of their groundwater. With the help of the Appalachian Research Defense Fund, they negotiated a settlement in which the mine operator agreed to drill new deep wells and install water purification systems in their homes.

Petitions to Have Land Designated as Unsuitable for Surface Coal Mining

Another way to protect groundwater from future mining operations is to petition the regulatory agency to designate areas susceptible to contamination as unsuitable for surface coal mining. Such a designation will prevent any coal mining in the area. A petition must be filed before mining operations begin and, in the case of an ongoing permit process, before the public comment period on an application is complete. One ground for such a petition is that the mining will cause a substantial decrease in the quantity or quality of water supplies—including aquifers and recharge areas.

Though this method is truly preventative, it is very difficult to succeed with such petitions because proof that damage to important water resources will occur is required. One petition was granted in Kentucky to protect lands adjacent to a public water supply reservoir. Pennsylvania leads the states in granting petitions, with 11 areas designated as unsuitable for mining as of this writing. Nationwide, only about 25 to 30 petitions have been approved.

Contact the regulatory agency for petition forms and for the exact procedure used by the state to make this designation. Some states, such as Kentucky, have handbooks to assist with petition preparation.

REFERENCES

MINING

Center for Law and Social Policy, Environmental Policy Institute. *The Strip Mine Handbook.* Brophy Associates, Inc. Washington, D.C. 1978. To order, contact The Environmental Policy Institute, 218 D Street, S.E., Washington, D.C. 20003. (202) 544–2600.

Galloway, L. Thomas, and Tom FitzGerald. *The Surface Mining Control and Reclamation Act of 1977: The Citizen's 'Ace in the Hole.'* 8 N. Ky. L. Rev. 259 (1981).

Kentucky Fair Tax Coalition. *Citizen's Water Handbook.* Prestonsburg, Ky. July 17, 1986. To order, contact: Kentucky Fair Tax Coalition, P.O. Box 864, Prestonsburg, Ky. 41653.

Morgan, Mark L., and Edwin A. Moss. *Citizens' Blasting Handbook.* Appalachia—Science in the Public Interest. Corbin, Ky. 1978.

Student Environmental Health Project (STEHP), Center for Health Services, Vanderbilt University. *Water Problems and Coal Mining: A Citizen's Handbook.* Nashville, Tenn. 1984. To order copies, contact: STEHP, Center for Health Services, Station 17, Vanderbilt University, Nashville, Tenn. 37232.

Low-cost groundwater testing for mine-related contaminants may be available at:
William Hoover Laboratory Project
West Virginia Mountain Streams Monitors
P.O. Box 170
Morgantown, West Virginia 26505

LAWS AND REGULATIONS

U.S. Code Citations

Surface Mining Control and Reclamation Act, 30 U.S.C.A. §§ 1201–1328 (West 1986 & Supp. 1988).

Clean Water Act, 33 U.S.C.A. §§ 1251–1387 (West 1986 & Supp. 1988).

Code of Federal Regulations Citations

30 C.F.R. Parts 700–955 (Mining regulations).
40 C.F.R. Parts 100–125 (Clean Water Act NPDES regulations).

CHAPTER 22

Transportation of Hazardous Materials

GROUNDWATER IS THREATENED not only by the commercial and industrial users and producers of hazardous materials or the landfills, wells, and other facilities where waste products end up, but also by the trucks, railroad cars, and pipelines that carry hazardous materials from factory to factory to waste site. Spills from these carriers can be a major source of groundwater contamination. This chapter describes how to use federal law to gather basic information about transportation of hazardous materials in your area, respond to spills or accidents, and take action to ensure that carriers that may affect your groundwater quality are meeting basic safety standards.

OBTAINING INFORMATION ABOUT TRANSPORTATION AND POTENTIAL PROBLEMS

Except for those situations where you actually observe or learn about a transportation spill, the first step for action will be gathering basic information about the transportation of hazardous materials and other potential contaminants through your area, including:

- The types and quantities of materials transported
- The modes of transportation (rail, truck, etc.)
- The routes traveled
- The hazards associated with the transported materials

Listed below are several sources of information that should be available to every community.[1] You may have access to other helpful information through special state and local programs and, once identified, from carriers themselves.

- Local emergency response plans (they should have been in place around the country by October 1988) are required to identify likely transportation routes for hazardous materials in your region. Local response committees

---

[1] You can reach the Federal agencies and offices referred to in this chapter through the Department of Transportation, 400 7th Street, S.W., Washington, D.C. 20590.

will also have detailed information about facilities in the community that handle hazardous materials (location, type of materials, etc.) that you can request and use to deduce for yourself likely transportation routes to and from these facilities. These programs were established by the Emergency Planning and Community Right-to-Know Act (EPCRA), discussed in detail in Chapter 17.

- State hazardous waste agencies or regional EPA offices should have information about transportation of hazardous waste in the form of manifest records and biennial reports from generators of waste. (See Chapters 16 and 17 for more about hazardous waste management.) These records should specify the type of waste, where wastes start and finish their journey, and what type of carrier was used.
- The Federal Railroad Administration (part of the Department of Transportation [DOT]) can provide you with maps of major railroad routes.
- Maps identifying pipelines used to transport petroleum, petroleum products, natural gas, and anhydrous amonia may be available through the Office of Pipeline Safety at DOT regional offices (and in some states, the state geological survey or state Pipeline Safety Office).
- Highway routes for hazardous materials transportation are available through the Bureau of Motor Carrier Safety within DOT. This office should have information on hazardous materials shippers and carriers around the country and in your area, including information on violators of DOT regulations (ranked by severity) and inspection results.

RESPONDING TO TRANSPORTATION SPILLS AND OTHER RELEASES

The main threat to groundwater posed by transportation is the spill or other release of hazardous materials in areas susceptible to contamination, like aquifer recharge areas. You should make sure that any spills of suspected hazardous materials, including oil products, that you observe or discover are reported to the National Response Center and that necessary cleanup is completed. (See Chapter 13 for details about how to report spills and initiate cleanup.) Local emergency response plans, discussed in detail in Chapter 17, should also address transportation-related spills.[2] Make sure local response teams are notified of spills and other transportation incidents—they may not be alerted by the transporter.[3]

[2] Review the suggestions in Chapter 17 for participating in the development and/or amendment of local emergency response plans, and use those approaches to make sure the local plan in your area includes information about hazardous materials transportation routes and plans for responding to accidents along those routes.

[3] CHEMTREC, the Chemical Transportation Emergency Center run by the Chemical Manufacturers Association, can provide technical information about transported chemicals and suggest methods for responding to spills. In an emergency, call CHEMTREC at (800) 424-9300.

PREVENTING TRANSPORTATION-RELATED CONTAMINATION

You can act to help prevent groundwater contamination from transportation of hazardous materials by watching for potential violations of minimum federal and state safety requirements. On the federal level, trucks, railroads, pipelines, and other carriers of hazardous materials must meet safety standards set under the Hazardous Materials Transportation Act and other federal laws. The federal Department of Transportation is the main agency responsible for enforcing these standards.

Although these minimum federal safety requirements are important, the current federal law does not resolve the hazardous materials transportation problem. In addition, enforcement by DOT cannot always be counted on. Citizens therefore should become involved at the state and local level (see Chapter 34) while simultaneously working for federal remedies to the hazardous materials transportation problem.

Among the most important of the federal minimum safety requirements are:

FOR RAILROADS AND TRUCKS

- Specific restrictions on the types of hazardous materials that can be transported and types of containers that can be used for various kinds of materials. (See 49 C.F.R. Part 177, 178.)
- Limitations on the kinds of tanks that can be used with tank cars for various materials and containers. (See 49 C.F.R. Part 179.)
- Design and loading requirements to prevent rupture or shifting of the cargo. (See 49 C.F.R. § § 177, Subpart B; 174, Subpart C.)
- Hazardous materials labeling and placarding requirements. Examples of placards are the diamond-shaped signs seen on the outside of trucks. (See 49 C.F.R. § § 172, 177.823.)
- Special restrictions for highly radioactive material. (See 49 C.F.R. § 177.842.)

FOR PIPELINES

- Restrictions on pipe materials, design, construction methods, and components for hazardous liquid pipelines. (See 49 C.F.R. Part 195, Subparts C and D.)
- Requirements for pressure tests, corrosion control, and periodic inspection. (See 49 C.F.R. Part 195, Subparts E and F.)

It will probably be difficult for you to determine whether many of these important safety regulations are being followed just by observing the carrier itself. Once you learn about the locations of facilities that handle hazardous materials and the transportation routes that service those facilities, you can, however, look for signals of potential problems, such as:

- Rail cars, cargo tanks or trucks that are corroded or have external signs of leakage or spills
- Signs of leakage along known or suspected transportation routes of hazardous materials carriers
- Trucks carrying radioactive materials along routes not designated for transport of such materials (placards should warn of such cargo)
- Rail cars or trucks with visible safety features, such as safety valves or locks on cars or doors, missing or in poor condition

Obvious problems like these should be enough to support a request to the regional office of the DOT to conduct an inspection of the carrier to determine whether safety and spill prevention requirements are being met. Direct the request to the office with jurisdiction over the carrier—either the Federal Highway Administration, Federal Railroad Administration, or Office of Pipeline Safety.

In addition to working on the federal level, your involvement at the state and local level is important, especially where federal law and enforcement efforts are inadequate. For example, in most states a state agency will have parallel or delegated carrier inspection authority. You should also make an inspection request to the state agency. See Chapter 34 for a more detailed discussion of opportunities for involvement at the state and local level.

Particularly for information about pipeline problems, transportation through sensitive areas, or heavily used routes, you may also want to review agency inspection reports and accident report records. Reports of DOT's carrier inspections and roadside checks should be available for review by the public at the regulatory agency (the state transportation agency or DOT). If such records are available, they can tell you whether certain carriers are frequent violators. Accident reports can also disclose carriers with a higher than average accident rate. Follow up on reported violations by asking the agency for information about enforcement actions and the status of compliance by the carrier.

If inspections disclose violations of safety requirements, or if your review of agency records identifies uncorrected violations, press the DOT to initiate enforcement action. Contact the Office of Hazardous Materials Transportation (for truck and rail problems) or the Office of Pipeline Safety at regional DOT offices with information about the suspected violation. Urge DOT to use its authority to issue a compliance order to the carrier. DOT may first issue a warning letter asking for voluntary compliance, or it may issue to the carrier a Notice of Probable Violation, and/or a proposed compliance order, or an assessment of penalties. If the carrier requests a hearing before the agency to review a proposed compliance order, you can request to participate to make sure your interests are

Watching for Transportation-Related Contamination

The Alabamians for a Clean Environment keep a watchful eye on trucks leading to the world's largest hazardous waste landfill in Sumter County, Alabama. They photographed numerous spills washing into roadside ditches and in one case discovered a trucker hosing down a truck placarded "Danger—PCBs" at a car wash adjacent to a large commercial fish pond. Several days later, most of the fish in the lake were dead. Members are using this information to try to force cleanup action and to raise awareness about the hazards associated with landfills and incinerators.

---

protected (you may need the assistance of a lawyer for hearings involving truck or rail problems because the hearing may involve witnesses, cross-examination, and other trial-type procedures; pipeline hearings are less formal).

Where you can identify potentially serious problems before they occur, you can also ask DOT to take action to prevent an imminent hazard. DOT can seek a court order to restrict or suspend the transport of hazardous materials by truck or rail. If a pipeline facility poses a hazard to life or property, you can ask DOT to issue an emergency corrective-action order, which could call for restricted or suspended use of the pipeline.

Finally, you can check to make sure transportators of hazardous waste comply with manifest requirements. Under RCRA, transporters of hazardous waste must have an EPA identification number, and for each shipment of waste, must obtain a manifest from the generator or shipper of the waste, which describes what the waste is (albeit, usually by general category only) and where it is to be transported for treatment or disposal. The manifest must accompany the shipment until it reaches its destination. You can ask the EPA (or EPA-approved state agencies) to review manifest records for specific transporters (the agency should have copies of manifests). Enforcing these requirements will help ensure that wastes are properly disposed of, rather than dumped along the roadside, and may enhance the speed of emergency response by informing emergency personnel about the identity of spilled wastes. Obtaining full compliance with manifest requirements will also assist you in developing information about the wastes going through your area.

REFERENCES

TRANSPORTATION OF HAZARDOUS MATERIALS

Cole, Leslie. "Hazardous Materials Emergency Response Training: The Colorado Training Institute." *Innovations*, September 1984. To order Innovations Reports, contact the Council of State Governments, P.O. Box 11910, Iron Works Pike, Lexington, Ky. 40578. (606) 252-2291.

ICF, Incorporated. *Lessons Learned: A Report on the Lessons Learned from State and Local Experiences in Accident Prevention and Response Planning for Hazardous Materials Transportation*. Prepared for the U.S. Department of Transportation. Washington, D.C. December 1985.

Maio, Domenic J., and Tai-Kuo Liu. *Truck Transportation of Hazardous Materials: A National Overview*. U.S. Department of Transportation, Office of Hazardous Materials Transportation. Washington, D.C. 1987.

U.S. Congress, Office of Technology Assessment. *Transportation of Hazardous Materials*. Washington, D.C. July 1986.

U.S. Congress, Office of Technology Assessment. *Transportation of Hazardous Materials—Summary*. Washington, D.C. July 1986.

U.S. Department of Transportation. *Community Teamwork: Working Together to Promote Hazardous Materials Transportation Safety: A Guide to Local Officials*. Washington, D.C. May 1983.

U.S. Department of Transportation. *1987 Emergency Response Guidebook for Initial Response to Hazardous Materials Incidents*. Washington, D.C. 1987. (Check for annual updates.)

U.S. Department of Transportation. *A Guide to the Federal Hazardous Materials Transportation Regulatory Program*. Washington, D.C. January 1983.

U.S. Department of Transportation. *Hazardous Materials Transportation, DOT Penalty Actions Resulting from Violations of Hazardous Materials Regulations*. Washington, D.C. July 1986.

All of the above DOT publications can be ordered from DOT Headquarters, Research and Special Programs Administration, Office of Hazardous Materials Transportation, 400 7th Street, S.W., Washington, D.C. 20590.

LAWS AND REGULATIONS

U.S. Code Citations

Hazardous Materials Transportation Act, 49 U.S.C.A. §§ 1801–1813 (West 1976 & Supp. 1988).

National Transportation Safety Board, 49 U.S.C.A. §§ 1901–1907 (West 1976 & Supp. 1988).

Natural Gas Pipeline Safety Act, 49 U.S.C.A. §§ 1671–1686 (West 1976 & Supp. 1988).

Hazardous Liquid Pipeline Safety Act, 49 U.S.C.A. §§ 2001–2014 (West 1976 & Supp. 1988).

Code of Federal Regulations Citations

49 C.F.R. Part 107 (Hazardous Materials Program Procedures).

49 C.F.R. Part 171 (General Information, Regulations, and Definitions for Hazardous Materials Regulation).

49 C.F.R. Part 172 (Hazardous Materials Communications Regulations and Hazardous Materials Tables).

49 C.F.R. Part 173 (General Requirements for Shipments and Packaging).

49 C.F.R. Part 174 (Rail Shipment Regulations).

49 C.F.R. Part 177 (Public Highway Hazardous Materials Shipment Regulations).

49 C.F.R. Part 178 (Shipping Container Specification Regulations).

49 C.F.R. Part 190 (Pipeline Safety Program Procedures).

49 C.F.R. Part 195 (Transportation of Hazardous Liquids by Pipeline Regulations).

49 C.F.R. Part 212 (State Safety Participation Regulations under Federal Railroad Laws).

PART IV

STATE
AND
LOCAL PROGRAMS
FOR
GROUNDWATER
PROTECTION

This part of the book is designed to serve a different purpose than the preceding chapters describing federal programs. The volume and variety of state and local laws relating to groundwater protection precludes us from providing detailed step-by-step advice for using specific programs, as we did in Part III. This part will instead highlight some of the programs being developed and implemented by state and local governments. It is not comprehensive, but does discuss some of the common and some of the unusual approaches now in use.

You should review the material in these chapters with an eye toward finding out whether similar programs exist in your state. Many of the programs discussed in this part are enforceable by citizens in the same way that some of the federal programs are. For example, you can always press administrative agencies to take action. In some cases, direct citizen enforcement in the courts may be possible. If you discover that your state or local governments have not enacted programs like those discussed in the following chapters, you should consider organizing to lobby for their adoption.

This part is organized to parallel the federal law chapters in Part III. General programs that could affect groundwater quality, but which are not directed toward particular contamination sources, are discussed at the outset in Chapters 23–26. Following is a series of chapters directed at particular contamination sources.

CHAPTER 23

Comprehensive State Groundwater Protection Programs

Contamination of groundwater has not been adequately addressed by the federal government due to a lack of comprehensive federal groundwater protection laws, and due to lack of funds, staff, and information. As the scope of the problem becomes more apparent, some states are developing strategies of their own to protect groundwater. This chapter discusses state groundwater protection programs and how citizens can be involved in their planning and implementation.

GENERAL STATE GROUNDWATER POLICIES AND STRATEGY

An important area for citizen involvement is in developing and implementing state groundwater policies and strategy. Every state has received a groundwater strategy grant from EPA for groundwater program development. As part of their groundwater strategies and protection programs, some states are also preparing aquifer maps and you should be involved in that process. (See Chapter 5.) States have used the groundwater strategy grants to:

- Develop state groundwater protection strategies
- Coordinate the responsibilities of various state agencies involved in groundwater protection
- Create a regulatory and legislative framework to accommodate state groundwater programs
- Develop technical activities designed to map and characterize groundwater resources, develop groundwater classification systems, and implement water quality standards

• Develop data bases to compile and distribute groundwater information
within the state

It is important for citizens to become involved in establishing the state
groundwater policy. Although this policy may not impose legal requirements, it sets goals for groundwater protection. In any state regulatory
decision where there are no specific laws or regulations regarding
groundwater that apply, you should use the groundwater policy to argue
for stringent controls to protect groundwater quality.

Goals for groundwater protection vary among states, ranging from non-
degradation, which protects groundwater quality at existing levels; to
limited degradation, which allows some contamination of groundwater
already deemed unpotable; to differential protection, which protects
aquifers or groundwater regions at different levels based on current or
anticipated uses. Montana, for example, has a formal nondegradation
policy that requires ". . . that any state waters whose existing quality is
higher than the established water quality standards be maintained at that
high quality unless it has been affirmatively demonstrated to the board
that a change is justifiable as a result of necessary economic or social
development and will not preclude present and anticipated use of these
waters." If your state has such a policy, use it to advocate the strongest
possible protection of groundwater.

If your state is just developing and implementing a groundwater protection strategy, take advantage of the opportunities for interested citizens to
participate. To assist in developing a strategy, many states have appointed
groundwater coordinating committees or advisory panels that include
representatives from the government, engineers, citizen activists, and
water resources experts. Find out when these committees meet and how
you can be appointed. The Ohio Environmental Protection Agency developed its state groundwater strategy with the assistance of a Groundwater
Task Force, which has become a continuing Interagency Groundwater
Advisory Council that provides advice to the state on all aspects of
groundwater protection and management. In Utah, ten public hearings
were held throughout the state so citizens could participate in developing
a groundwater strategy and the state sent draft strategy plans to the
public, affected industries, and government agencies to solicit suggestions from different perspectives.

Often, as part of the groundwater strategy, a memorandum of understanding between agencies within a state is entered to define each
agency's responsibilities in implementing and enforcing the groundwater
program. Request that this process be open to the public to ensure that
inspection or enforcement duties do not fall between the cracks in the
developing program. Obtain copies of these memoranda.

GROUNDWATER CLASSIFICATION SYSTEMS

Types of Classification Systems

An important element in many state groundwater protection strategies is the groundwater classification system. This is also an important area for citizen participation. A groundwater classification system accords differing levels of protection for geographic areas, aquifers, or portions of aquifers. The classification of an aquifer can result in restrictions on the siting of an industrial facility or ban it entirely; can regulate certain types of development; and can dictate the level of cleanup should the groundwater become contaminated.

Groundwater classifications are based on present and future use, existing quality, vulnerability to contamination, yield and availability of water, and economic and social considerations. The classes are defined by numerical standards for water quality, or by nonnumerical descriptions related to use, or by a combination of both. Contact the state environmental agency to find out how your state's groundwater classification system functions.

EPA formulated a groundwater classification system which divides groundwater into three classes:

I. Special groundwaters: Irreplaceable sources of drinking water and/or ecologically vital groundwater, which is highly vulnerable to contamination
II. Current and potential sources of drinking water and waters having other beneficial uses
III. Groundwaters not considered potential sources of drinking water and of limited beneficial use

Some states have adopted this system and others have adopted only certain aspects of it. It is important to work to prevent your state from adopting a system that, like EPA's system, designates some groundwater as unpotable (EPA's Class III designation). For Class III waters EPA would allow further degradation, without accounting for future needs and the possibility of improved treatment technologies. Work to persuade your state to consider options that do not classify groundwater as unpotable or not a potential source of drinking water.

Community Involvement in Groundwater Classification

You can help initiate a classification system or assist in classifying groundwater not yet covered by an existing system. Contact the state environmental agency to determine the process by which groundwater is classified. If your state holds public hearings as part of the classification process, attend the hearing and make comments (and submit a written copy), indicating the extent to which your community relies, and will in the future rely, on groundwater for drinking and domestic uses; the current and future availability of potable water in your community; and the vulnerability of the aquifer to point and nonpoint sources of contamination.

In some states, such as Montana and Florida, classification for a particular aquifer is not assigned until an operating or discharge permit is requested. In Montana, a permit applicant conducts an analysis of the existing groundwater quality and hydrology of the site. After the agency reviews this analysis, it classifies the aquifer and then reviews the permit application to determine if the facility will have a permissible impact on groundwater. Due to the large number of applicants, many states rely on the applicant's good faith in providing accurate information. If you believe the groundwater analysis is inaccurate or incomplete, notify the agency and request that they conduct their own analysis.

Some states revise their groundwater classifications periodically. Connecticut revises both its classification system and the accompanying maps once every three years. The state Department of Environmental Protection notifies chief elected officials and any interested parties when scheduled map revisions will take place.

Some states, such as Massachusetts, allow petitions for reclassification at any time. Typically, only those interested in degrading groundwater quality petition for reclassification and they have the burden of demonstrating why an aquifer should be degraded. Ask the agency to notify you when a petition to reclassify is received so you can participate. Use the reclassification process to upgrade a classification if circumstances have changed, for example, if the population using the aquifer has increased, if alternative sources have become contaminated, or if filtration techniques have been improved to purify degraded waters to acceptable drinking-water standards.

GROUNDWATER QUALITY STANDARDS

How Groundwater Quality Standards Work

Another element of a state groundwater protection strategy is the use of groundwater quality standards. These may apply in addition to drinking water standards, discussed in Chapter 11, that apply to groundwater that is a drinking-water source. Groundwater quality standards specify maximum levels of particular contaminants or restrict the design and operation of facilities that have the potential to contaminate groundwater. They are used to measure facility performance, trigger enforcement actions, and establish cleanup goals. They can be either numerical, indicating the maximum concentration of a contaminant allowed, or narrative, describing the appropriate level of protection without using numbers. To find out what type of standards your state uses and which contaminants are regulated, contact the state environmental office and request a list of the state's groundwater quality standards.

Standards can be used to protect groundwater uniformly throughout a state or diversely according to location. New Jersey applies standards diversely, with special standards for the delicate ecosystem of the Central Pine Barrens region. Wisconsin applies its two-tiered standards uniformly throughout the state. The first tier is the stringent "preventive action limit" that represents a low concentration of pollutants and is used as a warning signal and in facility design. The second, less stringent "enforcement standard" represents a higher concentration of pollutants. When the enforcement standard is exceeded, a facility is subject to immediate corrective action to bring it into compliance.

Many states use numerical groundwater quality standards in the discharge permits required by facilities that discharge contaminants from point sources. These standards limit the concentration of contaminants allowed to percolate into groundwater from a source of contamination. In many programs, monitoring wells are placed at the boundaries of the facility to detect contaminants in groundwater.

Narrative standards describe the level of protection afforded groundwater from generic types of activities but set no numerical levels of quality. For example, a narrative standard might state "discharges of any pollutants and disposal of any wastes to groundwater shall not cause a public health hazard." Find out if any permits have been denied based on the inability to meet narrative standards, as denial would indicate what types of permit challenges are likely to be successful and which aspects of the standards are strictly interpreted.

Citizen Participation in Standard Setting

Ask the state groundwater agency to place your name on a mailing list for notification of when standards will be set or altered. States may enact new standards for additional contaminants or alter existing standards based on recent health studies, the discovery of the contaminant in groundwater, or other factors. Almost all states have a public comment period or hold a public hearing to allow for input from interested parties. Participation at this juncture is your best opportunity to influence these standards, as once established, they will be extremely difficult to challenge. Setting groundwater quality standards involves scientific analysis of health effects and hydrogeology; unless you have expertise in these areas, your objective should be to locate information and experts that support a stringent standard.

Before commenting on proposed standards, determine if any other state, the EPA, or any scientific organizations have conducted studies on the health effects of the contaminant. If other states have established more stringent standards for this contaminant, find out why and obtain copies of the studies used as evidence. Contact the individuals who conducted the studies and if they believe the proposed standard is insufficient to protect public health, ask them to submit written or oral testimony to that effect for the public hearing.

You can test groundwater in your community (see Chapter 4), access the state's monitoring records for suspect facilities, and match monitoring records with allowable concentrations to determine if groundwater standards have been violated. If you discover evidence of a violation, petition the agency to take enforcement action. If the agency does not take action, you may have to consider suing the polluter or the agency for enforcement of the standard, if the state law gives you that right.

PROTECTING GROUNDWATER THROUGH CONTROL OF NONPOINT POLLUTION SOURCES

Another element of a state groundwater strategy is a program to control nonpoint sources of pollution. Under the Clean Water Act (CWA), states are required to develop programs to control nonpoint sources of pollution, such as runoff from farms, mining sites, construction, and urban areas. Since these are major causes of groundwater pollution, encourage your state to apply for grants from EPA for programs to control this contamination. It is particularly important for citizens to encourage states to develop these programs, because if a state does not submit its own program, as is required by the CWA, EPA is not required to implement a program for that state.

REFERENCES

COMPREHENSIVE STATE GROUNDWATER POLLUTION PROGRAMS

EPA, Office of Groundwater Protection. *Groundwater Protection Strategy.* Washington, D.C. 1984.

EPA, Office of Groundwater Protection. *Overview of State Ground-Water Program Summaries.* Vol. 1. Washington, D.C. 1985.

EPA, Office of Groundwater Protection. *State and Territorial Use of Ground-Water Strategy Grant Funds (Section 106, Clean Water Act)*. Washington, D.C. 1987.

EPA, Office of Groundwater Protection. *State Groundwater Program Summaries.* Vols. 1 and 2. Washington, D.C. 1985.

Henderson, T., J. Trauberman, and T. Gallagher. *Groundwater: Strategies for State Action* Environmental Law Institute. Washington, D.C. 1984, pages 88–95.

National Academy Press. *Ground Water Quality Protection: State and Local Strategies,* page 114. Washington, D.C. 1986.

LAWS AND REGULATIONS

U.S. Code Citations

Clean Water Act § 106, 33 U.S.C.A. § 1256 (West 1986).

Water Quality Act of 1987 § 319, 33 U.S.C.A. § 1329 (West Supp. 1988).

State Statutes

Mass. Regs. Code tit. 314, § 6.04 (1986).

Mont. Code Ann. § 75–5–103 (9), -303(1) (1987).

State Regulations

Mont. Admin. R. § 16.20.1013(4)(g) (1982).

CHAPTER 24

State and Local Land Use Controls

Effective land use controls can substantially lessen a number of threats to groundwater and recharge areas. By controlling the location of industrial facilities, underground injection wells, pesticide use, and the like, local governments can promote groundwater protection. State and local provisions vary, but the three principal areas of control are local land use regulation, state development laws, and sensitive area protection laws. Wellhead protection plans also serve as an important land use control, and are discussed in Chapter 12.

LOCAL LAND USE CONTROLS

Local governments may incorporate protections for underground drinking-water supplies into their land use controls, either pursuant to state direction or on their own initiative. The two basic elements of community planning and local land use regulation are the preparation of a comprehensive planning document (usually known as a "general plan" or "master plan") and regulation through zoning ordinances, subdivision regulations, site plan regulations, and building codes. Provisions can range from those that specifically mention groundwater criteria and restrict development that may harm groundwater resources to looser controls, which make no mention of groundwater or environmental protection but include language requiring decisionmakers to further the general public well-being. In between are zoning and planning requirements that include environmental protection language but do not specifically mention groundwater resources. Although controls that make specific mention of groundwater protection are the most desirable, citizens can also use the more general requirements to ensure that groundwater resources are considered.

322

Comprehensive Planning

The master plan is a general planning document that sets out an overview of a town or county's philosophy concerning development, growth, and preservation, as well as official maps identifying areas that are appropriate for different types of uses. In some communities, the comprehensive plan is only advisory; in others, zoning and permit decisions must adhere strictly to the master plan. Whether advisory or mandatory, the planning document is nevertheless important, because courts will rely on it, although to varying degrees, in reviewing land use disputes.

Ideally, local land use plans should identify sensitive areas such as groundwater recharge areas and prohibit, or at least restrict, activities harmful to groundwater in those areas. Make sure your local government considers aquifer recharge areas in its master plan and incorporates groundwater protection into subsequent regulation of development, regardless of whether such consideration is required by the state. Although such a master plan is not typical, it should be. If the general plan fails to consider aquifer recharge areas or the plan map places an industrial zone over a recharge area, work with planning officials and the public to amend the plan.

Zoning and Other Land Use Controls

Through the zoning process, local governments decide which areas of their community should be dedicated to particular uses. One of the principal goals of zoning is to minimize the negative impacts from incompatible activities. For example, zoning is an effective means of setting up buffers between industrial areas and residential areas. In this way, careful planning can reduce a variety of negative impacts on groundwater quality by controlling uses on the surface, even if groundwater is not specifically mentioned in the town's zoning ordinance.

You should be involved in your community's zoning process from the beginning. The first step is the enactment of a zoning ordinance, which allocates different areas to particular uses. Once ordinances are adopted, they can be difficult to overturn or alter, as courts generally defer to the judgment of local planning authorities. Early participation in the development of ordinances and regulations is thus important, and you should take advantage of the public hearings that are usually required.

Find out who decides zoning issues in your county or municipality. Contact your local government's public information office for the names of the agencies responsible for land use planning (various zoning and planning decisions may be handled by separate entities). Ask to be put on the mailing list so that you will be notified of proposed changes in your community's zoning ordinance.

Once a zoning ordinance is in place, uses must conform to the requirements for that particular area. Zoning records are available to the public and you can gather information about proposed developments by visiting the zoning office. Applicants who want to build a facility that may threaten groundwater might have to ask for a rezoning or a variance, which allows the developer to use the property in a manner not permitted by existing zoning regulations. Make sure you are on the zoning offices's mailing list so that you will be notified when zoning applications are filed. Find out the zoning designation requested by the applicant, and obtain a copy of the ordinance establishing current zoning designations. Your investigation also should include the following types of information:

- The name and address of the applicant
- The date of any public hearings
- The specific type of development proposed
- A copy of the area's zoning facility map, which will show zoning designations for nearby properties
- The procedures used by the zoning board to conduct the public hearing
- The standards used by the zoning board to make its decision in approving or denying the application
- Information on agency review procedures. For example, the local planning commission will often review applications and make recommendations to the zoning board. Find out who has what responsibilities. If the initial decision will be made by the planning commission, request an opportunity to address the commission
- The procedures to appeal an adverse decision by the zoning board

Next, you should contact the local planning commission to obtain a copy of the area's current master plan. Review the development proposal and the plan to see if they conform to each other.

You can also research and investigate the would-be developer of the property. Find out if the developer has similar projects elsewhere. See if there have been problems at those other facilities, looking particularly for past violations of environmental regulations and laws. Look at the company's financial status and the kind of insurance coverage it has. You can also investigate the principal people involved in the company to see what kind of environmental track record they have.

With this information in hand, you should then be prepared to go before the zoning board and make your case. If the board approves the project anyway, you may be able to appeal the decision to a reviewing body or take the board to court.

In addition to zoning restrictions, your city may use other land use mechanisms to protect sensitive aquifer areas. For example, mandatory clustering provisions require a certain percentage of subdivided property to be kept as open space. Because the permitted development is clustered around a centralized area, clustering helps to protect recharge areas by reducing the size and number of fertilized lawns and paved surfaces.

Even where a particular use is allowed in a given area, your community's regulations may require conditional use permits and site plan reviews for activities that pose special threats, including the risk of groundwater contamination. For example, Massachusetts allows certain uses and structures only upon issuance of a special permit, even if the area is otherwise zoned for the specified type of activity. Site plan reviews, such as those established by several towns on Cape Cod, require detailed information about the types and quantities of chemicals to be used on a site, design measures to prevent and contain chemical spills, and the availability of proper means of disposal of hazardous wastes generated on the site.

The permitting and review processes give citizens an opportunity to learn about and comment upon activities that may threaten water quality. These processes can also be used to suggest mitigation measures to be included as conditions in a permit. Ideally, both permit and site plan reviews should be public processes and should complement any health, building, and fire code permit requirements. Encourage officials from related state and local agencies to participate in the review process and to lend their expertise on particularly sensitive projects.

Active citizen involvement can be extremely effective in bringing groundwater concerns to the attention of community planners and in ensuring that protective measures are incorporated into local land use controls. When Connecticut adopted a law requiring local development plans to consider protection of underground drinking-water supplies, the Housatonic Valley Association's Groundwater Action Project (GAP) began to assist towns in developing sound groundwater protection strategies. GAP formed a group of local, state, and staff experts to help town selectmen and the general community understand the impacts of specific land uses, evaluate the quality of local groundwater, and focus on detailed groundwater protection planning. Each community then developed local action committees with representatives from the town commission and local concerned citizens. Similar efforts might prove useful in your community.

Although many local governments include drinking-water protections in their land use controls, state regulations sometimes preempt local controls. For example, preemption may occur if your state has a particularly detailed program to protect groundwater. Often preemption depends upon the type of activity being regulated. For example, most states in the Midwest do not preempt local zoning or health ordinances relating to solid waste disposal, but do preempt local regulation of hazardous waste facilities.

STATE DEVELOPMENT LEGISLATION

Some states have decided to play an even greater role in influencing local development decisions by passing site location and development laws. These laws control the location of large developments and force planning to avoid or mitigate the many potential environmental problems that large developments cause.

State development laws generally require developers to submit their plans to a permit review and public hearing process. In Hawaii, for example, county planning commissions have responsibility for issuing permits for unusual uses within agricultural and rural districts. Other states, such as Maine and Vermont, divide the state into districts and create district review boards.

Some state development laws specifically recognize the need to protect recharge areas. For example, Vermont prohibits the issuance of a development permit in areas supplying significant amounts of recharge waters to aquifers. In Maine, developers must provide detailed information on the hydrogeologic aspects of underlying aquifers if the Department of Natural Resources has determined that the proposed development is a hazardous activity that is likely to discharge pollutants into a significant groundwater aquifer and is located above a primary recharge area. Although not the exclusive factor, one of the issues that must then be considered in deciding whether to grant a permit is whether the development will have an adverse effect on the environment, including, of course, water quality.

If your state requires a permit application and recognizes the potential impacts of locating large developments over recharge areas, you can scrutinize the developer's profile of area hydrogeology. What are the qualifications of the firm that prepared the report? Did the firm ask for assistance from state experts or use available state information? Are there adequate provisions to protect natural drainage? If the state has a non-degradation groundwater policy, the developer may have to prove that the project will not cause any adverse impacts on the quality of area groundwater. If your state has a similar policy, carefully scrutinize the developer's evaluation of background groundwater quality, site geology, and hydrologic and soil conditions. You should also review the developer's plans for chemical and waste management and the proposed plan of action if the development is found to contaminate groundwater.

SENSITIVE AREAS PROTECTION

Recharge Areas Protection

Because recharge areas can underlie large areas of land already under extensive use, enactment of strict land use controls over these areas has been politically difficult. Nevertheless, some jurisdictions are beginning to recognize the need for special protection of aquifer recharge areas.

In addition to permitting provisions in state development laws, local governments can protect sensitive groundwater resources in other ways as well. Some states have added recharge areas to the list of restricted areas covered by source-specific controls, prohibitions on discharge, and stringent classification grading. For example, San Antonio, Texas, regulates the recharge area for the Edwards Aquifer by using three different zones, in which development is regulated according to the zone's sensitivity to potential contamination.

Your state may require local governments to consider the need for protecting these areas or may offer incentives—in the form of funding or technical aid—to local governments working on protective measures. Massachusetts, for example, provides a cash incentive for local governments to develop hydrogeologic information on their recharge areas, learn about the impacts of existing land use activities within these areas, and adopt innovative land use controls. Under this program, the state gives money to selected cities, towns, and districts to acquire property to protect sensitive aquifers and recharge areas.

In Washington, local governments may apply for state funding to establish and protect special "groundwater management areas." These areas may include sole source aquifers and primary source aquifers, as well as any geographic area where groundwater quality is threatened by or susceptible to contamination by land use activities or seawater intrusion or overdraft. Citizens who belong to an established association of holders of groundwater rights located within a proposed management area may request special protection.

Critical Areas Protection

States sometimes protect recharge areas under broad land management laws designed to protect ecologically sensitive areas. Often called "critical areas" laws, these statutes typically prohibit or control development within certain types of sensitive areas, particularly coastal and wetlands areas. Although these laws may not have been designed with groundwater protection specifically in mind, coastal and wetland areas are often important for groundwater recharge, so that protecting these areas results in protection for groundwater too.

To find out if an area is protected, contact the county clerk or registry of deeds office (where wetlands inventories may be filed). The state agency that protects wetlands in your state is also a good source of information.

If an area near you is protected by a coastal or wetlands protection law (or you think it ought to be), learn more about how your state's critical areas law operates. Review the act's criteria for critical area selection. If you think an area qualifies but has not yet been designated, you can begin the designation process by petitioning the state or local agency in charge to consider nominating the area for inclusion under the critical areas protections.

Your state's act may require the responsible state agency to periodically review the state's coastal and wetlands areas for possible critical areas designation. The act may also require regional development commissions or local governments to recommend areas that they believe the state should consider. Verify that the state environmental agency has performed its statutory duties.

Wetlands and coastal acts in Maryland, South Carolina, and New York, among other states, require or authorize the state environmental agency to compile an inventory of sensitive areas and to design an environmental management strategy for their protection. Some states, such as Louisiana, provide financial or technical assistance to help local governments establish a program for critical area protection. Alabama encourages counties and municipalities to assume coastal area responsibilities by developing and implementing programs to achieve wise use of land and water resources in coastal and wetlands areas. Check to see if these inventories are complete. Has the agency enacted programs within a reasonable amount of time? Has it taken steps to help or encourage local governments to implement protective plans?

Some states, such as Mississippi, may require permits and/or environmental impact statements prior to development within specially designated areas. Several states allow members of the public to sit on the state permit review board. Find out who is on the board and make sure that board members have adequate information on the potential impacts the proposed development might have on area groundwater. If you are interested, you may want to try to become a member of such a board.

REFERENCES

STATE AND LOCAL LAND USE CONTROLS

Babcock, Richard F. *The Zoning Game: Municipal Practices and Policies.* University of Wisconsin Press. Madison, Wis. 1966.

DiNovo, Frank, and Martin Jaffe. *Local Groundwater Protection: Midwest Region.* American Planning Association. Chicago, Ill. 1984.

Gordon, Noah J., ed. *1987 Zoning and Planning Law Handbook.* Clark Boardman Co., Ltd. New York, N.Y. 1987.

Healy, Robert G. *Land Use and the States.* The Johns Hopkins University Press. Baltimore, Md. 1976.

Housatonic Valley Association, Inc. *A Final Report on the Activities of the Groundwater Action Project.* Cornwall Bridge, Conn. February 1988.

Mandelker, Daniel R. *Land Use Law.* Mitchie Co. Charlottesville, Va. 1982.

Moss, Elaine, ed. *Land Use Controls in the United States: A Handbook on the Legal Rights of Citizens.* Dial Press. New York, N.Y. 1977.

National Research Council. *Ground Water Quality Protection—State and Local Strategies.* National Academy Press. Washington, D.C. 1986.

Sherry, Susan. *High Tech and Toxics: A Guide for Local Communities.* Golden Empire Health Planning Center. Sacramento, Calif. 1985.

Water Resources Action Project. *Tools for Community Water Supply Protection.* Natural Resources Forum. Charlestown, N.H. 1985.

Witten, Jon D. *Tools for Ground and Surface Water Protection: A Summarized Discussion for Municipal Officials.* Prepared for: Landlaw's 1988 Land Use Conference. Sandwich, Mass. 1988.

LAWS AND REGULATIONS

State Statutes

Ala. Code § 9–7–12 (1987).
Fla. Stat. Ann. §§ 380.045, 380.050 (West 1988).
Haw. Rev. Stat. § 205–6 (1985).
La. Rev. Stat. Ann. § 49:213.9 (West 1987).
Mass. Gen. Laws Ann. ch. 40A, § 9 (West 1979 & Supp. 1988).
Me. Rev. Stat. Ann. tit. 38, c.3, § 483 (1987).
Md. Nat. Res. Code Ann. §§ 9–301–302 (1983).
Miss. Code Ann. § 49–27–11 (Supp. 1987).
N.H. Rev. Stat. Ann. § 674:1–2 (1986).
N.Y. Envtl. Conserv. Law §§ 24–0301, 24–0903 (McKinney 1984).
S.C. Code Ann. §§ 48–39–40, 48–39–80 (Law. Co-op. 1986).
Vt. Stat. Ann. tit. 10, §§ 6026, 6083, 6086 (1984 & Supp. 1988).
Wyo. Stat. §§ 35–12–101 to -119 (Supp. 1987).

State Regulations

Maine Land Use Regulations (1975).
Wash. Admin. Code § 173–150 (1986).

CHAPTER 25

State Superfund Programs

Areas contaminated by hazardous materials present serious threats to groundwater. The federal Superfund program focuses cleanup efforts on those contaminated sites around the country that present the greatest danger. (See Chapter 13.) The remaining problem sites are left to the responsibility of the individual states, almost all of which have some type of hazardous substances cleanup program. Although there are a few significant differences, these state programs generally parallel the federal Superfund law. The citizen's role in state cleanups is also similar: instigating action and advocating thorough cleanups.

IDENTIFYING AND RANKING SITES

If you learn the process used under your state cleanup statute for investigating and setting priorities among sites that may require remedial action, you will be more effective in influencing the agency to conduct cleanups at sites near you. State agencies identify potential sites in need of cleanup through their own investigations, reports filed by facility operators disclosing a leak or other problem, citizen complaints, and other sources. In New York, for example, each county annually surveys its land to locate suspected inactive hazardous waste sites. The state Department of Environmental Conservation then investigates and determines the relative need for action at each site.

Louisiana requires a similar hazardous waste assessment by the secretary of the Department of Environmental Quality. The assessment includes the location of the site, type of waste, and estimated quantity, as well as possible methods for reducing, recycling, neutralizing, and disposing of the waste. The risk of each site and the estimated remedial cost are then measured against available funding. A five-year plan ranks sites that need immediate attention and can be cleaned up with present funding. Semiannual updates are required for certain developments, including technological and regulatory changes.

Missouri has a typical ranking system to determine which identified sites need priority cleanup:

Ranking	Classification	Type of Action Required
1	Presenting an imminent danger of irreversible or irreparable harm to the public health and welfare.	Immediate action required.
2	Significant threat to the environment.	Action required.
3	Does not present a significant threat to the public health or environment.	Action may be deferred.
4	Site properly closed.	Requires continued management.
5	Site properly closed, no evidence of present or potential adverse impact.	No further action required.

State programs involve citizens in this site identification and ranking process in several ways. Massachusetts, for example, requires its Department of Environmental Quality Engineering to release annually a list of suspected contaminated sites. Once a site is listed, state officials have two years to complete an investigation (if warranted) and to publish in local newspapers a notice summarizing the results of the investigation.

After you have learned how your state evaluates sites for cleanup, your investigations of the specific site that concerns you should focus on the criteria the state uses to determine whether to conduct a cleanup action. Many of the factors relevant to Superfund cleanups will also be considered by state agencies: types of substances and their toxicity, potential for migration in soil and water, proximity of residents and water supplies, and the like. Be sure of your facts and obtain the best possible evidence, then work with agency officials to convince them of the importance of action at sites threatening groundwater.

PARTICIPATING IN CLEANUP DECISIONS

Once the agency decides a site needs cleanup, its action may depend on how it perceives the severity of the threat. Ask the agency about the requirements for citizen participation in cleanup decisions. Influencing the agency to do the right kind of cleanup is as important as initiating action. In Massachusetts, ten or more residents may petition for a public meeting, and the agency must then present a plan for public participation in decisions regarding actions at the site. Affected communities may apply for technical assistance grants.

Like Superfund, some states authorize two basic levels of response: (1) emergency action at sites presenting imminent threats to health or the environment and (2) long-term remedial action. Arkansas, for example, maintains two separate funds: one for responding to emergency threats, the other for long-term remedies. Refer to Chapter 13 for more about types of cleanup approaches and the issues to consider in developing your position on the best cleanup method.

GOING AFTER RESPONSIBLE PARTIES

State cleanup money is usually used only when the state cannot find a responsible party who is solvent enough to pay for cleanup. The state Superfund agency usually searches for potentially responsible parties but, as in many other agency tasks, citizens can provide valuable assistance. Because state funds are limited, finding a responsible party may make cleanup more likely.

You must learn how your state determines which parties are potentially responsible. The pool of potential responsible parties is normally extensive. Massachusetts looks to the owners and operators of the site, the owners and operators of the trucks or other vessels used to transport the wastes, the persons who arranged for the transportation, the transporter himself, and anyone who caused or assisted in causing any release or threat of imminent release.

To find the potentially responsible parties, you might, for example, interview citizens living near the site, past or present employees of the facility, and employees of waste haulers who may have used the site. You also might review the files of the state or local agencies that issue business licenses or that otherwise regulate businesses.

Some states provide incentives for responsible parties to voluntarily become active in the cleanup process. North Carolina encourages voluntary cooperation from responsible parties by establishing a maximum financial responsibility. No responsible party who voluntarily participates in the implementation of a remedial action program may be required to pay more than $3 million.

SPECIAL LIABILITY RULES

State cleanup statutes sometimes impose broader liability than the federal Superfund in ways that directly benefit citizens. California's Hazardous Substance Account Act, for example, is intended to compensate individuals, under certain circumstances, for costs of out-of-pocket medical expenses, lost wages, or loss of business income resulting from injuries caused by a release of hazardous substances. New Jersey allows citizens to make claims against the state fund for property or health damage.

Minnesota has established a $2 million victim compensation fund for people injured by toxic contamination. The governor appoints a five-member board to determine whether an individual claiming personal or property injuries is eligible for a damage payment and how much that payment should be. The Minnesota compensation fund is an innovative solution, though skeptics are critical of the small size of the fund and the $250,000 limitation on individual payments. (Minnesota formerly had a more liberal citizen compensation system. In 1983 the state enacted a sweeping toxics liability law that made it easy for victims to sue for personal losses resulting from toxic contamination. In 1985, however, in response to pressure from industry and the insurance companies, the state repealed these provisions of the law and substituted the compensation fund.)

REFERENCES

STATE SUPERFUND PROGRAMS

EPA, Office of Groundwater Protection. *Survey of State Groundwater Quality Protection Legislation, 1985.* Washington, D.C. 1987.
Hazardous Materials Control Research Institute. *Management of Uncontrolled Hazardous Waste Sites.* Silver Spring, Md. 1980.
Kirkland, Janis L., and James A. Thornill. *Federal and State Remedies to Clean Up Hazardous Waste Sites.* 20 U. Rich. L. Rev. 379 (Winter 1986).

LAWS AND REGULATIONS

State Statutes

Ark. Stat. Ann §§ 8–7–401 to 8–7–519 (1987 & Supp. 1987).
Cal. Health & Safety Code §§ 25370–25382 (West 1984 & Supp. 1988).
La. Rev. Stat. Ann. § 30:1149.7 (West Supp. 1988).
Mass. Gen. Laws Ann. ch. 21E, §§ 3A, 5, 14 (West Supp. 1988).
Minn. Stat. Ann. §§ 115B.25–.37 (West 1987 & Supp. 1988).
Mo. Ann. Stat. § 260.445 (Vernon Supp. 1988).
N.J. Stat. Ann. §§ 13:1E–100 to –108 (West Supp. 1988).
N.Y. Envt'l. Conserv. Law § 27–1305 (McKinney 1984 & Supp. 1988).
N.C. Gen. Stat. § 130A–310.9 (Supp. 1987).

State Regulations

N.J. Admin. Code tit. 7, §§ 1I–1.1 to 6.3 (1988).

CHAPTER 26

State Environmental Policy and Protection Acts

State environmental policy acts (SEPAs) are another useful tool for citizens to use to protect groundwater. Like the National Environmental Policy Act (NEPA), discussed in Chapter 15, these state laws set up an environmental impact statement process and require thorough consideration of environmental effects, including effects on groundwater, before governmental officials proceed with actions that might harm the environment. Some states also have environmental *protection* acts, which can be used to protect groundwater. Unlike SEPAs, which require only analysis of impacts, state environmental protection acts actually limit the degree of environmental degradation that can be caused by projects.

STATE ENVIRONMENTAL POLICY ACTS

Whereas NEPA only addresses federal agency actions, SEPAs govern the actions of state and local officials, and occasionally of private parties. Generally, however, SEPAs operate in much the same way as NEPA: if a state or local agency is proposing to take an action that might have a significant effect on the environment, the agency cannot proceed until it first prepares and circulates an EIS, also known as an environmental impact report, or EIR, in some states. As with NEPA, the public has a right to comment on the EIS and the agency must respond to any comments received. Although in most states the agency may proceed with the proposed action once the required procedural steps have been followed, the action can still be challenged in state court if the agency has not followed all the necessary steps, has failed to prepare an EIS where one is required, or has not based its decision to go ahead with the project on a supportable record. In one particularly pertinent example, New York's SEPA was used to force that state's Department of Environmental Conservation to study the impact of a proposed sand and gravel mining operation on an underlying aquifer.

Although modeled after NEPA, many of these state laws provide even greater protection than their federal counterpart. In California and Minnesota, for example, the environmental policy acts apply not only to state and local agency action but to private development projects as well. Washington State's Environmental Policy Act not only establishes an environmental impact statement process but also provides a basis for denying project approval when an EIS shows that the impacts will violate formally designated policies for environmental protection. And South Dakota and New York have statutes that expressly require feasible mitigation measures to be utilized.

ENVIRONMENTAL PROTECTION ACTS

In addition, several states have environmental protection acts that allow citizens to challenge environmentally harmful actions regardless of whether an EIS is prepared or even required. Under the Michigan Environmental Protection Act, citizens can bring a court action to halt activities that threaten the destruction or impairment of natural resources. Other states, including Connecticut, New Jersey, Minnesota, Indiana, South Dakota, and Florida have similar statutes. Like common-law nuisance suits (see Chapter 8), these actions can be brought on the basis of potential harm to the environment, regardless of whether an EIS has been prepared or whether it is adequate. These statutory claims provide broader relief than nuisance actions, however, because citizens do not have to show that the action is unreasonable or that they have suffered property damage, only that there is a sufficient likelihood of environmental damage.

State environmental quality laws offer a valuable tool for groundwater protection by allowing citizens to participate in agency decisionmaking, by forcing governmental officials to consider the environmental consequences of their actions, and in some cases, by providing an opportunity for citizens to go to court to stop environmentally destructive actions. Ask your state environmental protection agency for a copy of your state's environmental protection act and any implementing regulations.

REFERENCES

STATE ENVIRONMENTAL POLICY AND PROTECTION

Anderson, Frederick, Daniel Mandelker, and A. Dan Tarlock. *Environmental Protection: Law and Policy.* Little, Brown & Co. Boston, Mass. 1984.
Renz, Jeffrey T. *The Coming of Age of State Environmental Policy Acts,* 5 Pub. Land L. Rev. 31 (1984).
Robinson, Nicholas A. *SEQRA's Siblings: Precedents from Little NEPA's in the Sister States,* 46 Alb. L. Rev. 1155 (1982).
Slone, Daniel K. *The Michigan Environmental Protection Act: Bringing Citizen Initiated Environmental Suits into the 1980s,* 12 Ecology L. Q. 271 (1985).

LAWS AND REGULATIONS

U.S. Code Citation

The National Environmental Policy Act of 1969, 42 U.S.C.A. §§ 4321–4335, 4341–4347 (1976).

State Statutes

California Environmental Quality Act, Cal. Pub. Res. Code §§ 21000–21177 (Deering 1987 & Supp. 1988).
Connecticut Environmental Policy Act of 1973, Conn. Gen. Stat. Ann. §§ 22a–1 to 27 (West 1985 & Supp. 1988).
Environmental Protection Act of 1971, Fla. Stat. Ann. § 403.412 (West 1986).
Hawaii Rev. Stat. §§ 343–1 to –8 (1985 & Supp. 1987).
Ind. Code Ann. §§ 13–1–10–1 to –8 (West 1981).
Maryland Environmental Policy Act of 1973, Md. Nat. Res. Code Ann. §§ 1–301 to 305 (1983 & Supp. 1986).
Massachusetts Environmental Policy Act, Mass. Gen. Laws Ann. ch. 30, §§ 61, 62–62H (West 1979).
Michigan Environmental Protection Act of 1970, Mich. Comp. Laws Ann. §§ 691.1201–1207 (West 1987).
Minnesota Environmental Policy Act of 1973, Minn. Stat. Ann. §§ 116D.01–07 (West 1987).
New York State Environmental Quality Review Act, N.Y. Envtl. Conserv. Law §§ 8–0101–0117 (McKinney 1984 & Supp. 1988).
North Carolina Environmental Policy Act of 1971, N.C. Gen. Stat. §§ 113A–1 to – 10 (1983 & Supp. 1985).
South Dakota Environmental Policy Act, S.D. Codified Laws Ann. §§ 34A–9–1 to –13 (1986).
Virginia Environmental Quality Act, Va. Code Ann §§ 10–177 to –186 (1985 & Supp. 1987).
Washington State Environmental Policy Act, Wash. Rev. Code Ann. §§ 43.21C.010 to .910 (1983 & Supp. 1988).
Wisconsin Environmental Policy Act of 1971, Wis. Stat. Ann. § 1.11 (West 1986 & Supp. 1987).

CHAPTER 27

State Regulation of Solid and Hazardous Waste Disposal

Inadequate or poorly sited waste disposal facilities create a serious threat to groundwater. The U.S. General Accounting Office believes that this country may have as many as 425,000 potential Superfund sites, the bulk of which are municipal and industrial landfills managed by the states. State and federal regulations distinguish between hazardous and solid waste disposal systems, with RCRA specifying the hazardous waste system and the states addressing the majority of solid waste management issues. The details of both solid and hazardous waste disposal vary from state to state.

SOLID WASTE DISPOSAL

State Program Approaches to Solid Waste

Pursuant to subtitle D of RCRA, states are the primary regulators of solid waste disposal sites, including municipal waste landfills, industrial waste surface impoundments, oil-sludge land application, and mining and agricultural waste impoundments. State management plans vary in approach and comprehensiveness. Your state's plan may use a number of substantive requirements to effectively control solid waste. Plans often require permits that outline restrictions on disposal, including surface or groundwater-monitoring requirements, liner requirements, and standards for leachate collection systems. Alternatives to disposal, such as resource recovery or encouragement for recycling, as in New Jersey, also might be a component of a state plan. Some states mandate financial assurances as part of any closure provision required in the solid waste permit.

You can help protect your groundwater by being alert to public hearings, community education programs, and other opportunities for citizen involvement during the permitting, enforcement, and other stages of a waste management plan. For example, under federal RCRA regulations, plans that are EPA-approved must provide for public participation in development and implementation of state and local plans, in regulatory development, and in permit issuance and renewal proceedings. Public participation also must be part of the drafting of the annual state work program, which verifies that the state is carrying out its plan correctly. Mississippi law provides that any person claiming an interest in the facility property or project, and situated such that they may be affected by the project, shall be granted a formal evidentiary hearing upon request to the permit board. Appeals of final decisions of the permit board may subsequently be taken to court.

Recycling to Reduce Solid Waste

Implementing and enforcing recycling requirements can reduce waste flow to landfills. Reduced waste lessens the strain on established landfills and lowers the demand for new facilities, decreasing the likelihood that groundwater will be exposed to solid waste contaminants.

Find out if your state has recycling requirements for disposal of solid waste. In Connecticut, no landfill or refuse-handling facility will be allowed to accept recyclable items after 1991. Recycling efforts also are initiated at the local level. Seattle, for example, has an extensive recycling program. Residents are given containers for paper, glass, aluminum, and tin; the containers are collected for recycling free of charge. Recycling of plastics also is being considered.

HAZARDOUS WASTE DISPOSAL

Siting a Hazardous Waste Disposal Facility

Although the federal government regulates the management of hazardous waste, your state law largely governs the location of hazardous waste disposal facilities. This siting process is an opportunity for you to help ensure that waste facilities do not threaten your groundwater with contamination. If you disagree with a final siting decision, appeals are generally available either before your state siting board or in court.

There are two basic approaches to siting decisions, each presenting opportunities for public involvement through a variety of local advisory committees, public hearings, representation on state facility siting boards, and participation in the site designation process. Some states use a preselection system, assessing the hazardous waste problem as a whole, and then selecting *potential* sites where facilities may be constructed as needed. Other states use an "ad hoc" system, selecting a site once a permit or site application is received. Sometimes both processes are used, especially when a particular location is under review from a preselected group of sites.

New Jersey has a preselection system, which presents several opportunities for public involvement. The New Jersey Hazardous Waste Facility Siting Commission works with an Advisory Council to develop a Major Hazardous Waste Facilities Plan and to propose and adopt site designations for new major hazardous waste facilities. The Commission is responsible for a public information program addressing the necessity and opportunities for public participation in the siting process. When the Commission proposes a site, the affected municipality is notified and given a grant to conduct a site suitability study. After the suitability study is completed, a hearing is held by an administrative law judge. Final Commission review of the judge's recommendations is conducted with the addition of temporary voting members appointed by the governing bodies of the affected county and municipality.

Ad hoc systems also provide opportunities for public involvement. In New York, a certificate of environmental safety and public necessity is required in order to construct or operate a new industrial hazardous waste facility. The application for this certificate is reviewed by a facility-siting board, which includes two members from the host community. Notice is given to the Chief Executive Officer of each affected municipality. The final decision of the Board is based on the record of a public adjudicatory hearing held after public notice.

Some states have increased public participation in their siting system by providing for negotiations between the community and the facility developer. In states where negotiated agreements are required prior to the construction of a facility, an impasse in negotiations may block or delay the facility pending resolution of disputed issues through arbitration. Negotiations focus on ways to make the facility more acceptable to the community. For example, Massachusetts requires that a Local Assessment Committee, made up of members of the host community, negotiate terms of a siting agreement with the builder. Such an agreement may address the community's health, safety, and environmental concerns, including special benefits and compensation for demonstrable adverse impacts caused by the facility.

Issues to Focus on in the Siting Process

- Is the facility genuinely needed? Is it in the best interest of the community and the state to solve their waste problems using this site for a facility? Wisconsin requires the developer to document the need for the facility during the state permitting process.
- Where will the waste come from? States cannot ban wastes from out of state, but certain controls on disposal of such wastes can be imposed at the state or local level. (See page 341.)

- What are the anticipated health effects?
- What opportunities will there be for access to operating and monitoring records?
- What is the background of the party or parties running the facility? Ask for references. Your state may require disclosure of criminal or civil charges related to hazardous or solid waste disposal involving the applicant.

State and Local Hazardous Waste Management Regulations

In addition to participating in the siting process, you can help protect groundwater by ensuring that state and local hazardous waste management regulations are enforced. States must be as stringent as the RCRA in implementing and enforcing its state hazardous waste management plans, and so most states follow the RCRA model for their own programs. States and localities can have stricter substantive regulations if they are not inconsistent with the federal requirements; check your state laws for these additional mandates. For example, states, unlike the federal system, may adopt a nondegradation policy, pursuant to which any increase in toxics in groundwater triggers enforcement action. Some states, like Illinois and Kansas, ban land disposal of certain wastes or certain disposal methods altogether, subject to narrow exceptions.

Disposal of Household Hazardous Waste

Although it is not illegal for municipal solid waste facilities to accept household hazardous waste, the accumulation of this waste from households in your state may create a serious threat to your groundwater, even if your municipal or state facility has leachate and monitoring controls.

Find out if your state or community has a household hazardous waste collection program. For example, Florida's Amnesty Days program designated certain days for citizens throughout the state to dispose of collected hazardous waste at no charge. The waste was then shipped to a federally approved hazardous waste facility. Always find out which facility will be accepting the wastes and make sure it does not have a history of leaks into groundwater or any current operating problems.

Local organizations or companies may have a hazardous waste collection program you can use. For example, the League of Women Voters promotes household hazardous waste collection around the country, often on a county by county basis. Some chemical companies, such as Dow Chemical, have also developed household hazardous waste collection programs. Check with the major companies in your area to find out if they have a program in place. If you find there are no programs in your area, you should initiate one at the state and local level.

LOCAL AND STATE ATTEMPTS TO RESTRICT WASTE DISPOSAL BY ORIGIN

Government entities, both state and local, may try to restrict or prohibit waste disposal from outside their boundaries by law. While some state restrictions on the disposal of waste from areas outside the relevant government jurisdiction are valid, states cannot prohibit the importation of waste material generated out-of-state without risking a violation of the United States Constitution as an improper restriction on interstate commerce.

Localities also may try to limit disposal of waste originating outside the local area. Local prohibitions against nonlocal wastes may be tailored to avoid the constitutional problems that state restraints encounter. Nevertheless, local restrictions, including city ordinances, zoning laws, local hazardous waste management laws, or laws governing waste transportation may be rendered unenforceable by state laws. For example, some states have enacted preemption provisions that prevent the localities from enforcing their local laws concerning siting hazardous wastes facilities. State preemptions vary in type, with Maryland generally preempting local laws entirely and Florida authorizing the governor and cabinet to issue variances to local land use laws instead of preempting local laws outright.

More Information on State Hazardous Waste Programs

- Information on state hazardous waste programs may be available from the community relations coordinator or public information office at your state regulatory office or from environmental groups who have worked on this issue within your state.
- Regional hazardous waste projects, formed through attorneys general's offices, share reports and general information among the states. For more information, contact National Association of Attorneys General, 444 North Capitol Street, Washington, D.C. 20013, (202) 628-0435.

REFERENCES

STATE REGULATION OF SOLID AND HAZARDOUS WASTE DISPOSAL

Association of State And Territorial Solid Waste Management Officials. *State Programs for Hazardous Waste Site Assessments and Remedial Actions.* Washington, D.C. June 1987.

Citizen's Clearinghouse for Hazardous Waste. *How to Deal with a Proposed Facility.* Arlington, Va. 1986.

DiNovo, F., and M. Jaffee. *Local Groundwater Protection: Midwest Region.* American Planning Association. Chicago, Ill. 1984.

Doyle, Paul. *Hazardous Waste Management: An Update.* National Conference of State Legislatures. Washington, D.C. 1987.

EPA, Office of Solid Waste. *Report to Congress: Solid Waste Disposal in the United States.* Washington, D.C. 1988. For a copy of the Executive Summary (EPA/530–SW–88–011A), call the RCRA Hotline (see Appendix D).

Holznagel, Bernard. *Negotiation and Mediation: The Newest Approach to Hazardous Waste Facility Siting.* 13 Envtl. Aff. 329 (1986).

Lord, John. *Hazardous Wastes from Homes.* Enterprise for Education. Santa Monica, Calif. 1986.

Lennett, David J., and Linda E. Greer. *State Regulation of Hazardous Waste.* 12 Ecology L. Q. 183 (1985).

Muir, Dr. Warren R., and Joanna Underwood. *Promoting Hazardous Waste Reduction: Six Steps States Can Take.* Inform, Inc. New York, N.Y. 1987.

Seldman, Neil. "The Rise and Fall of Recycling." *Environmental Action,* Vol. 18, page 12. January/February 1987.

U.S. General Accounting Office. *Superfund: Extent of Nation's Potential Hazardous Waste Problem Still Unknown.* GAO/RCED–88–44. Washington, D.C. December 1987.

For more information about collection programs and advice about how to establish such a plan in your area, order *A Survey of Household Hazardous Wastes and Related Collection Programs* PB–87–108–072/AS by calling National Technological Information Service at (800) 336-4700.

LAWS AND REGULATIONS

Code of Federal Regulations Citation

40 C.F.R. §§ 256.60–.63 (1987).

State Statutes

Connecticut Public Act No. 87–544, § 6 (not codified but see Conn. Gen. Stat. Ann., Appendix Pamphlet at 346, 348) (West 1988).

Fla. Stat. Ann. §§ 403.723, 403.7264 (West 1986 & Supp. 1988).

Ill. Ann. Stat. ch. 111 ½, para. 1021 (Smith-Hurd Supp. 1988).

Kan. Stat. Ann. § 65.3458 (Supp. 1987).

Md. Nat. Res. Code Ann. § 3–705(d) (Supp. 1988).

Mass. Ann. Laws ch. 21D, §§ 5, 12 (Law. Co-op. 1988).

Miss. Code Ann. §§ 17–17–43, –45, 49–17–28, –29 (Supp. 1988).

N.J. Stat. Ann. §§ 13:1E–52(c), –54, –55(a), –58(d)(4), –59(a), –92 et seq. (West Supp. 1988).

N.Y. Envtl. Conserv. Law §§ 27–1105(1), (3)(c)–(3)(g) (McKinney Supp. 1988).

Wisc. Stat. Ann. §§ 144.44(2)(f)(6), (2)(n)(4), (2)(nm), (2)(nr), (2)(om) (West Supp. 1988).

State Regulations

N.Y. Comp. Codes R. & Regs. tit. 6, §§ 361.1 et seq. (1988).

CHAPTER 28

State Regulation of Commercial and Industrial Facilities

A variety of state and local programs address groundwater pollution caused by commercial and industrial facilities. In addition to state plans required under the Emergency Planning and Community Right-to-Know Act (EPCRA) (see Chapter 17), state controls may include discharge permit programs, regulations governing storage and management of hazardous materials, community awareness and response programs, recycling requirements, and cleanup plans. Your state and local government also can use zoning and other land use controls to protect groundwater, as discussed in Chapter 24.

STATE POLLUTION DISCHARGE PERMIT PROGRAMS

States often use permit systems to control commercial and industrial discharge that could contaminate groundwater. These programs generally parallel the federal National Pollutant Discharge Elimination System (NPDES) permit program. (See Chapter 17.) Participation in the state program is one of the most direct ways to control industrial contamination of groundwater. Under the New Jersey Pollution Discharge Elimination System permit program, all facilities that discharge to groundwater are required to receive a permit. (This goes beyond the federal requirement that dischargers to navigable waters obtain an NPDES permit. See Chapter 17.) The New Jersey requirement applies to any discharge with the potential to violate state ambient groundwater standards, including landfills, injection wells, feedlots, and land treatment units. Other New Jersey laws require permits for the storage of any significant amount of hazardous liquids or pollutants. These permits impose standards designed to keep pollutants from entering surface waters or groundwater.

Permit conditions are often based on potential harm determined by the classification of the groundwater which the pollutants might contaminate. (Groundwater classification systems are described in Chapter 23.) Arizona requires any facility with a point source disposal that could affect groundwater to obtain a Department of Health Services permit setting discharge limits that conform to state groundwater quality classification standards. Missouri and Pennsylvania have similar restrictions on industrial wastes.

A state may also implement a discharge permit program in conjunction with land use controls designed to protect certain classifications of groundwater. Aquifers falling into certain classifications may be protected by land use controls that place additional restrictions on industrial discharges. In Connecticut, the siting of facilities that require an industrial waste discharge permit is not allowed in watershed areas that are protected pursuant to the state's classification system (watershed areas now used or potentially usable as drinking water).

HAZARDOUS MATERIALS CONTROLS

Another method to control contamination from commercial or industrial sources is to regulate the storage and management of hazardous or toxic materials. Under a model by-law drafted by the Cape Cod Planning and Economic Development Commission, the town of Barnstable, Massachusetts, requires all nonresidential users of toxic or hazardous materials, who store 50 gallons or 25 dry pounds of such materials, to register these substances with the Board of Health. Requirements are also established for the storage of materials and spill containment.

Genoa Township in Livingston County, Michigan, developed a policy plan for groundwater protection by holding public meetings, hiring mapping experts, and distributing a business questionnaire on toxic and hazardous materials. After learning the type, quantity, location, and disposal method of these types of substances in the community, the township put together a policy to advise commercial and industrial facilities how to protect groundwater. This program included advice on the safest disposal methods, what materials should not be discharged through septic systems, and what materials could easily be recycled or reused.

Control of hazardous substances can also come through statewide referenda. California voters overwhelmingly approved the precedent-setting Proposition 65, a law that prohibits the discharge of any carcinogen or reproductive toxin into state drinking water. The governor was required to issue a list of chemicals subject to this prohibition. Listing triggers strict labeling requirements for consumer products containing these substances and mandatory posting of warnings in workplaces where these materials are handled. Currently, 230 chemicals are subject to Proposition 65, with more being proposed by scientists and environmentalists. In addition, Proposition 65 requires state employees to make public any contamination of drinking water as defined by the law. Designated government workers are required to notify the relevant county officials of illegal discharges likely to cause injury to public health within 72 hours of the release. The local health officer must immediately alert the local news media and make the information public.

State hazardous waste regulations may apply to more or different wastes than federal law, increasing control of hazardous waste within your state. Regulations in some states, for example, apply to the small-quantity generators of hazardous waste (less than 100 kilograms per month) that are not necessarily regulated under federal law. Some other states treat PCBs as hazardous waste, unlike the federal law which regulates these substances less strictly under TSCA.

EMERGENCY PREPAREDNESS
AND RESPONSE PROGRAMS

Though all states are required to develop emergency response plans and community right-to-know materials pursuant to the Emergency Planning and Community Right-to-Know Act (EPCRA), these programs may not be operational immediately. (See Chapter 17 for a discussion of EPCRA.) In addition to programs developed by your state or EPA, the Chemical Manufacturers Association has devised the Community Awareness and Emergency Response Program (CAER), which encourages plant officials to make the public aware of hazards in a community. CAER programs facilitate the drafting of Material Safety Data Sheets on certain chemicals and codevelopment of response plans with local communities in the case of an accident. For more information on this program, contact CAER Coordinator, Chemical Manufacturers Association, 2501 M Street, N.W., Washington, D.C. 20037, (202) 887-1150.

REDUCING WASTE AT THE SOURCE

You can work with your state, local government, or area industries to prevent or reduce industrial waste through a variety of programs. Prevention of industrial waste is vital to the preservation of your groundwater and is more efficient than trying to clean up groundwater once it is contaminated.

At a basic level, industry can adopt better "housekeeping" practices—

such as reducing the amount of spillage that goes into drainage systems and then becomes hazardous waste. Industry also can find alternatives to the use of some chemical products. (See Figure 28.1.) For example, a 3M plant began using pumice rather than 40,000 pounds of toxic chemicals per year to clean metal circuits. A General Motors automobile plant reduced paint wastes by over 50 percent through a process modification that caused the paint to adhere better to metal. There are obvious benefits to industry from waste reduction (also known as "source reduction"), including lower cost for raw materials, lower cost for disposal, and better public relations.

Recycling raw waste materials within the manufacturing process (in-process recycling) can similarly benefit industry and help prevent unnecessary waste. By recycling toxic organic chemicals back into the production process, a North Carolina pesticide-manufacturing facility saved disposal and raw material costs (a great incentive to industry), reduced the need to transport and dispose of highly toxic materials, and thereby lessened threats to groundwater. EPCRA reporting requirements (see Chapter 17) can help you determine which facilities in your area are most "suspect" and would benefit most from waste reduction or recycling.

Many state or local programs encourage or require source reduction. Minnesota, for example, has a technical and research assistance program run by the Minnesota Waste Management Board, designed to assist generators of hazardous waste to identify and apply methods of reducing the generation of hazardous waste, to improve management of hazardous waste, and to comply with hazardous and industrial waste regulations. The Board may also make grants to generators of hazardous and industrial waste for studies to determine the feasibility of applying specific methods and technologies to reduce the generation of hazardous and industrial waste. Rhode Island law also provides for a program to research, develop, and demonstrate hazardous waste reduction, recycling, and treatment technologies and to provide grants to universities, government entities, and private organizations for research, development, and commercial demonstration of such technologies.

PROPERTY TRANSFER AND CLEANUP STATUTES

One effective way to ensure cleanup of industrial contamination that has already occurred can be found in New Jersey's Environmental Cleanup Responsibility Act (ECRA), which calls for cleanup of industrial or hazardous waste sites before they can be sold or transferred. The owner or operator must notify the New Jersey Department of Environmental Protection in writing of any conveyance or closure and submit an environ-

FIGURE 28.1
INDUSTRY WASTE REDUCTION: SOME SUCCESS STORIES

	3M Corporation Saint Paul, MN	Stanley Furniture Stanleytown, VA	The Cooper Tool Group Lufkin Division Apex, NC
Savings in dollars per year	$50,000,000.	$140,000.	$10,000.
Savings as % of net profit	5.44%	0.51%	0.25%
Payback time	2—3 years	5 weeks	5 weeks
Cost to install program/ process (one-time cost)	$250,000,000.	$15,000.	$1,000.
Amount spent as % of total expenses	0.27% (over past 12 years)	0.01%	0.00062%
Description of program/process	The company uses a life cycle approach to reducing the amount of hazardous waste generated. Over the last 12 years the program has involved over 2,200 individual projects—including product reformulation, process modification, equipment redesign, housekeeping, and recycling.	This modification of the company's tray booth spraying system allows them to mix the sludge with wood chips, put them in hazardous waste bags, and burn them. They use the fuel, or sell it.	The company modified their manufacturing process to recover nickel from their process wastewater.

Stories of successful waste reduction programs abound.

• A leading chemical company established a program in 1987 that reduced waste generated at the company's facilities by more than 100,000 tons, saving an estimated $250 million.

• A Texas chemical manufacturer of a nylon intermediate (adiponitrile) developed a new process that improves product yield, cuts the amount of wastewater generated in half, and saves $10 million a year.

• A company that makes furnaces recovered calcium fluoride from its sludge, saving more than $170,000 per year.

• A Pennsylvania die manufacturer used a solvent recovery unit to reclaim 1,1,1-trichloroethane, which was then used as a degreasing agent, producing a savings of nearly $5,000 per year.

The range of savings in these programs highlights the fact that waste reduction can work in businesses of all sizes and shapes.

Courtesy of the Water Pollution Control Federation.

mental evaluation of the site. If the DEP determines the site still contains contamination, the company must develop and carry out a state-approved cleanup plan. Violation of ECRA can mean stiff penalties or voiding of the sale or transfer.

If your state has such a law, ask to be notified of any proposed cleanup plan or site approval in your community. Your knowledge of the local residents and businesses may help to ensure that the owners or operators are identified and held liable. Compelling a solvent owner/operator to clean the site will help prevent the site from becoming a Superfund problem.

REFERENCES

STATE REGULATION OF COMMERCIAL AND INDUSTRIAL FACILITIES

Alternative Technology and Policy Development Section. *Alternative Technology for Recycling and Treatment of Hazardous Wastes.* Toxic Substances Control Division, Dept. of Health Services. Sacramento, Calif. 1986. Free copies available by calling (916) 324-1807.

Corbett, Judy, with Susan Sherry. *Waste/Source Reduction: A Key to Prevention: Part I: Setting Official Policy.* Local Government Commission. Sacramento, Calif. 1986. To order, contact: Local Government Commission, 909 12th Street, Sacramento, Calif. 95814. (916) 448-1198.

DiNovo, Frank, and Martin Jaffee. *Local Groundwater Protection: Midwest Region.* American Planning Association. Chicago, Ill. 1984.

Fogarty, David. *Great Lakes Toxic Hotspots: A Citizen's Action Guide.* Lake Michigan Federation. Chicago, Ill. 1985. To order, contact: Lake Michigan Federation, 59 East Van Buren Street, Suite 2215, Chicago, Ill. 60605. (312) 939-0838.

Lennett, David J., and Linda E. Greer. *State Regulation of Hazardous Waste,* 12 Ecology L.Q. 183 (1985).

Newberry, Rick. *Integrated Waste Management.* West Michigan Environmental Action Council. Grand Rapids, Mich. October 1986. To order, contact: WMEAC, 1432 Wealthy Drive, SE, Grand Rapids, Mich. 49506. (616) 451-3051.

Oldenburg, Kirsten, and Joel Hirschhorn. "Waste Reduction—A New Strategy to Avoid Pollution." *Environment.* March 1987.

For more information on how to prevent industry contamination through discharge controls, response plans, recycling, or legislation, contact INFORM, Inc., a research organization that gives practical advice on how to reduce hazardous wastes created by industry. Their book, *Cutting Chemical Wastes: What 29 Organic Chemical Plants Are Doing to Reduce Hazardous Wastes,* by Dave Sarokin, Warren Muir, Catherine Miller, and Sebastian Sperber (published by INFORM, Inc., New York, N.Y. 1986), describes the initiatives taken by industries in three states to reduce their waste stream. Their address is 381 Park Avenue South, New York, N.Y., 10016. (212) 689-4040.

LAWS AND REGULATIONS

U.S. Code Citation

42 U.S.C.A. § 6921(d)(4) (West Supp. 1988).

Code of Federal Regulations Citation

40 C.F.R. § 761.1(b).

State Statutes

9 Minn. Stat. Ann. §§ 115A.152, .154 (West Supp. 1988).
N.J. Stat. Ann. §§ 58:10A–21 *et seq.*, 13:1E–1 *et seq.*, 13:1K–6 *et seq.* (West 1979 & Supp. 1988).
Pa. Stat. Ann. tit. 35, § 691.307(a) (Purdon Supp. 1988).
R.I. Gen. Laws §§ 23–19.10–1 *et seq.* (Supp. 1988).

State Regulations

Ariz. Comp. Admin. R. & Regs. 9–20–201 to –226 (1987).
Conn. Water Quality Standards § IV, part 36 and § V; *see,* [1 State Water Laws] Env't Rep. (BNA) 731:1024–731:1030 (June 3, 1988).
Minn. R. 9200.9500 (1987).
Mo. Code Regs. tit. 10, §§ 20–6.010, –7.031 (1987).
N.J. Admin. Code tit. 7, §§ 14A–1.2(d), (e), 14A–6.15(d)(2)(i), 26B– (1988).

CHAPTER 29

State Regulation of Waste Disposal Wells

As discussed in Chapter 18, the federal Underground Injection Control (UIC) program provides opportunities for citizens to help protect groundwater from waste disposal wells. State laws often contain further requirements. In particular, state regulations may be more stringent on such issues as the siting of new facilities, waste characterization (testing the composition of injected wastes), monitoring, and inspection of wells, and can therefore be used as an additional avenue for citizen action.

SITING

Citizen participation in siting is critical to ensure that new waste disposal wells do not pose a threat to groundwater. Promoting safe design and proper location of disposal wells from the outset is an important preventative step to avoid future groundwater contamination from well failure.

Under the federal UIC program, hazardous and industrial waste injection wells (Class I wells) can go through an underground source of drinking water (USDW), but cannot inject into or above a USDW. Some states have additional requirements that are even more stringent. In California, for example, no point along the length of a hazardous waste injection well, as measured vertically or horizontally, can be located within half a mile of a potential source of drinking water.

Learn what requirements your state has and then locate existing drinking-water sources to see if they will be closer to a proposed injection well than state or federal regulations permit. (See Chapter 5 for information on mapping.) If there are potential drinking-water sources too close to a proposed well, petition your state environmental agency to stop the well. Depending on the requirements of state law, your petition may include, among other things, an analysis of existing groundwater quality, specific information about the hydrogeology of the area, and an estimate of the number of people dependent on groundwater for drinking (taking into account the area's likely future growth).

Even when the site of the proposed injection well is the required distance from USDWs, the geology of the site may not adequately confine injected wastes, and thus might result in contamination of an USDW. To challenge a well on this basis, you probably will need a professional hydrogeologist to analyze the site and prepare testimony, which you can submit at the hearing on the well permit.

WASTE CHARACTERIZATION

Citizens can also help ensure that requirements for testing the composition and compatibility of injected wastes are enforced. Wastes must be tested for compatibility with the well equipment, with the formation into which they will be injected, and with other wastes injected into the same well. Incompatibility with well equipment can corrode the casing and result in leakage. Incompatibility with the formation can compromise the integrity of the confining layers above or below the well, resulting in waste migration. Mixing incompatible wastes within a well can lead to chemical reactions that explode or plug up the well.

Although the federal UIC program requires that injected hazardous wastes must be sampled before the initial injection, no further analysis is required to determine compatibility with well equipment or the formation. Your state, however, may have requirements of its own. California, for example, requires monthly sampling and analysis for hazardous wastes injected on-site. Hazardous wastes injected at commercial injection facilities must be sampled prior to injection every time the composition differs from the previous discharge. In Texas, operators of hazardous and industrial waste injection wells must obtain approval from the Water Commission before injecting a waste stream that has not already been approved.

Examine your state's regulations to establish exactly what the operator must do to comply with state characterization requirements. Request a copy of the well's waste characterization records and pay particular attention to whether the wastes are corrosive or reactive. Have a chemical engineer review the list of wastes to determine whether they are compatible with each other. You can also obtain a copy of the original permit application to determine the composition of the formation into which the wastes are injected. These records should include the agency's finding that the injected wastes are compatible with the underground formation. If you discover that incompatible wastes are being injected into the well, petition the state agency to stop the operation until injected wastes are separated or treated.

MONITORING

Monitoring of disposal wells is one way to ensure that wastes do not migrate from the well and contaminate groundwater. Although the federal program generally does not require groundwater-monitoring wells, some states do. California, for example, requires the use of monitoring wells to determine changes in drinking-water quality within one-half mile of the well. Florida requires monitoring wells above the injection zone near Class I (hazardous and industrial and municipal waste) wells.

Determine if your state has more stringent monitoring and reporting requirements than the federal UIC program. Obtain information about the design of the well's monitoring system and copies of monitoring records to determine if the system is effective. If wastes have migrated in the past, find out what corrective action was taken. Notify the state agency responsible for regulating underground injection of any previously undetected migration. You can petition the agency to enforce its regulations if necessary.

INSPECTION

Disposal wells should be inspected regularly to assess their mechanical integrity. You can help ensure that operators comply with inspection requirements and that unsafe wells are shut down.

Determine the inspection and testing requirements your state has for disposal wells. The frequency and other aspects of testing may be stricter than the federal requirements. The federal program requires mechanical integrity testing every five years. California requires mechanical integrity testing of well bores at least once a year and pressure testing at least once every six months. EPA requires certain wells that are used in the oil and gas development process (Class II wells) to be pressure tested to 200 psi (pounds per square inch), but states such as Oklahoma require a much stricter standard (in some cases up to 1,000 psi).

Examine mechanical integrity test results to check for equipment failure and leakage. If you find that the mechanical integrity of an injection well is compromised, petition the agency to conduct a thorough inspection and to determine if the well should be repaired or shut down.

CLASS V INJECTION WELLS

Class V wells include 32 subcategories of wells that do not fit into any other EPA classification. Examples include agricultural drainage wells, stormwater drainage wells, cesspools, industrial waste disposal wells, aquifer recharge wells, and radioactive-waste disposal wells. Estimates of the numbers of Class V wells range from 173,159 to as high as 1 million. Not all of these wells pose a serious threat to groundwater, but many do, particularly agricultural drainage wells, raw-sewage waste disposal wells, industrial process water and waste disposal wells, and automobile service station waste disposal wells.

EPA is currently working on technical requirements for Class V wells. Until these regulations are completed, Class V wells are authorized by federal rule and are subject to closure and other enforcement actions if found to be contaminating a USDW or adversely affecting public health. While there are states and localities that have strict regulations governing subcategories of Class V injection wells, many impose inadequate regulations or none at all.

To find out about Class V wells that may be in your area, the first step is to identify the agency responsible for regulating these wells. This can be done by contacting the EPA regional office with jurisdiction for your state and asking to speak to someone in charge of the regional UIC program. Ask for a copy of the report your state filed for EPA's *Inventory of Class V Wells*. States such as New York, New Jersey, and California have computerized listings of the types and locations of wells around the state, the types of fluids injected, and the state's regulatory structure respecting each type of well.

With this information in hand, you should then determine the appropriate subcategory of the well that concerns you and the regulatory requirements that apply. In Iowa, for example, owners of agricultural drainage wells must register with the state and develop alternatives to be phased in by July 1991. Ohio requires a permit for new Class V wells under a variety of circumstances. Other states, such as Florida, prohibit injection well approval if groundwater quality standards might be violated.

If you are concerned about a particular well, ask the regulating department to identify the type of the well in question and what construction and operation requirements apply. Take advantage of any opportunities for citizen participation in permit approval decisions. Request and examine any required operation and maintenance records to ensure that the operator is complying with state regulations.

CLASS II INJECTION WELLS

Class II wells are related to oil and gas operations and are often regulated stringently at the state level, usually by the same agency that regulates oil and gas activities generally.

Class II wells are often abandoned as the oil or gas resources are depleted. Abandoned wells can act as conduits for migrating fluids, creating a threat to groundwater. The federal UIC program requires these wells to be plugged to prevent the movement of fluids into or between underground sources of drinking water. Some states have more stringent requirements. If you suspect problems with an abandoned well, you should obtain documentation relating to the date and method of plugging. You can petition the agency to inspect the plug to ensure that there is no migration of wastes.

Another thing you can do is check the permitting status and operating requirements for wells in your area that are still active. Contact the state agency responsible for regulating Class II wells to establish the location of these wells in your community. Determine what permitting requirements your state has, including procedures for closure and plugging. Check the permitting status and operating regulations for the specific wells you are concerned about. Most states with Class II wells have some sort of notice requirement for the permitting process. You should also take advantage of inspection and reporting procedures. California requires mechanical integrity testing of Class II wells once a year.

ADDITIONAL REGULATIONS

Some states impose additional requirements for alternative disposal or pretreatment of wastes. Kansas, for example, requires that wastes meet minimum pretreatment requirements set by the Secretary of Health and Environment. Examine your state's pretreatment requirements and determine if the injected wastes can meet these requirements. Some states also prohibit the use of injection wells except where no feasible alternatives exist. If your state has such a limitation, determine what other treatment and disposal methods can be used for the particular types of wastes, and see if you can demonstrate that these wastes can be disposed of or treated economically by means other than underground injection. This type of evidence should be presented during testimony at the permit hearing.

REFERENCES

STATE REGULATION OF WASTE DISPOSAL WELLS

EPA, Office of Drinking Water. *Class V Injection Wells—Current Inventory, Effects on Groundwater, and Technical Recommendations* (Executive Summary). Washington, D.C. 1987. Copies can be ordered by calling toll free 1-800-336-4700, or by writing to the U.S. Dept. of Commerce, National Technical Information Service, 5285 Port Royal Road, Springfield, Va. 22161. The document number is EPA 570/9–87–007 or PB 88136817.

EPA, Office of Groundwater Protection. *State Groundwater Program Summaries.* Vols. 1 and 2. Washington, D.C. 1985.

Fortuna, Richard, and David Lennett. *Hazardous Waste Regulation: The New Era.* New York, N.Y. McGraw-Hill. 1987.

Gordon, Wendy, and Jane Bloom. *Deeper Problems: Limits to Underground Injection As a Hazardous Waste Disposal Method.* Natural Resources Defense Council. New York, N.Y. 1985.

Pettyjohn, Wayne A. *Protection of Public Water Supplies from Ground-Water Contamination.* Center for Environmental Research Information. Cincinnati, Ohio. 1986.

Underground Injection Practices Council, Inc. *International Symposium on Class Five Injection Well Technology Proceedings.* Oklahoma City, Okla. 1987.

LAWS AND REGULATIONS

Code of Federal Regulations Citation

40 C.F.R. § 146.13(b)(3).

State Statutes

Cal. Health & Safety Code §§ 25159.15(a)(1), 25159.17 (Deering Supp. 1988).
Iowa Code Ann. § 159.29 (West Supp. 1988).
Kan. Stat. Ann. § 65–3439(c) (1985).
Ohio Rev. Code Ann. § 6111.043 (Page Supp. 1987).

State Regulations

Florida Underground Injection Control Rule, Fla. Admin. Code Ann. r. §§ 17–28.25(1)(f), 17–28.61(2, 3) (Harrison 1985).
Kan. Admin. Regs. 28–46–24(e) (1986).
Ohio Underground Injection Control Program, Ohio Admin. Code § 3745–34–13 (1984).
Oklahoma Corporation Comm'n, Oil & Gas Cons. Div'n. General Rules and Regulations, Rule 3–305(A)(2)(f) (1987).

CHAPTER 30

State Regulation of Underground Storage Tanks

State and local regulation of underground storage tanks is an important area for citizen action. The federal program, described in Chapter 19, is relatively new and incomplete. It alone does not provide adequate protection against groundwater contamination from leaking underground storage tanks (USTs). Many states have UST programs that predate the federal law; others are developing new programs that meet or exceed minimum federal requirements. Citizen action is essential to ensure that these state and local programs are aggressively enforced and that the many gaps left by federal law are filled.

LOCATING UNDERGROUND STORAGE TANKS IN YOUR COMMUNITY

The Information You Need

The first step in preventing groundwater contamination from underground storage tanks is determining where they are. Under federal law, owners of a regulated operating underground storage tank, or a tank taken out of service after 1974, must notify the state of that tank's location, age, size, type, and uses (see Chapter 19). By contacting the proper state agency (see Appendix E) and asking for this information you can locate most USTs covered by the federal law; in many cases the agency will have a map available.

Some states have imposed stricter registration requirements to make up for lack of compliance with this federal notification requirement. Florida regulations, for example, require that a petroleum UST must be registered

357

before it can be filled by a supplier, a requirement that is improving the effectiveness of the registration process.

The location of older tanks, those taken out of service before 1974, is harder to determine, in part because federal notification requirements do not apply to these tanks. Even though no longer in service, they still may contain hazardous substances that can leak. To find information on these older tanks, check to see if your state has a registration process that covers tanks abandoned before 1974.

If abandoned USTs have not already been mapped by the local fire department or other group, consider a community mapping project. By involving residents in locating abandoned USTs, you increase their awareness of the potential for groundwater contamination. A community mapping project in West Oakland, Michigan, worked with long-time residents who knew the location of old gasoline pumps, and with building inspectors, fire chiefs, and local officials. They also received suggestions and telephone calls from residents in response to newspaper articles about abandoned USTs, and they conducted door-to-door surveys in industrial areas.

Where to Find the Information

Depending on your state's regulations, UST records may be found at places ranging from state UST offices or the state public safety, public health, or environmental agencies to the local fire department or the town clerk's office. Some records are organized by tank location, some by purpose and contents, and others by size. Records for underground tanks storing gasoline for service stations, for example, may be maintained by the local fire department and organized by location, whereas records for underground home-heating oil tanks may be maintained by the department of health and organized by size. The key is to approach all possible sources. Consider urging your state to centralize and cross reference their records.

REGULATIONS ADDRESSING
UST LEAKS AND CLEANUP

If you discover that a UST is leaking, you must act quickly to ensure that contamination is abated and the tank is removed for repair or replacement. First, determine if state or local agencies have procedures for operators to follow in the event of a spill or leak. States may require the operator to immediately report such occurrences to the local fire department, with substantial penalties for failure to report. Specific follow-up actions by the owner/operator of the leaking facility also may be required. Connecticut requires a commercial owner/operator to immediately stop using and empty the leaking part of the tank system, recover and properly dispose of the discharged liquid and anything contaminated by it, and restore the environment. In the event of a spill or leak, contact your local fire department (or other responsible agency) and monitor cleanup activities to ensure that the operator complies with abatement requirements.

Once the spill or leak has been controlled, the tank must be repaired or replaced and any contamination cleaned up. States such as Connecticut allow only one relining or repair of petroleum storage tanks. In states such as Maine and Florida, contractors must be certified to repair or install these tanks.

Under many state mini-superfunds, cleanup can be done by state agencies. Tank owners are generally ultimately responsible for leaking UST cleanup costs, however, and must reimburse the fund for any cleanup expenditures made by the state. (See Chapter 25.) The Massachusetts Superfund makes tank owners legally responsible for up to three times the cleanup cost. Some states, such as Florida and Maine, create special funds to be used for cleanup of leaking USTs.

Many states have special tank-cleanup statutes, under which the owners who fail to clean up the wastes or to remove the leaking tank may have to pay cleanup costs, as well as civil and criminal penalties. New Hampshire imposes strict liability for cleanup costs on any owner or operator of a UST that discharges oil into groundwater. Maine requires any person who contaminates groundwater with discharges of oil or petroleum products from USTs to report and clean up the spill and to restore water supplies that were contaminated by the spill or face civil penalties. Likewise, Ohio makes owners/operators responsible for cleanup costs for discharges of petroleum from USTs. A person who does not comply with a cleanup order is subject to a civil penalty of $10,000 per day for each day that the violation continues.

Florida has created a unique system to encourage cleanup that relies on the carrot of funding rather than the stick of penalties. The state Underground Petroleum Environmental Response Act entitles underground petroleum storage tank owners to reimbursement for cleanup costs from petroleum discharges if they have complied with state regulations, not intentionally damaged tanks or falsified records, and reported the discharge within a certain grace period.

REGULATIONS GOVERNING UNDERGROUND STORAGE TANK OPERATION

States have developed a variety of UST regulations that go beyond the minimum federal requirements discussed in Chapter 19. If a tank violates applicable requirements in your state, report it to the agency in charge of enforcing the regulations. Enforcement of UST regulations may be the responsibility of one or more state or local agencies, ranging from special state UST offices to local fire departments. The key is to find the right agency and press for action. If your state does not have some of the following types of regulations, suggest that they be considered.

Permits

Check to see if your state has permit requirements for the operation of tanks in addition to the federal registration requirements. Permits may be required for more than one phase of UST operation (and from more than one office). Ohio, for example, requires a permit to install, remove, repair, or alter in any way a tank that stores flammable or combustible liquids, or to change or replace any pipeline associated with the tank. Suffolk County requires a permit to operate before a tank may be filled with toxic or hazardous substances. In Massachusetts the fire department may deny permits or impose conditions on permits for replacement or modification of tanks if the proximity of the tank to a public or private well, aquifer, recharge area, or surface water endangers a public or private water supply.

Design and Construction

Find out what state regulations require for design and construction of USTs. Overfill protection, leak detection, or secondary containment systems (such as a double-walled tank) often are required for new USTs. New Hampshire requires all three for petroleum storage tanks. Florida suggests overfill protection consisting of an impervious containment system or a tight fill device. California requires secondary containment for new tanks containing hazardous substances that are nonmotor vehicle fuels.

Installation

Faulty installation of tanks is a major cause of leakage. Some states train and certify UST installers, and some require testing of new equipment. The Conservation Law Foundation's model program recommends testing every new or replacement tank and its piping separately, with air pressure between three and five pounds per square inch. Industry standards prepared by the American Petroleum Institute encourage testing according to manufacturers' instructions, which recommend the safest air pressure level for the type of tank and piping.

Monitoring

Monitoring requirements are designed to detect leaks before damaging chemicals escape into the ground. In some cases you can check compliance with these requirements by asking to review monitoring records at the facility. Techniques include sensors monitoring the level of liquid in the tank; detectors monitoring the piping between the tank and the dispensor; sensors to detect vapors in the soil; and monitoring wells to test the surrounding groundwater for contamination. For double-walled tanks, visual inspection, pressure loss detectors, or automatic product sensors may be used to monitor the space between primary and secondary containers.

One common type of monitoring is inventory control. This involves keeping a detailed record of the amount of tank product used and received, to ensure that none is unaccounted for. Because it is difficult to measure precisely the volume of liquid in a tank, and because even small leaks can cause huge problems, this method of leak detection is usually inadequate.

Testing for Tightness

Find out how often USTs should be tested for leaks. Most states require periodic testing of USTs, with the frequency increasing as the tank gets older. In South Dakota, if the tank owner elects to use testing and inventory control as the release detection mechanism (see above), the tightness test must be conducted twice a year and be capable of detecting a 0.1 gallon per hour leak rate with a probability of detection of 0.99 and a probability of false alarm of 0.01 from any part of the UST system. Many tank tightness tests are not this accurate.

Determine how the tests are conducted in your state and ask for the results for nearby tanks. If you suspect a nearby UST is leaking, contact the appropriate agency and ask them to determine if the tank should be tested.

Abandoned or Out-of-Service Tanks

Your state may require USTs that are no longer being used be removed, or drained and sealed, to prevent future leaking. In Ohio, a permit must be obtained from a fire official to remove, abandon, or temporarily take out of service any flammable or combustible liquid tank. If removal is necessary, it must be approved by the fire official. States or cities may have special abandoned tank removal programs. The fire department of Three Rivers, Michigan, held workshops and provided assistance to owners of

Model Laws

If your state does not have adequate UST laws, you may initiate some action by developing a model UST law. For instance, the Passaic River Watershed Association (PRWA) drafted a model ordinance for UST regulation, which generated widespread interest from other communities. Once the major oil companies realized that this effort could lead to similar ordinances in the other 567 localities of the state, they approached the PRWA to work on a statewide bill. This became the New Jersey Underground Storage Tanks Law.

A model UST ordinance prepared for Massachusetts cities and towns can be obtained from the Conservation Law Foundation (see the references for this chapter).

abandoned tanks, and less than a year later, 50 tanks were removed, one of which was discovered to be leaking. Determine what the procedure is in your state for taking a tank out of service and when it must be removed from the ground. If the operator does not comply, petition the agency to take enforcement action.

State Regulation of Tanks Exempt from Federal Regulations

Action at the state and local levels is especially important for tanks that are exempt from federal UST regulations, such as underground heating oil and farm fuel tanks. Some states and localities have begun to realize that these tanks pose almost as great a threat to groundwater supplies as commercial and industrial tanks and are imposing restrictions to prevent groundwater contamination from these tanks. Suffolk County on Long Island, for example, tests underground heating oil tanks over 1,100 gallons ten years after installation and every five years after that. Check to see if your community regulates underground heating oil tanks and other tanks exempt from federal law. If not, and if your community has a large number of such tanks over an aquifer or in groundwater recharge zones, meet with local officials to suggest regulations requiring registration and periodic tank testing.

REFERENCES

STATE REGULATION OF UNDERGROUND STORAGE TANKS

Conservation Law Foundation of New England, Inc. *Massachusetts Prototype: Model Bylaw/Ordinance for Regulating Underground Hazardous Material Storage.* Boston, Mass. To order, contact: Conservation Law Foundation of New England, 3 Joy St., Boston, Mass. 02108. (617) 742-2540.

East Michigan Environmental Action Council. *Community Involvement in Underground Fuel Storage Tank Mapping*. Birmingham, Mich. 1985.

Environmental Defense Fund. *State and Local Underground Storage Tank Regulations*. Washington, D.C. 1987.

EPA, Office of Groundwater Protection. *Overview of State Groundwater Program Summaries*. Vol. 1. Washington, D.C. 1985.

EPA, Office of Groundwater Protection. *State Groundwater Program Summaries*. Vol. 2. Washington, D.C. 1985.

Massachusetts Audubon Society, Community Groundwater Protection Project. *Underground Storage Tanks And Groundwater Protection*. Lincoln, Mass. December 1984. Revised and Reprinted July 1986.

LAWS AND REGULATIONS

State Statutes

Cal. Health and Safety Code § 25284 (West 1984 & Supp. 1988).
Fla. Stat. Ann. §§ 206.9915 *et seq.*, 376.30 *et seq.* (West Supp. 1988).
Me. Rev. Stat. Ann. tit. 38, § 561 *et seq.* (Supp. 1987).
Mass. Gen. Laws Ann. ch. 21E, § 5(e) (West Supp. 1988).
N.H. Rev. Stat. Ann. § 146–C: 11 (Supp. 1987).
N.J. Stat. Ann. § 58: 10A–21 *et seq.* (West Supp. 1988).
Ohio Rev. Code Ann. § 3737.01 *et seq.* (Baldwin 1982).

State Regulations

Conn. Agencies Regs. § 22a–449d–1(j)(3) (1986).
Fla. Admin. Code Ann. r. §§ 17–61.060(2), 17–61.050(1)(a) (1988).
Mass. Regs. Code tit. 527, § 9.24 (1986).
Ohio Admin. Code § 1301:7–7–28 (1985).
S.D. Admin. R. 74:03:28 (1987).

Local Regulations

Suffolk County Sanitary Code, Art. 12, §§ 1207, 1210(d).

CHAPTER 31

State Regulation of Agricultural Sources

Pesticides, fertilizers, and animal feedlots can contaminate groundwater in agricultural areas. The contamination generally is the result of the leaching of agricultural chemicals from crops and animal wastes from feedlots. Because the chemicals from several different sources combine in rainwater and are often washed far away from their origins, pinpointing any one source of contamination is difficult. Consequently, those state and local governments that do regulate chemical applications do not regulate the sources directly, but instead they impose or encourage controls on the application and storage of pesticides and fertilizers to minimize the potential for groundwater contamination from agricultural runoff and leaching.

In addition to the federal government's regulation of the sale and manufacture of agricultural chemicals (see Chapter 20), some states have comprehensive legislation to protect groundwater from agricultural contamination. Other states rely on a patchwork of laws that regulate agricultural chemicals.

FINDING THE INFORMATION

In addition to the information provided on pesticide labels and packaging, your state may have "right-to-know" laws that give you the right to know the date, type, and amount of a pesticide applied at a particular site. Texas law allows any member of the community to find out which pesticides were applied to a particular area if the applicator stores or uses over a certain amount per year. Washington state law enables any person to request data on government use of pesticides at a given site but limits requests regarding commercial use to chemically sensitive individuals. Washington also has a Hazardous Substances Information and Education Office that disseminates information collected by the state. Some state right-to-know laws specifically exempt pesticide application, so read your state laws carefully. Once you find out where and when pesticides are used, you can work to limit the use of harmful pesticides in areas where they may contaminate groundwater.

PRIVATE AGRICULTURAL SOURCES OF GROUNDWATER CONTAMINATION

Pesticides

Before it can be used in a state, a pesticide must meet EPA's registration requirements and any additional requirements imposed by the state. A state may register pesticides that EPA has not yet registered if they have makeup and use patterns similar to those of a federally registered pesticide. If EPA finds that the state registration plan does not meet EPA's requirements, it may remove the state's authority to register pesticides.

The state charges a fee for registration and citizens should encourage their state to increase registration fees and use the money to fund programs that will protect groundwater and reduce pesticide use. The Pesticide Act of Iowa imposes fees on farmers, pesticide manufacturers, and retailers, in part to fund research on improved agricultural practices. Programs like this in your state may provide information for farmers.

In the registration process, most states rely primarily on the health and environmental data submitted to EPA, because they lack the expertise to conduct extensive testing. Some states request additional information specific to conditions in that state, such as the pesticide's reaction to different temperatures and levels of acidity. California's registration procedure, for instance, is more thorough than any program in the country, including EPA's. In California and a number of other states there is a public comment period, during which citizens can bring their concerns about groundwater to the attention of the regulators before a final decision about a pesticide registration is made.

States typically have authority (although it may be seldom used) to impose restrictions on the method, site, timing, and rate of application of certain pesticides and these must be printed on the pesticide label. Use restrictions may be imposed due to health risks, past incidents of contamination, or the vulnerability of groundwater in certain areas of the state. If the pesticide is not compatible with the geology of a certain area, the state may restrict its application in that area.

After the initial registration, a state can modify or revoke a pesticide's state registration if evidence indicates it is causing groundwater contamination. For example, Wisconsin has adopted restrictions on the use of aldicarb as part of its groundwater protection program. Farmers must file a "report of intended application" with the state Department of Agriculture. Based on a soil and groundwater analysis, the Department may permit the use, require a monitoring well to be maintained downgradient, or prohibit the use. If the concentration of aldicarb reaches the state's groundwater enforcement standard, the state will prohibit its use within one mile of where the standard is exceeded. In addition to the above restrictions, Wisconsin limits aldicarb application to two pounds of active ingredient per acre, applied once every other year. (Aldicarb is discussed further in Chapter 20.)

Some states have pesticide review procedures that provide extensive opportunity for public participation. In Massachusetts, when restrictions are placed on a pesticide's use by the Pesticide Bureau Subcommittee, aggrieved parties can appeal the Subcommittee's decisions. Citizens can seek permission to intervene in these appeal proceedings. A magistrate holds a hearing and the state Pesticide Board then issues a final decision, which can be appealed to the state courts.

California has a strong program for reconsidering problem pesticides when the pesticide or its harmful degradation products are found in groundwater, or below specific soil zones. The registrant must request a hearing before a committee of representatives of several state agencies concerned with groundwater in order to avoid immediate cancellation. After the hearing, the pesticide must be canceled unless the committee determines continued use does not present a health risk or is an economic necessity.

Crop Fertilizer

Groundwater contamination from fertilizers generally is addressed through education and local demonstration projects. The U.S. Department of Agriculture is expanding its efforts in groundwater protection. Check with the Agricultural Stabilization and Conservation Service (ASCS) county committee, or if there is no county committee, the ASCS state committee. These committees authorize federal cost-sharing for individual conservation projects that use environmentally beneficial agricultural practices. The ASCS county committee must meet publicly once a year. Since public comments are not accepted on individual cost-sharing proposals, they must be made when the program is developed or modified. Check your state department of agriculture to see if it also has a program to help farmers reduce fertilizer use, or programs for education and training.

Irrigation Water

"Chemigation"—the practice of putting chemicals directly into irrigation water—can contaminate groundwater. Contamination occurs when the irrigation water flows back through the pump and into the well that the water came from or when excess irrigation water is applied. Some states require one-way check valves between the irrigation pump and the point of chemical injection to prevent back-flow. The responsible state agency may keep inspection records of the chemigation equipment that will indicate if the operator is in compliance. You may request this agency to inspect the equipment if you believe improper operation may be causing groundwater contamination. Nebraska and Kansas require a permit for chemigation, in addition to recordkeeping and periodic inspections. When applying for a renewal permit, the operator must list the names and amounts of chemicals used during the previous year.

To protect groundwater from the buildup of salts that may result from irrigation, some states have adopted an irrigation scheduling policy. This saves water and reduces leaching of salt and other minerals from over-irrigation.

Animal Wastes

Leaching and disposal of animal wastes may cause groundwater quality standards for nitrates and bacteria to be violated. These violations can be reduced by state regulations restricting lot size and number of animals, as well as by collecting and containing waste to assure proper disposal. North Dakota requires any operator of an animal feedlot that contains over 200 animals, and is located in a flood plain or within a certain distance of state water, to obtain a permit from the state health department. If the permit application indicates there may be a pollution problem, the operator must submit a plan for water pollution control. The department may inspect the site and withdraw the permit. If you believe feedlot owners in your state are not in compliance with existing regulations, request an inspection by the state department of agriculture or health.

Some localities require a conditional use permit under zoning ordinances for feedlots or poultry operations with more than a certain number of animals. Jefferson County, Wisconsin, requires that the applicant have sufficient acreage and waste disposal methods to keep soil nitrate levels within the U.S.D.A. Soil Conservation Service technical guidelines. If your locality does not have such a requirement, contact the Minnesota Project, by telephoning 507-765-2700 and request a copy of their Model Ordinance for Ground Water Protection. This ordinance contains standards for animal feedlots and manure storage areas.

GOVERNMENT APPLICATION OF PESTICIDES

Another potential threat to groundwater is presented by use of pesticides on golf courses, rights-of-way, or parks by state and local governments. These control efforts are subject to different laws and review standards than are private applications of pesticides. By participating in the various review procedures, including those under state NEPAs, you can ensure that these programs take into account the vulnerability of groundwater supplies in your community. (See Chapter 26.)

Many states have environmental policy acts that may be used to require a formal environmental review of the impacts of proposed pest- and weed-control programs. Such a review requires the state to study the area to be sprayed, the potential health effects of the pesticides, and their potential for affecting nontarget species and causing contamination of groundwater.

Massachusetts has used generic environmental impact reports (GEIRs) to evaluate state pest eradication programs, which on an individual project basis might go unregulated but when taken in total may cause widespread human exposure or some cumulative damage to the environment. This process can be used to determine impacts of statewide programs such as gypsy moth spraying, mosquito abatement, and weed control on railroad rights-of-way.

Because ongoing local pest- or weed-control programs must have funds authorized each year, the appropriations process might be the best place to achieve review. Whether review of pest control activities is part of the appropriations process or takes place as part of a periodic program review, you should participate in the review proceedings.

Some communities have a practice of examining alternatives before automatically approving the use of potentially dangerous pesticides. For example, Berkeley, California, has mandated that pesticides are only to be applied on city property after examining other possible options. Encourage your state to consider funding alternatives to use of pesticides as part of its groundwater protection strategy.

REFERENCES

STATE REGULATION OF AGRICULTURAL SOURCES

Concern, Inc. *Pesticides.* Washington, D.C. 1985.
DiNovo, Frank, and Martin Jaffe. *Local Groundwater Protection: Midwest Region.* American Planning Association. Chicago, Ill. 1984.

EPA, Office of Groundwater Protection. *Overview of State Groundwater Program Summaries.* Vol. 1. Washington, D.C. 1985.

EPA, Office of Groundwater Protection. *Pesticides in Groundwater: Background Document.* Washington, D.C. 1986.

EPA, Office of Groundwater Protection. *State Groundwater Program Summaries.* Vols. 1 and 2. Washington, D.C. 1985.

EPA, Office of Groundwater Protection. *State Program Briefs/Pesticides in Groundwater.* Washington, D.C. 1986.

EPA, Office of Groundwater Protection. *Survey of State Groundwater Quality Protection Legislation, 1985.* Washington, D.C. April 1987.

Holden, Patrick W. *Pesticides and Groundwater Quality: Issues and Problems in Four States.* National Academy Press. Washington, D.C. 1986.

The Minnesota Project. *Model Ordinance for Groundwater Protection.* Preston, Minn. July 1984.

Morandi, Larry. *Protecting Groundwater from Farm Chemicals: State Legislative Options.* 18 Env't Reptr. 1941 (December 18, 1987).

Pye, V. I., R. Patrick, and J. Quarles. *Groundwater Contamination in the United States.* University of Pennsylvania Press. Philadelphia, Pa. 1983.

Raymond, Lyle S., Jr. *Chemical Hazards in Our Groundwater: Options for Community Action.* Cornell University. Ithaca, N.Y. 1986.

LAWS AND REGULATIONS

U.S. Code Citations

7 U.S.C.A. §§ 136v(a), (c)(4) (1976).
16 U.S.C.A. § 3434 (1982).

State Statutes

Cal. Food & Agric. Code § 13141–13152 (West 1988).
Idaho Code § 22–3402(5)(c), (10)(b) (1977).
Iowa Code Ann. § 206 *et seq.* (1987 & Supp. 1988).
Kan. Stat. Ann. §§ 2–3301–3316 (Supp. 1987).
Neb. Rev. Stat. § 46–612.01 (1984), § 46–1121 (1984 & Supp. 1987).
Texas Agric. Code Ann. § 125.005(i) (Vernon Supp. 1988).

State Regulations

Cal. Admin. Code tit. III, §§ 6172, 6181, 6182, 6186 (1988).
Mass. Regs. Code tit. 33, § 8.08 (1986).
N.D. Admin. Code § 33–16–03–05(1)–(3) (1978).

Local Regulations

Jefferson County, Wisconsin, Zoning Ordinance § 11.03–11.05 (1975).

CHAPTER 32

State Regulation of Mining

Mining activities can be a threat to groundwater quality, depending on the method, size of operations, and the substance being mined. Strict controls placed on mining operations *before* they begin are critical, as the substances mined create highly toxic waste when mixed with water or oxygen. An example is the Bunker Hill mining and smelting site in northern Idaho, where soil, groundwater, and surface water are contaminated by a deadly combination of arsenic, cadmium, fluoride, chromium, lead, mercury, and PCBs. This is one of the nation's largest Superfund sites, covering 21 square miles. At the Anaconda Copper Corporation site near Butte, Montana, contamination from mining wastes was so severe that a citizen coalition forced the corporation to buy out 30 homeowners living near the site. Citizens should work for effective state mining regulation and enforcement, in order to prevent future disasters like these from occurring.

GROUNDWATER PERMITS

One of the most direct ways to control mining contamination is through involvement in the groundwater-permitting process if there is one in your state. These permits generally apply to activities, including mining, that have an impact or could potentially have an impact on groundwater quality. For example, Arizona requires all mining activities with potential groundwater impacts to obtain a groundwater discharge permit.

Groundwater permit applications typically must include information on the design of the discharge facility, the flow and quantity of pollutants, and a hydrogeologic assessment of the proposed site, including analysis of existing groundwater quality and vulnerability to contamination.

Some states impose substantive requirements in the groundwater permit for liners, leak detection systems, and monitoring wells for pits, ponds, and lagoons associated with mining activities. Additionally, some states are developing groundwater protection regulations for heap leaching—a technique for leaching minerals from low-grade ores by applying acids or cyanide, allowing them to percolate through the heap, and separating the minerals from the acids or cyanide. Some 14 cases of cyanide leaking from the surface impoundments used to collect the wastes produced by this process have been reported in the western states. To address this problem, Idaho issued new regulations for heap leaching in 1988 and regulations are now being considered for Nevada and South Dakota.

Some states require a permit applicant to ensure that proposed operations will not violate state groundwater quality standards. Most states with standards use the national primary drinking-water standards. (See Chapter 11.) Due to the extent of mining operations in the state, New Mexico has set standards for additional contaminants that result from mining activities. For example, New Mexico has standards for cyanide, uranium, chloride, and sulphate, which are not covered by a national primary drinking-water standard.

Citizens should take part in the groundwater permit process at the responsible state agency. When an applicant requests a permit most states require notice to be placed in a general circulation newspaper. You can place your name on the issuing agency's mailing list to receive individual notice of permit applications. A public comment period follows—usually 30 days—during which the agency will accept written comments from all interested parties. During this time, you can request that the agency hold a hearing. In Colorado a hearing must be held upon a showing of "good cause," while in New Mexico it requires a finding of "substantial public interest." The New Mexico standard involves, for example, showing that the issues are too complex or important to be reviewed through written comments alone, without opportunity for discussion between the commenters and regulators.

When reviewing a permit application, make note of the existing water quality, the hydrogeology of the site, the substance being mined, the method of mining, associated processes to extract the mined substance from ores, and any plans for discharges. Because the information is technical, and because some permit requirements may be developed by the agency staff and thus may not appear in the laws and regulations, you should consult with agency staff members when reviewing groundwater permits.

You may be able to appeal a permit decision if it does not adequately protect groundwater. The appeals process varies widely among the states. In some states the agency head may appoint an administrative law judge or hearing officer to hold an administrative hearing. Other states, such as Nevada, provide for appeals to a state environmental board. Contact the entity issuing the permit (the agency director) or the state attorney general's environment or natural resources staff to find out how to appeal the permit decision. Assistance from an attorney may be necessary.

STATE MINING PERMITS

In addition to the laws on groundwater discharge, many states also have laws requiring broader mining permits. In some states, a groundwater permit is required as part of the mining permit, and a detailed hydrologic assessment of the site is required. In other states, the permitting agency seeks a recommendation from the state environmental agency regarding a mining operation's potential impacts on groundwater. In Montana, the Department of State Lands' Hard Rock Minerals division issues permits for large minerals mining operations (on state lands), while the Water Quality Bureau reviews the permit to evaluate the operation's potential impact on groundwater and makes recommendations for additional restrictions to protect groundwater from contamination. For smaller mining operations, the Water Quality Bureau handles all aspects of the groundwater-permitting process.

Ohio's procedure for permitting operations involves two hydrologic assessments, one from the Division of Reclamation and the other from the Department of Environmental Protection (DEP). DEP issues a Permit to Install (PTI) for the aspects of the operation that produce wastewater that may contaminate groundwater. The Division of Reclamation conducts a cumulative hydrologic impact assessment that evaluates the potential for runoff from the mining operation.

You can participate in these permit reviews in the same ways as described for groundwater permits on page 500. As with those permits, the amount of public input into the formal review process varies from state to state. Some states have a comment period and hold a public hearing if there is sufficient concern. If your state's minerals permit review has a hydrologic component, submit written comments explaining in detail why the mining operation, as proposed, does not adequately protect groundwater from contamination.

The appeals process for mineral-mining permits also varies from state to state. In Arizona, you must submit a request to the Lands Commissioner to reconsider the issuance of a permit. If your concerns are valid, the Commissioner may hold a hearing in which interested parties may voice their concerns. In Ohio the Chief of Reclamation makes permit decisions, which can be appealed to the Reclamation Board of Review.

Generally, the appeals process involves a hearing before a reclamation or minerals board and then appeal to a state court if you are still not satisfied with the conditions of the permit.

LAND USE CONTROLS

Some states may control surface-mining operations through laws that regulate all land use within the state. These laws may be implemented by the state government or delegated to counties or municipalities. Maine, for example, has a Site Location Act, which requires many activities, including mining, to be reviewed by the State Board of Environmental Protection or municipal planning board, if the municipality has authority to implement the Act. At a public hearing, the operator must show that his activity will not create adverse effects on the environment or public health, with particular attention given to potential effects on aquifers that may serve as a drinking-water source. Land use controls are discussed further in Chapter 24.

STATE ENVIRONMENTAL POLICY AND PROTECTION ACTS

Your state may have an environmental policy or protection act that can be used to protect groundwater from mining contamination. For a discussion of these laws and how they are important to groundwater protection, see Chapter 26.

REFERENCES

STATE REGULATION OF MINING

Citizen's Clearinghouse for Hazardous Wastes. *Action Bulletin.* Arlington, Va. May 1987.

EPA, Office of Groundwater Protection. *State Groundwater Program Summaries.* Vols. 1 and 2. Washington, D.C. 1985.

Gulf Resources Submits Draft Plan to EPA for Cleanup of Bunker Hill Mining Waste Site. 17 Environment Rep. (BNA) 1720 (February 6, 1987).

Slone, Daniel K. *The Michigan Environmental Protection Act: Bringing Citizen Initiated Suits in the 1980s.* 12 Ecol. Law Q. 271 (1985).

"South Dakotans May Vote on Heap-Leaching." 18 *Env.* Rep. (BNA) 20 (May 1, 1987).

Student Environmental Health Project, Center for Health Services, Vanderbilt University. *Water Problems and Coal Mining: A Citizen's Handbook*. Nashville, Tenn. 1984.

LAWS AND REGULATIONS

State Statutes

Ariz. Rev. Stat. Ann. § 49–241, –243, –261, –321 (1987).
Colo. Rev. Stat. § 25–8–501,–502 (1982).
Me. Rev. Stat. Ann. tit. 38, §§ 481–490 (Supp. 1987).
Environmental Protection Act of 1970, Mich. Comp. Laws Ann. §§ 691.1201–1207 (West Supp. 1984).
Mont. Code Ann. § 82–4–227 (1979).
Nev. Rev. Stat. § 40–445.274 (1988).
N.M. Stat. Ann. § 74–6–5 (1987).
Ohio Admin. Code Ann. §§ 1513.07(A)(4)(d), 1513.13(A)(1), 6111.03(g), 6111.035 (1988).

State Regulations

Idaho Dep't of Health and Welfare, Rules and Regulations, Div'n of Environment, tit. 1, ch. 2, § 01.2800 (1988).
Montana Water Pollution Control Regulations, Mont. Admin. R. 1620.904, 1620.1001 (1985).
Nevada Water Pollution Control Regulations, Nev. Admin. Code ch. 445, §§ 140–174 (1987).
New Mexico Water Quality Control Comm'n Regulations, As Amended, Part 3–103A (February 27, 1987).

CHAPTER 33

State Regulation of Septic Systems

Septic systems can cause serious contamination of groundwater. There are more than 22 million such systems operating today, serving a third of the nation's population, and discharging about 1 trillion gallons of effluent into soils each year. To minimize the risk of groundwater contamination, septic systems must be properly designed, constructed, sited, and maintained, and they must not be used to dispose of wastes that are toxic or those that prevent the biological decomposition of the human wastes such systems are designed to handle.

Be aware that to begin to take action it may be necessary to approach several local and state agencies to determine which one regulates septic systems, and, if more than one agency has such regulatory authority, to identify the different areas that each agency regulates. Frequently, the design, construction, siting and, to some degree, maintenance of septic systems are regulated by the state Department of Public Health. Requirements for septic maintenance, as well as for waste transportation and disposal, are often imposed by the state environmental agency.

SITING

Many states regulate the siting of septic systems. These regulations typically specify: the types of soils where septic systems may or may not be placed; the percolation rate (the amount of time it takes effluent to travel through the soil) that is necessary to treat effluent adequately; and the distance required between the septic tank and the groundwater. Minnesota's regulations require that each site be evaluated as to:

- Depth to the highest known or calculated groundwater table or bedrock
- Soil conditions, properties, and permeability

- Slope of the terrain at the proposed site
- The existence of lowlands, local surface depressions, and rock outcrops
- All legal setback requirements—for example, from property lines, water supply wells, buried water pipes and utility lines, and existing and proposed buildings
- Surface water-flooding probability

Many states use similar criteria in evaluating a septic system permit.

Some states, such as Minnesota, also permit local governments to establish standards that are stricter than state regulatory standards. Wisconsin has a different approach; it requires each county to adopt private sewage system ordinances that meet the minimum standards set forth in the state regulations. Counties then administer and enforce the ordinances.

If you suspect that septic systems are installed in your community in areas that pose a threat to groundwater, check the applicable state and local siting regulations to determine whether such systems have been improperly sited. You can also review proposed developments to insure that septic systems meet local requirements. If you find inadequate siting criteria, work with the regulators to have these criteria upgraded.

INSPECTIONS AND MAINTENANCE

Some states and many local governments have regulations regarding required inspection and maintenance schedules for septic systems. Florida, for instance, requires all owners to check their septic tanks at least once every three years, or once a year if garbage grinders are discharging to the tank, to determine if sludge needs to be removed. The Stinson Beach County Water District in California inspects every septic system within its authority on a biennial basis to determine the need for pumping, to examine the absorption field for surfacing effluent, and to determine whether an odor investigation is necessary. In Ottawa County, Michigan, the Health Department conducts inspections of septic systems before a residential sale is completed to certify that the septic system meets the applicable requirements.

Some states and communities regulate the use of organic chemical solvents that are marketed commercially to owners of septic systems and purport to declog or degrease septic systems. EPA reports that there is little evidence that these cleaning solvents perform any of their advertised functions. They may actually hinder effective septic system operation by destroying bacteria needed to degrade wastes. Additionally, some of these solvents are highly toxic and pass through the septic system into groundwater aquifers. Florida and other states prohibit the advertisement, sale, or use of such solvents for the purpose of cleaning septic systems.

Also check to see if a grant program exists in your state for owners of faulty septic systems. These grant programs may pay all or part of the cost to repair or replace failing septic systems. States usually establish maximum grant amounts and impose eligibility requirements. In Wisconsin, for instance, grant recipients may not have an annual income exceeding $32,000, and in no case may the combined grants received by a residence or establishment exceed $3,000 for each septic system repair or replacement. Maryland has a similar grant program available to owners whose systems fail a soil percolation test. Maryland's grants are available only for the purpose of constructing certain statutorily defined "innovative and alternative on-site sewage disposal systems."

SEPTAGE DISPOSAL

You may also be able to take steps to ensure that septage that is pumped from the holding tank during regular tank maintenance is disposed of properly. Some states require those who service septic systems to have a permit. For instance, Wisconsin, Maryland, and Massachusetts require every person involved in septic tank pumping and transportation of the waste to have a state license, and also to have their vehicles and equipment inspected and approved by state authorities.

Look for state and local government regulations pertaining to proper methods of septage disposal. In some instances the untreated septage may be disposed of directly into a solid waste landfill. This risks groundwater contamination as harmful bacteria from the decomposing septage seep into aquifers. Some states, therefore, require the septage to be disposed of at a "publicly owned treatment work" where it is treated along with municipal sewage before disposal in a landfill.

States and communities also may allow land application of septage over specified types of agricultural or dormant lands. Land application is generally preferable to landfilling because land application allows for the assimilation and recapture of nutrients by plants, nutrients which otherwise might pollute groundwater if the waste were discarded in a landfill. Land application has its hazards, however, primarily resulting from over-application of septage and sewage to the land. States that allow land application often regulate the types of lands where wastes may be spread, the amount of waste spread, and whether the waste must be treated before application. States that allow land application include Wisconsin, Minnesota, and Maryland.

SEPTIC DISPOSAL WELLS

Septic systems sometimes rely on wells or seepage pits for treatment of waste in the soil. Seepage pits have uncased sidewalls and bottoms and are usually dug five to ten feet above the water table. The potential for groundwater contamination from these wells is high, particularly if they are improperly designed, sited, or operated.

In addition to the general septic system regulations that apply to systems using the well disposal method, other state or local regulations, specifically applicable to disposal wells, also may govern these systems, depending on their size and the location and the type of wastes that they receive. (See Chapter 29.)

LARGE RESIDENTIAL SEPTIC SYSTEMS

Many areas without access to municipal sewage systems rely on large centralized septic systems to dispose of wastes (generally between 2,500 and 5,000 gallons per day) from groups of homes, apartment complexes, motels, or other types of residences. An advantage of large centralized septic systems is that operation and maintenance are required at only one facility. On the other hand, because many residences may share the system, it is difficult to assign responsibility for repairs. If a problem has been chronically neglected by the system's users, you might want to suggest that the local government assume daily management of the system.

You also should determine whether your state or locality has special design criteria for large residential septic systems, and if so, whether those criteria are met by the system in question. Some states, such as Minnesota, realize that larger systems present unique problems to groundwater (including the limited ability of the soils to treat and dispose of large volumes of waste over a long time), and in response have enacted special statutes concerning large system design and operation.

COMMERCIAL AND INDUSTRIAL SEPTIC SYSTEMS

Septic systems used by commercial and industrial establishments pose a significant threat to the quality of nearby groundwater. Septic systems are not designed to treat many of the toxic wastes businesses may place in them. In some instances the toxic substances may prevent human wastes from properly decomposing in the septic system. In addition, many of the toxic substances pass through the septic tank with the effluent, are not absorbed by the soil, and leach into groundwater aquifers. Any business that produces 100 kilograms (about 220 pounds, or one 55-gallon barrel) or less per month of hazardous wastes is exempt from federal hazardous waste rules (see Chapters 16 and 17), and could be using septic systems for such waste. Examples of industrial waste substances that septic systems are *not* designed to treat are:

- Organic solvents and metal degreasers discarded by printers
- Organic and inorganic chemicals discarded by the photoprocessing industry
- Soil and stain removers discarded by laundries and laundromats
- Solvents such as trichloroethylene and perchloroethylene used by dry cleaners
- Waste oils, degreasers, and automotive fluid discarded by service stations
- The large volumes of grease and cleansers discarded by restaurants
- Dyes discarded by beauty salons

Due to increasing incidents of groundwater contamination by the disposal of toxic commercial and industrial wastes, some states now regulate the types of wastes that may be placed into septic systems. States also may regulate the amount of wastes that may be discarded, and may impose strict permitting, monitoring, and reporting requirements on business establishments using septic systems.

Massachusetts, for instance, requires a permit before certain pollutants may be discharged into groundwater from septic systems. The state issues individual permits that set pollutant-specific discharge limits, which are designed to keep the total discharges to the aquifer from all sources either at or below what is needed to achieve minimum groundwater quality criteria. In addition to permit requirements, Massachusetts also imposes monitoring, recordkeeping, and reporting requirements on the amount and types of pollutants discharged from each facility.

New York requires a State Pollutant Discharge Elimination System (SPDES) permit for all septic systems that are designed to discharge 1,000 gallons or more per day of sewage effluent, and for all systems that discharge any industrial wastes. Both Massachusetts and New York allow local governments to set even more stringent requirements for waste disposal. Other states that have enacted special regulations for commercial and industrial septic systems include Florida, Maryland, and New Hampshire.

If you believe that a commercial or industrial septic system is contaminating groundwater, find out what state and local regulations apply and pressure responsible agencies to enforce them. Obtain a copy of the facility's operating permit. Determine whether the permit is in compliance with applicable regulations and whether the establishment is operating in accordance with the permit specifications. For instance, is the establishment exceeding its total allowable volume flow discharge? Are the amount and type of constituents discharged into the system consistent with the permit? Is the system maintained according to permit specifications? If you find violations, you should request the responsible state agency to inspect the system and bring enforcement action. In some states, you may be able to take enforcement action in the courts yourself.

REFERENCES

STATE REGULATION OF SEPTIC SYSTEMS

DiNovo, Frank, and Martin Jaffe. *Local Groundwater Protection: Midwest Region.* American Planning Association. Chicago, Ill. 1984.

EPA, Office of Ground Water Protection. *Septic Systems and Groundwater Protection: A Program Manager's Guide and Reference Book.* Washington, D.C. 1986.

EPA, Office of Ground Water Protection. *Septic Systems and Ground-Water Protection: An Executive's Guide.* Washington, D.C. 1986.

EPA, Office of Ground Water Protection. *State Ground-Water Program Summaries.* Vol. 2. Washington, D.C. 1987.

LAWS AND REGULATIONS

State Statutes

Md. [Health-Envtl.] Code Ann. § 9–1401 *et seq.* (Supp. 1986).

Wis. Stat. Ann. § 144.245 (West Supp. 1987).

State Regulations

Fla. Admin. Code Ann. r. 10D–6.041 *et seq.* (1986).

Md. Regs. Code tit. 10, §§ 17.02 *et seq.* (1987), 17.16 *et seq.* (1986).

Mass. Regs. Code tit. 310, § 15.01 *et seq.;* tit. 314, §§ 5.01 *et seq.,* 6.01 *et seq.* (1986).

Minn. R. 7080.0010 *et seq.* (1985).

N.H. Code Admin. R. Ws 1000.01 *et seq.* (1984).

N.Y. Comp. Codes R. & Regs. tit. 6, § 751.1 *et seq.* (1985).

Wis. Admin. Code ILHR 83.01 *et seq.* (February 1985), NR 113.01 *et seq.* (September 1987), NR 124.01 *et seq.* (June 1986).

CHAPTER 34

State Regulation of Transportation of Hazardous Materials and Highway Runoff

Two major threats to groundwater are related to transportation. First, hazardous materials may leak or spill during transportation and seep into groundwater. Second, salt and oil washed off roads by rain or snow can contaminate groundwater. The Department of Transportation sets some federal safety standards for the transportation of hazardous materials. (See Chapter 22.) In addition, these potential contamination sources can be addressed effectively at the state and local level. For example, federal law establishes a framework for local emergency response plans that can address transportation-related hazards at the state and local level. (See Chapter 17.) In addition, your state or community may have regulations to protect the public health from spills and accidents not addressed by an emergency response plan. Because transportation and highway runoff are primarily locally controlled, grass-roots action is particularly effective in reducing this threat to groundwater.

HAZARDOUS MATERIALS TRANSPORTATION

Federal law established minimum safety standards for trucks, railroads, pipelines, and other carriers of hazardous materials. (See Chapter 22.) In addition, states and communities impose a wide variety of hazardous

How Citizens Are Helping

In Chickasaw, Alabama, citizens learned that Waste Management, Inc., planned to store toxics and other wastes at a nearby port while awaiting pickup by their incinerator ships. To reach the port, the wastes had to travel through the town of Chickasaw. The citizens organized the Chickasaw Community Affairs Group (CCAG) and worked with the town council to pass ordinances placing restrictions on the transport of hazardous materials through the town, including:

• Pre-notification of the waste trucks' route and time of travel to allow for a police escort;
• Maintenance of 150 feet of distance between the truck and the nearest vehicle;
• Bans on travel during adverse weather conditions; and
• Lowered weight restrictions on some streets and bridges.

The CCAG also succeeded in working with the nearby city of Mobile, Alabama, to implement further restrictions on hazardous waste transportation and storage.

---

materials transportation controls. Colorado requires transporters of hazardous materials to obtain an annual permit. Transporters passing through once must obtain a single trip permit, which requires proof of liability insurance and is valid for a period of no more than 72 hours. New York City has rush-hour curfews for transportation of hazardous materials and bans transporters of petroleum products and pressurized gas from city streets unless they are making pickups or deliveries. Portland, Oregon, used a planning model devised by the Federal Highway Administration to calculate the probability of an accident involving hazardous materials along its roadways and, on the basis of this data, banned hazardous materials along certain routes. Cities may close selected bridges, roads, tunnels, or waterways to hazardous materials transportation and conduct transporter education programs to inform shipping companies and operators of safer alternative routes. Your state department of transportation or your community's fire department or department of health may have information about restrictions on times, routes, weather conditions, and speeds as well as requirements for ensuring financial responsibility.

Although state and local regulations exist, your own involvement in preventing transportation-related contamination is important. Some important issues to be aware of include:

• What hazardous materials are stored in or transported through your community? Reports on stored materials exceeding a certain quantity must be

filed with your local planning committee. (See Chapter 17.) Smaller quantities of toxics or those exempt from reporting may be investigated based
on your knowledge of local industrial and commercial facilities.

- What routes are used to transport the hazardous materials? Your local
  health or fire department may have maps of approved routes for these
  materials. Accident data from these sources, as well as your state DOT or
  Materials Transportation Bureau, may help pinpoint where materials have
  been transported and where problems are likely. Your community also
  might conduct a road survey to observe trucks bearing warning labels.
- What are the potential hazards of the materials being transported? Obtain
  the DOT Emergency Response Guidebook and the profiles of toxic materials available from EPA's Office of Pesticides and Toxic Substances.
- Are local officials actively promoting hazardous material transportation
  safety? DOT's *Community Teamwork* (see references section for this chapter) provides ideas for developing a hazardous material transportation
  safety program at economical costs.

HIGHWAY RUNOFF

Salt, oil, gasoline, and other substances wash off roads after rainfall and
can pollute your groundwater. The main concern is groundwater contamination from sodium, a component of salt applied to roadways for de-
icing.

Your attention to local road-salting policies may help ensure that
groundwater is protected from sodium contamination. Information on
these policies is available through the Department of Public Works at the
state and local levels. Your involvement may include petitioning local
officials to test salting alternatives such as calcium magnesium acetate
and salt/sand mixtures, to designate sensitive areas for reduced salting,
or to require road salt storage sheds designed to prevent leaching. Massachusetts and Connecticut have designated reservoirs and groundwater
recharge zones as reduced salting areas. If the salt distribution equipment
in your area is operated manually, suggest the use of automatic equipment, which can reduce the potential for misapplication.

In addition to local policies addressing highway runoff, the Clean
Water Act requires states to develop remedial programs for highway
runoff as a source of "nonpoint" (i.e., non-source-specific) contamination.
These programs may employ preventative measures such as ditch or
drainage systems that block surface runoff from reaching groundwater.
States receive grants to implement these programs. Your state may have a
citizen advisory committee that will allow you to participate in the
development of more effective management programs.

REFERENCES

State Regulation of Transportation of Hazardous Materials and Highway Runoff

ICF, Incorporated. *Lessons Learned: A Report on the Lessons Learned from State and Local Experiences in Accident Prevention and Response Planning for Hazardous Materials Transportation.* Report prepared for U.S. Department of Transportation and U.S. Environmental Protection Agency. Washington, D.C. December 1985.

Pioneer Valley Planning Commission. *The Road Salt Management Handbook: Introducing a Reliable Strategy to Safeguard People and Water Resources.* West Springfield, Mass. November 1986. To order, contact the Commission at 26 Central Street, West Springfield, Mass. 01089.

U.S. Department of Transportation. *Community Teamwork: Working Together to Promote Hazardous Materials Transportation Safety: A Guide for Local Officials.* Washington, D.C. May 1983.

U.S. Department of Transportation. *Guide for Preparing Hazardous Materials Incident Reports.* Washington, D.C. 1985.

U.S. Department of Transportation. *A Guide to the Federal Hazardous Materials Transportation Regulatory Program.* Washington, D.C. 1983. (An updated version should be available from the DOT, Office of the Secretary, Technology Sharing Program by spring of 1989.)

U.S. Department of Transportation. *1987 Emergency Response Guidebook for Initial Response to Hazardous Materials Incidents.* DOT P 5800.4. Washington, D.C. 1987. (Check for annual updates.)

All of the above DOT publications can be ordered from either your DOT regional office or DOT Headquarters, Research and Special Programs Administration, DHM–51, Office of Hazardous Materials Transportation, 400 7th Street S.W., Washington, D.C. 20590.

The Environmental Policy Institute's model local ordinance on the transportation of hazardous materials can be obtained by calling (202) 544-2600.

EPA Regional Offices

Region 1
John F. Kennedy Federal Building
Room 2203
Boston, MA 02203
(617) 565-3715

Region 2
26 Federal Plaza
New York, NY 10278
(212) 264-2525

Region 3
841 Chestnut Street
Philadelphia, PA 19107
(215) 597-9800

Region 4
345 Courtland Street, N.E.
Atlanta, GA 30365
(404) 347-4727

Region 5
230 South Dearborn Street
Chicago, IL 60604
(312) 353-2000

Region 6
1201 Elm Street
Dallas, TX 75270
(214) 767-2600

Region 7
726 Minnesota Avenue
Kansas, KS 66101
(913) 236-2800

Region 8
1 Denver Place
999 18th Street
Suite 1300
Denver, CO 80202
(303) 293-1603

Region 9
215 Fremont Street
San Francisco, CA 94105
(415) 974-8071

Region 10
1200 Sixth Avenue
Seattle, WA 98101
(206) 442-5810

APPENDIX B

Government Sources for Reference Materials

EPA Offices: Several offices within the EPA offer guidance materials and fact sheets on programs, regulations, and enforcement strategy, including the Office of Groundwater Protection, The Office of Solid Waste and Emergency Response, the Office of Pesticide Programs, the Office of Drinking Water, the Office of Pesticides and Toxic Substances.

You should contact these offices either at headquarters, 401 M Street, SW, Washington, DC, 20460, or through the regional offices listed in Appendix A to inquire about their publications or about materials prepared by other agencies, which might be helpful to you. EPA publications are mailed to you at no cost.

The Federal Register: The Federal Register is an important source of information for you, as it contains proposed and final agency action for nearly all of its regulatory activities. The Code of Federal Regulations contains the regulations formulated by the agencies, the rulebook for nearly all of the contamination sources we have described in this book. To order copies of the *Federal Register* or Code of Federal Regulations (CFR), contact:

> Superintendent of Documents
> U.S. Government Printing Office
> Washington, D.C. 20402
> (202) 783-3238

All orders must be accompanied by payment.

General Accounting Office: The U.S. General Accounting Office (GAO) publishes numerous reports at the request of Congressional committees. Its many comprehensive studies include the status of regulatory requirements, EPA enforcement of environmental protection laws, and the progress of federal programs. The GAO publishes an annual index of reports for each subject matter it has investigated, and provides free computer readouts on all titles published under a given topic, providing information on the background of the document, GAO's findings on the subject investigated, and recommendations to the agencies. Use this resource to identify GAO reports of value to your investigation.

These documents are free and can be ordered by calling or writing the GAO. Forward your requests to:

U.S. General Accounting Office
P.O. Box 6015
Gaithersburg, Maryland 20877
(301) 275-6241

Ordering Congressional Documents: Reviewing a proposed bill, act, report, or public law prepared by Congress is essential to understanding its ramifications. To order Congressional documents, send your request with a self-addressed mailing label to:

Senate Document Room
Hart Office Building B-04
Washington, D.C. 20510

The first copy of each document requested is free. There is a charge for additional copies or requests for more than six documents.

APPENDIX C

Organizations Providing Information and Services

The following is by no means an exhaustive list of national and regional organizations that may be of help to you; exclusion or inclusion of any group is not meant as a criticism or an endorsement. We do hope that at least some of these groups will be of help to you.

National Organizations

American Council of Independent Laboratories, Inc. 1725 K Street, N.W., Suite 412, Washington, D.C. 20006, (202) 887-5872. The Council is the trade association of over 300 testing laboratories, of which about half test drinking water for contaminants. Publishes a directory of members, provides referrals, and has published the "Accreditation of Environmental Testing Labs: An Independent Lab Perspective."

American Public Works Association. 1313 E. 60th Street, Chicago, Illinois 60637, (312) 667-2200. The APWA assists utilities in delivering services to the public, informs them about legislation related to drinking-water issues, and is devoted to keeping water technology up to date. Has published a directory of public works.

American Water Works Association. 6666 West Quincy Avenue, Denver, Colorado 80235, (303) 794-7711. The AWWA develops and shares information about drinking-water technologies and water utility management. Publishes manuals, handbooks, and magazines, with over 250 titles offered for sale.

Association of State Drinking Water Administrators. 1911 North Fort Myer Drive, Arlington, Virginia 22209, (703) 524-2428. Coordinating body and clearinghouse for state drinking-water officials and drinking-water policies.

Atlantic States Legal Foundation, Inc. 658 W. Onondaga Street, Syracuse, New York 13204, (315) 475-1170. ASLF is a nonprofit environmental organization dedicated to bringing legal, technical, and organizational support to aid in the enforcement of environmental statutes, in particular the Clean Water Act. It relies on membership support to fund its legal attacks against a variety of threats to the environment.

Citizen's Clearinghouse for Hazardous Wastes. Box 926, Arlington, Virginia 22216, (703) 276-7070. CCHW specializes in community organizing, particularly in fighting the threats caused by hazardous wastes. They have also pub-

388

lished over 50 citizen action guides on subjects ranging from how to organize and hire a lawyer to strategies for recycling. CCHW also provides referrals to local groups.

Clean Water Action Project. 317 Pennsylvania Avenue, S.E., Washington, D.C. 20003, (202) 547-1196. CWAP is a citizens' organization working for clean and safe water at an affordable cost, control of toxic chemicals, and the protection of our nation's natural resources. CWAP is active in lobbying, research, organizing, and coalition-building.

CONCERN. 1794 Columbia Road, N.W., Washington, D.C. 20009, (202) 328-8160. CONCERN has published several citizen guides on groundwater, pesticides, and drinking water, which give an overview of the issues, state and federal legislation, and protection guidelines. CONCERN networks information on the various issues and functions as a clearinghouse.

Environmental Action. 1525 New Hampshire Avenue, N.W., Washington, D.C. 20036, (202) 745-4870. EA has fact packets on drinking water, groundwater pollution, and RCRA issues. EA works in Congress and other public forums, but particularly with grassroots environmental organizations to promote groundwater safety, and publishes the bi-monthly magazine *Environmental Action*. EA has also published a handbook on how to sue polluters.

Environmental Defense Fund. 1616 P Street, N.W., Suite 150, Washington, D.C. 20036, (202) 387-3500. Contact: Brian Day or David Boose. EDF is a national environmental advocacy organization, focusing primarily on environmental policy on the federal level. Scientists and lawyers in EDF's N.Y. and D.C. offices are active in the pursuit of federal groundwater protection legislation. EDF has developed the Environmental Information Exchange, a computer system whereby activists in various states can share groundwater technical, scientific, and legal strategies and expertise, and also find out the latest national developments in groundwater legislation, litigation, or policy.

Environmental Research Foundation. P.O. Box 3541, Princeton, New Jersey 08543-3541. ERF is a privately funded information center that obtains technical information on landfills, incinerators, and hazardous waste, and makes such information accessible to the average person. Upon request, ERF makes available at no charge their "RACHEL" computer databank (Remote Access Chemical Hazards Electronic Library). Almost all personal computers are compatible with RACHEL. ERF also has a weekly newsletter, *RACHEL's Hazardous Waste News*, compiled with information from RACHEL.

Environmental Law Institute. 1616 P Street, N.W., Washington, D.C. 20036, (202) 328-5150. ELI specializes in environmental education, training, and publication, and is the publisher of *Environmental Reports*, a looseleaf service on the latest developments in environmental litigation, policy, and legislation. It conducts national conferences on environmental issues and has published several books on groundwater-related issues.

Environmental Policy Institute. 218 D Street, S.E., Washington, D.C. 20003, (202) 544-2600. EPI was active on the reauthorization of the Safe Drinking Water Act and is now tracking the development of new drinking-water regulations. EPA has sponsored citizens conferences on groundwater protection and actively participates in formulating groundwater protection policy.

Environmental Task Force. 1525 New Hampshire Avenue, N.W., Washington, D.C. 20036, (202) 745-4870. (Now merged with Environmental Action.) ETF acts as a liason between local and state environmental groups working on similar issues, and provides organizational support services to grass-roots efforts, including technical and legal advice. ETF publishes the *RE:SOURCES* bulletin and other publications.

Friends of the Earth. 218 D Street, S.E., Washington, D.C. 20003, (202) 543-4312. FOE lobbies for several groundwater-related statutes, and is working for the enactment of a comprehensive groundwater law. They provide technical information relating to drinking water, groundwater, and pesticides.

Fund for Renewable Energy and the Environment. 1001 Connecticut Avenue, N.W., Washington, D.C. 20036, (202) 466-6880. FREE is a non-profit organization dedicated to environmental education. It published the *State of the States Report*, a guide to state environmental legislation and policy.

Government Accountability Project. 25 E Street, N.W., Suite 700, Washington, D.C. 20001, (202) 347-0460. GAP is a non-profit public interest group, which provides legal support to whistleblowers. GAP is currently researching contamination of the aquifers at the Department of Energy's Fernald, Ohio, nuclear production plant. They are comparing official DOE and contractor claims with internal documents obtained under the Freedom of Information Act.

Greenpeace. 1436 U Street, N.W., Washington, D.C. 20009, (202) 462-1177. Greenpeace uses direct action, a canvassing program, publications, and community organizing to publicize the sources of toxic contamination of water. They are focusing on the halting of illegal dumping and source reduction of toxins.

League of Women Voters. 1730 M Street, N.W., Washington, D.C. 20036, (202) 429-1965. The LWV has embarked on a Drinking Water Education Project, which will combine citizen education and community involvement with local research by individual League chapters on the status of local drinking water systems and monitoring programs. Contact them for information on household hazardous waste collection programs and guides to groundwater protection.

The Local Government Commission. 909 12th Street, Suite 205, Sacramento, California 95814, (916) 448-1198. The Commission produces guidebooks and provides technical assistance to local governments. They focus primarily on the minimization and recycling of hazardous and solid wastes.

National Audubon Society. 801 Pennsylvania Ave, S.E., Washington, D.C. 20003, (202) 547-9009. The Audubon Society and its many regional chapters are active in promoting conservation legislation, litigation, and grass-roots action. With over a half million members, it is one of the largest environmental organizations in the country.

National Toxic Campaign. 37 Temple Place, Fourth Floor, Boston, Massachusetts 02111, (617) 482-1477. The campaign works with grass-roots groups across the country on a variety of toxics issues through its 40 affiliates. It is concerned about drinking-water contamination from Superfund sites and stresses prevention and cleanup as methods to stave off tomorrow's Superfund sites. They published the *Citizens Toxic Protection Manual.*

National Cancer Institute. 9000 Rockville Pike, Bethesda, Maryland 20892, (301) 496-5615. NCI conducts cancer research and issues guidance on cancer-causing substances.

National Coalition Against the Misuse of Pesticides. 530 Seventh Street, S.E., Washington, D.C. 20003, (202) 543-5450. NCAMP focuses on and has reports on the issues of pesticides in drinking water, the lack of federal standards for bottled water, and the impacts of agriculture on groundwater, as well as monitoring state and federal legislation relating to pesticides.

National Conference of State Legislatures. 1050 17th Street, Suite 2100, Denver, Colorado 80265, (303) 623-7800. The NCSL publishes and provides advice on hazardous waste facility-siting legislation, Superfund assessments and remedial action, RCRA-related issues, and groundwater supply questions.

National Demonstration Water Project. 602 South King Street, Suite 402, Arlington, Virginia 22209, (703) 478-8652. NDWP works through a network of affiliates in rural areas and small communities to address water quality issues, and also works on the issues of field sanitation for migrant workers, and wastewater.

National Institute for Occupational Safety and Health. Robert A. Tafts Laboratory, 4676 Columbia Parkway, Cincinnati, Ohio 45226, 1-(800) 35-NIOSH. NIOSH has information on the health and safety risks posed by various chemicals and production processes.

National Institute of Environmental Health Sciences. P.O. Box 12233, Research Triangle Park, North Carolina, 27709 (919) 541-3345. The Institute conducts research on environmental hazards and the resulting health effects.

National Solid Waste Management Association. 1120 Connecticut Avenue, N.W., Washington, D.C. 20036, (202) 659-4613. NSWMA represents the major hazardous waste disposal firms and is a good source for material on disposal technology. They also have background information on the major waste firms.

National Water Well Association. 6375 Riverside Drive, Dublin, Ohio 43017, (614) 761-1711. The NWWA is a professional society and trade association representing all segments of the groundwater industry.

National Wildlife Federation. 1400 16th Street, N.W. Washington, D.C. 20036, (202) 797-6800. NWF has written a state guide to groundwater and has published several books on toxics, waste reduction, and other groundwater-related issues. With a staff of lawyers, scientists, and researchers, they also specialize in environmental litigation and lobbying. They publish annually the *Conservation Directory*, which lists organizations, agencies, and individuals involved in protection of natural resources.

Natural Resources Defense Council. 1350 New York Avenue, N.W., Suite 300, Washington, D.C. 20005, (202) 783-7800. NRDC has worked for the implementation of several groundwater-related statutes, monitoring of EPA's enforcement of environmental laws, the restriction of deepwell injection and other issues. They published *A Citizen's Guide to Groundwater* and *Deeper Problems: Limits to Underground Injection as a Hazardous Waste Disposal Method.*

Rachel Carson Council. 8940 Jones Mill Road, Chevy Chase, Maryland 20815, (301) 652-1877. Named for the famous conservationist who was one of the first to recognize in her book *Silent Spring* the dangers of pesticide contamination, this organization provides grass-roots support, educational materials, and lobbying efforts on behalf of pesticide issues.

Sierra Club. 730 Polk Street, San Francisco, California 94109. (415) 981-8634. The Sierra Club advocates conservation of natural resources through lobbying, publication, education, and legislative efforts. With over 300 chapters nationwide and over 500,000 members, it is one of the oldest and most influential environmental organizations in the country, responsible for the preservation of national parks, forests, and other wilderness areas, as well as wildlife preservation.

Trial Lawyers for Public Justice. 2000 P Street, N.W., Washington, D.C. 20036, (202) 463-8600. Trial Lawyers for Public Justice represents citizens injured by toxics, brings citizen suits under various environmental law statutes, and defends "whistleblowers" who have reported safety hazards in the workplace, among many other public interest cases.

Regional Organizations

California Action Network. Box 464, Davis, California 95617, (916) 756-8518. A statewide membership group involved in public education and lobbying on the issue of pesticide contamination of groundwater.

Center for Rural Affairs. Box 405, Walthill, Nebraska 68067, (402) 846-5428. They are concerned about groundwater in the western corn belt region.

Citizens' Action Coalition. 3951 N. Meridian, Indianapolis, Indiana 46208, (317) 921-1120. Provides information about water and groundwater resource issues, issues regarding reduction of toxic chemicals, and waste reduction legislation in Indiana.

Citizens for a Better Environment. Main office: 506 South Wabash, Suite 523, Chicago, Illinois 60605, (312) 939-1530. CBE is a federated organization attempting to influence regulation and legislation on the state level. CBE-Illinois is active on drinking water and the establishment of a state groundwater protection program. Regional offices are located in CA, MN, MI, WI, and IN.

Coalition for the Environment. 6267 Delmar Boulevard, St. Louis, Missouri 63130, (314) 727-0600. The 26,000-member Coalition organizes statewide on groundwater issues (32 percent of the state's drinking water is groundwater). They are active on the issues of contaminated wells, PCB's, radioactive dumps, and leaking landfills. They conduct public education and workshops on groundwater.

Conservation Law Foundation. 3 Joy Street, Boston, Massachusetts 02108-1497, (617) 742-2540. The Foundation's Groundwater Protection Project is involved in improving protection of New England's groundwater. Has published the booklet *Underground Petroleum Storage Tanks: Local Regulation of a Groundwater Hazard*, which provides a prototype ordinance and detailed information.

Del-Aware. 6 Stockton Avenue, New Hope, Pennsylvania 18938, (215) 862-9862. Dedicated to protection of resources, especially water, in the Delaware Valley region. Has worked on stopping the Limerick 2 nuclear plant and promoting sound water management, and other issues affecting water supplies and quality.

East Michigan Environmental Action Council. 21220 West 14 Mile Road, Birmingham, Michigan 48010, (313) 258-5188. EMEAC is a community action group educating the public to influence environmental policy. Has citizen guidebooks on waterwell testing and water management.

Ecology Center of Louisiana. Box 19146, New Orleans, Louisiana 70179, (504) 891-2447. A nonprofit membership organization that provides technical advice to citizens and lobbies on state legislation relating to "right-to-know" and groundwater.

Environmental Planning Lobby. 33 Central Avenue, Albany, New York 12210, (518) 462-5526. Lobbies New York State on water quality standards, toxic cleanup, and surface water issues. Involved in a Safe Drinking Water Initiative, and water quality standards for surface water throughout New York state.

Kootenai Environmental Alliance. 1909 Pine Hill Court, Coeur d'Alene, Idaho 83814, (208) 667-2229. Works on water quality issues in lakes and aquifers.

Lake Michigan Federation. 59 East Van Buren, Suite 2215, Chicago, Illinois 60605, (312) 939-0838. The only group dedicated to lake-wide action to protect and improve water quality. The Federation works on cleaning the lake's toxic hotspots, protection of wetlands, and halting shoreline erosion.

Life of the Land. 250 South Hotel Street, Honolulu, Hawaii 96813, (808) 521-1300. Life of the Land operates a canvass, performs some research, and is engaged in some lobbying to establish a state water code. They have asked the Governor to perform an inventory of the wells in Hawaii.

Manasota-88. 5314 Bay State Road, Palmetto, Florida 33561, (813) 722-7413. Manasota-88 is a statewide group active on the issues of radionuclides and chemicals in drinking water. They participate in rulemaking at the state level and at the EPA on groundwater standards. They publish information, and conduct public education and lobbying.

National Water Center. P.O. Box 264, Eureka Springs, Arkansas 72632, (501) 253-9755. The National Water Center gathers, distills, and disseminates information on the full spectrum of water issues. Has published a resource on wastewater treatment, *We All Live Downstream—A Guide to Waste Treatment That Stops Water Pollution.*

Natural Resource Council of Maine. 271 State Street, Augusta, Maine 04330, (207) 622-3101. Conducts education and advocacy on issues relating to water and river conservation. Has extensive experience in protection of rivers and groundwater.

New England Interstate Water Pollution Control Commission. 85 Merrimac Street, Boston, Massachusetts 02114, (617) 367-8522. The Commission provides a myriad of educational materials on groundwater as a resource, its major contamination problems, and effective management strategies, including brochures, slide shows, lectures, and other publications. (It is the publisher of *LUSTline*, a comprehensive newsletter on national and regional developments concerning underground storage tanks.)

Rural Water Resources. 218 Central Street, Box 429, Winchendon, Massachusetts 01475, (508) 297-1376. RWR provides free service to groups or communities and provides educational materials on the evaluation of local water supply needs and protection, technical review of engineering plans, and funding applications. (Component of Rural Housing Improvement, Inc.)

Silicon Valley Toxics Coalition. 361 Willow Street, Suite 3, San Jose, California 95110, (408) 287-6707. Focuses on the toxic contamination from high-tech industries that threaten groundwater and workers in Silicon Valley. Publishes a monthly newsletter, *Toxic News.*

Texas Center for Rural Studies. Box 2618, Austin, Texas 78768, (512) 474-0811. Researches and organizes on issues relating to toxic contamination. The Center's main interest in drinking water is in relation to toxic wastes and pesticides in water.

The Minnesota Project. 2222 Elm Street, S.E., Minneapolis, Minnesota 55414, (507) 765-2700. A statewide group working on a variety of rural issues. Their work focuses on public education on land management activities. They have worked on the issue of groundwater contamination for over four years, and have developed a model protection ordinance which is in various stages of adoption in eight counties. Concerned about nonpoint pollution of groundwater—fertilizers and pesticides—into the "Karst Area" bedrock, which does not filter contaminants before usage.

Utah Environment Center. Box 8357, Salt Lake City, Utah 84108, (801) 322-0220. Active on groundwater issues in Utah.

Vermonters Organized for Cleanup. 289 North Main Street, 2nd Floor, Barre, Vermont 05641, (802) 476-7757. VOC is concerned with toxic sites and their affects on drinking-water supplies.

Wisconsin's Environmental Decade. 14 West Mifflin Street, Suite 5, Madison, Wisconsin 53703, (608) 251-7020. A large membership-based organization, they conduct public education and some statewide lobbying on environmental issues.

Below is a list of Public Interest Research Group (PIRG) organizations around the country. PIRGs are devoted to representation of the public interest on a variety of concerns, including threats to the environment.

PIRG Office List

Alaska PIRG (AKPIRG)
Director: Jeff Bohman
430 W. 7th Ave.
Anchorage, AK 99501
(907) 278-3661

British Columbia PIRG (BCPIRG)
Contact: Dorrie Nagler
Simon Fraser University
Room TC-304-305
Barnaby, British Columbia V51S6
CANADA
(603) 291-4360

California PIRG (CALPIRG)
Director: Susan Birmingham
1147 South Robertson Blvd.
#203
Los Angeles, CA 90035
(213) 278-9244

Colorado PIRG (CoPIRG)
Director: Rich McClintock
1724 Gilpin Street
Denver, CO 80218
(303) 355-1861

Connecticut PIRG (ConnPIRG)
Director: James Leahy
University of Connecticut
Box U-8
Storrs, CT 06268
(203) 486-5002

Florida PIRG (FPIRG)
Director: Sylvia Ward
1441 East Fletcher Avenue
Tampa, FL 33612
(813) 971-7564

Illinois PIRG
Organizing Committee
Contact: Megan Seibel
845 Chicago Avenue, Suite 3C
Evanston, IL 60202
(312) 492-9828

Indiana PIRG (INPIRG)
Env. Proj. Dir.: Mick Harrison
Activities Desk, 1st Floor
Indiana Memorial Union
Bloomington, IN 47405
(812) 855-7575

Iowa PIRG (IPIRG)
Director: Jim Dubert
Iowa State University
Memorial Union, Room 36
Ames, IA 50011
(515) 294-8094

Maine PIRG (USMPIRG)
Contacts: Christie Johansen/Berla
   Stacey
94 Bedford Street
Portland, ME 04103
(207) 780-4044

Maryland PIRG (MaryPIRG)
Director: Shannon Varner
3110 Main Dining Hall
University of Maryland
College Park, MD 20742
(301) 454-5601

Massachusetts PIRG (MASSPIRG)
Director: Doug Phelps
29 Temple Place
Boston, MA 02111
(617) 292-4800

PIRG in Michigan (PIRGIM)
Director: Elise Jacques
212 South 4th Ave. #207
Ann Arbor, MI 48104
(313) 662-6597

Minnesota PIRG (MPIRG)
Director: Niel Ritchie
2512 Delaware Street, S.E.
Minneapolis, MN 55414
(612) 627-4035

Missouri PIRG (MoPIRG)
Director: Beth Zilbert
4144 Lindell Boulevard
Suite 410
Saint Louis, MO 63108
(314) 534-7474

Montana PIRG (MontPIRG)
Director: Brad Martin
2356 Corbin Hall
Missoula, MT 59812
(406) 243-2907

New Jersey PIRG (NJPIRG)
Director: Ken Ward
99 Baynard Street
New Brunswick, NJ 08901
(201) 247-4606

New Mexico PIRG (NMPIRG)
Contacts: Kevin Bean/Robyn Reed
University of New Mexico
Box 66 SUB
Albuquerque, NM 87131
(505) 277-2757

New York PIRG (NYPIRG)
Director: Jay Halfon
9 Murray Street
New York, NY 10007
(212) 349-6460

Ohio PIRG (OPIRG)
Director: John Meyer
Wilder Box 89
Oberlin, OH 44074
(216) 775-8137

Ontario PIRG (OPIRG)
Provincial Coord.: Lisa Kristianson
455 Spadina Ave.
Room #201
Toronto, Ontario M55G8
CANADA
(416) 598-1576

Oregon PIRG (OSPIRG)
Director: Joel Ario
027 SW Arthur
Portland, OR 97201
(503) 222-9641

Rhode Island PIRG (RIPIRG)
Director: Rich McClintock
4th Floor
265 Atwell's Ave.
Providence, RI 02903
(401) 331-7474

Texas PIRG
Contact: Heather Cusick
123 South LBJ
San Marcos, TX 78666
(512) 396-5020

United States PIRG (U.S. PIRG)
Director: Gene Karpinski
215 Pennsylvania Avenue S.E.
Washington, D.C. 20003
(202) 546-9707

Vermont PIRG (VPIRG)
Director: John Gilroy
43 State Street
Montpelier, VT 05602
(802) 223-5221

Washington PIRG (WashPIRG)
Director: Wendy Wendlandt
340 15th East, Suite 350
Seattle, WA 98112
(206) 322-9064

West Virginia PIRG (WVPIRG)
Director: Brian Gessner
Mountainlair, SOW
West Virginia University
Morgantown, WV 26506
(304) 293-2108

Hotlines, Emergency Assistance, and Other Important Phone Numbers

Below is a list of the toll-free numbers offered by EPA Headquarters:

Hotlines		*Hours*
RCRA/Superfund Hotline	800-424-9346	Monday–Friday
—in Washington, D.C.	202-382-3000	8:30–4:30
(Also serves as LUST hotline)		
Title III, Emergency Planning Community		
Right-to-Know Hotline	800-535-0202	Monday–Friday
—in Washington, D.C.	202-479-2449	8:30–4:30
Safe Drinking Water Hotline	800-426-4791	Monday–Friday
—in Washington, D.C.	202-382-5533	8:30–4:30
National Pesticides Telecommunications Network	800-858-7378	Seven days a week
—in Texas	806-743-3091	24 hours a day
Asbestos Hotline	800-334-8571	Monday–Friday
		8:15–5:00
National Response Center Hotline	800-424-8802	Seven days a week
—in Washington	202-426-2675	24 hours a day
(For reports of oil and hazardous substances accidents and spills)		
Inspector General's Whistle-Blower Hotline	800-424-4000	Monday–Friday
—in Washington, D.C.	202-382-4977	10:00–3:00
(For confidential reports of agency-related waste, fraud, mismanagement, or abuse from the public or EPA employees)		Callers can leave messages at all other times.
Emergency Information Numbers		
Federal Emergency Management Agency	202-646-2400	24 hours
Agency for Toxic Substances and Disease Registry	404-452-4100	24 hours
National Response Center in Washington, D.C.	800-424-8802	24 hours
	202-426-2675	
	202-267-2675	

Nuclear Regulatory Commission	301-951-0550	24 hours
U.S. Department of Energy Radiological Assistance	202-586-8100	24 hours
CHEMTREC	800-424-9300	24 hours
Association of American Railroads/Bureau of Explosives	202-639-2222	24 hours

Other Important Numbers

TSCA Assistance Line	202-554-1404	All operate during normal business hours.
Office of Groundwater	202-382-7077	
EPA Office of Community Liaison	202-382-4454	
EPA Office of Legislative Analysis	202-382-5414	
EPA Public Information Center	202-382-2080	
Chemical Referral Center	800-262-8200	

Agencies Designated to Receive Notifications of Underground Storage Tanks

Alabama (EPA Form)
Alabama Department of
  Environmental Mgmt.
Ground Water Section/Water Division
1751 Federal Drive
Montgomery, Alabama 36130

Alaska (EPA Form)
Department of Environmental
  Conservation
Pouch O
Juneau, Alaska 99811
907/465-2653

American Samoa (EPA Form)
Executive Secretary
Environmental Quality Commission
Office of the Governor
American Samoan Government
Pago Pago, American Samoa 96799
Attention: UST Notification

Arizona (EPA Form)
Attention: UST Coordinator
Arizona Department of Health
  Services
Environmental Health Services
2005 N. Central
Phoenix, Arizona 85004

Arkansas (EPA Form)
Arkansas Department of Pollution
  Control and Ecology
P.O. Box 9583
Little Rock, Arkansas 72219
501/562-7444

California (State Form)
Ed Anton
California Water Resources Control
  Board
P.O. Box 100
Sacramento, California 95801
916/445-9552

Colorado (EPA Form)
Kenneth Mesch, Section Chief
Colorado Department of Health
Waste Management Division
Underground Tank Program
4210 East 11th Avenue
Denver, Colorado 80220
303/320-8333 Ext. 4364

Connecticut (State Form)
Hazardous Materials Management
  Unit
Department of Environmental
  Protection
State Office Building
165 Capitol Avenue
Hartford, Connecticut 06106

Delaware (State Form)
Division of Air and Waste
  Management
Department of Natural Resources and
  Environmental Control
P.O. Box 1401
89 Kings Highway
Dover, Delaware 19903
302/736-5409

District of Columbia (EPA Form)
Attention: UST Notification Form
Department of Consumer and
  Regulatory Affairs
Pesticides and Hazardous Waste
  Management Branch
Room 114
5010 Overlook Avenue, S.W.
Washington, D.C. 20032

Florida (State Form)
Florida Department of Environmental
  Regulation
Solid Waste Section
Twin Towers Office Building
2600 Blair Stone Road
Tallahassee, Florida 32301
904/487-4398

Georgia (EPA Form)
Georgia Department of Natural
  Resources
Environmental Protection Division
Underground Storage Tank Program
3420 Norman Berry Drive
Hapeville, Georgia 30354

Guam (State Form)
James B. Branch, Administrator
Guam Environmental Protection
  Agency
P.O. Box 2999
Agana, Guam 96910
Overseas Operator (Commercial Call
  646-8863)

Hawaii (EPA Form)
Chief, Noise and Radiation Branch
Hawaii Department of Health
591 Ala Moana Boulevard
Honolulu, Hawaii 96801
808/548-4129

Idaho (EPA Form)
Underground Storage Tank
  Coordinator
Water Quality Bureau
Idaho Department of Health & Welfare
Division of Environment
450 W. State Street
Boise, Idaho 83720
208/334-4251

Illinois (EPA Form)
Underground Storage Tank
  Coordinator
Division of Fire Prevention
Office of State Fire Marshal
3150 Executive Park Drive
Springfield, Illinois 62703-4599

Indiana (EPA Form)
Division of Land Pollution Control,
  UST Program
Indiana State Board of Health
P.O. Box 7015
Indianapolis, Indiana 46207
317/243-5060

Iowa (State Form)
Iowa Department of Water, Air and
  Waste Management
900 East Grand
Des Moines, Iowa 50319
515/281-8692

Kansas (EPA Form)
Office of Environmental Geology
Kansas Department of Health &
  Environment
Forbes Field, Building 740
Topeka, Kansas 66620
913/862-9360 Ext. 221

Kentucky (State Form)
Natural Resources Cabinet
Division of Waste Management,
    Attention: Vicki Pettus
18 Reilly Road
Frankfort, Kentucky 40601
502/564-6716

Louisiana (State Form)
Patricia L. Norton, Secretary
Louisiana Department of
    Environmental Quality
P.O. Box 44066
Baton Rouge, Louisiana 70804
504/342-1265

Maine (State Form)
Attention: Underground Tanks
    Program
Bureau of Oil & Hazardous Material
    Control
Department of Environmental
    Protection
State House—Station 17
Augusta, Maine 04333
207/289-2651

Maryland (EPA Form)
Science and Health Advisory Group
Office of Environmental Programs
201 West Preston Street
Baltimore, Maryland 21201

Massachusetts (EPA Form)
UST Registry, Department of Public
    Safety
1010 Commonwealth Avenue
Boston, Massachusetts 02215
617/566-4500

Michigan (EPA Form)
Ground Water Quality Division
Department of Natural Resources
Box 30157
Lansing, Michigan 48909

Minnesota (State Form)
Underground Storage Tank Progam
Division of Solid and Hazardous
    Wastes
Minnesota Pollution Control Agency
1935 West County Road, B-2
Roseville, Minnesota 55113

Mississippi (EPA Form)
Department of Natural Resources
Bureau of Pollution Control
P.O. Box 10385
Jackson, Mississippi 39209

Missouri (EPA Form)
Gordon Ackley, UST Coordinator
Missouri Department of Natural
    Resources
P.O. Box 176
Jefferson City, Missouri 65102

Montana (EPA Form)
Solid and Hazardous Waste Bureau
Department of Health and
    Environmental Science
Cogswell Building, Room B201
Helena, Montana 59620

Nebraska (EPA Form)
Nebraska State Fire Marshal
P.O. Box 94677
Lincoln, Nebraska 68509-4677

Nevada (EPA Form)
Attention: Underground Storage
    Tanks
Division of Environmental Protection
Department of Conservation and
    Natural Resources
Capitol Complex
201 S. Fall Street
Carson City, Nevada 89710
800/992-0900 Ext. 4670

New Hampshire (EPA Form)
Water Supply and Pollution Control
  Commission
Hazen Drive
P.O. Box 95
Concord, New Hampshire 03301
Attention: UST Registration
603/271-3503

New Jersey (State Form)
Underground Storage Tank
  Coordinator
Department of Environmental
  Protection
Division of Water Resources (CN-029)
Trenton, New Jersey 08625
609/292-0424

New Mexico (EPA Form)
New Mexico Environmental
  Improvement Division
Ground Water/Hazardous Waste
  Bureau
P.O. Box 968
Sante Fe, New Mexico 87504
505/827-2933 or 505/827-2918

New York (EPA Form)
Bulk Storage Section
Division of Water
Department of Environmental
  Conservation
50 Wolf Road, Room 326
Albany, New York 12233-0001
518/457-4351

North Carolina (EPA Form)
Division of Environmental Mgmt./
  Ground Water Section
Dept. of Natural Resources &
  Community Development
P.O. Box 27687
Raleigh, North Carolina 27611
919/733-5083

North Dakota (State Form)
Division of Hazardous Waste Mgmt.
  and Special Studies
North Dakota Department of Health
Box 5520
Bismarck, North Dakota 58502-5520

**Northern Mariana Islands (EPA
  Form)**
Chief
Division of Environmental Quality
P.O. Box 1304
Commonwealth of Northern Mariana
  Islands
Saipan, CM 96950
Overseas Operator: 6984
Cable Address: GOV. NMI Saipan

Ohio (State Form)
State Fire Marshal's Office, UTN
Department of Commerce
8895 E. Main Street
Reynoldsburg, Ohio 43068
State Hotline 800/282-1927

Oklahoma (EPA Form)
Underground Storage Tank Program
Oklahoma Corporation Comm.
Jim Thorpe Building
Oklahoma City, Oklahoma 73105

Oregon
Underground Storage Tank Program
Hazardous and Solid Waste Division
Department of Environmental Quality
P.O. Box 1760
Portland, Oregon 97207
503/229-5788

Pennsylvania (EPA Form)
Pennsylvania Department of
  Environmental Resources
Bureau of Water Quality Management/
  Ground Water Unit
9th Floor, Fulton Building
P.O. Box 2063
Harrisburg, Pennsylvania 17120

Puerto Rico (EPA Form)
Director, Water Quality Control Area
Environmental Quality Board
Commonwealth of Puerto Rico
P.O. Box 11488
Santurce, Puerto Rico 00910
809/725-0717

Rhode Island (EPA Form)
UST Registration
Department of Environmental
  Management
204 Cannon Building
75 Davis Street
Providence, Rhode Island 02908
401/277-2234

South Carolina (State Form)
Attention: Susana Workman
Groundwater Protection Division
South Carolina Dept. of Health and
  Environmental Control
2600 Bull Street
Columbia, South Carolina 29201
803/758-5213

South Dakota (EPA Form)
Office of Water Quality
Department of Water and Natural
  Resources
Joe Foss Building
Pierre, South Dakota 57501

Tennessee (EPA Form)
Terry K. Cothron, Director
Division of Ground Water Protection
Tennessee Department of Health and
  Environment
150 Ninth Avenue, North
Nashville, Tennessee 37219-5404
615/741-7206

Texas (EPA Form)
Underground Storage Tank Program
Texas Water Commission
P.O. Box 13087
Austin, Texas 78711

Utah (EPA Form)
Kenneth L. Alkema
Division of Environmental Health
P.O. Box 45500
Salt Lake City, Utah 84145-0500

Vermont (State Form)
Underground Storage Tank Program
Vermont AEC/Waste Management
  Division
State Office Building
Montpelier, Vermont 05602
802/828-3395

Virginia (EPA Form)
Russell P. Ellison, III, P.G.
Virginia Water Control Board
P.O. Box 11143
Richmond, Virginia 23230-1143
804/257-6685

Virgin Islands (EPA Form)
205(J) Coordinator
Division of Natural Resources
  Management
14 F Building 111, Watergut Homes
Christianstead, St. Croix, Virgin
  Islands 00820

Washington (State Form)
Earl W. Tower, Supervisor
Department of Ecology, M/S PV-11
Management Division, Solid and
  Hazardous Waste
Olympia, Washington 98504-8711
206/459-6316

West Virginia (EPA Form)
Attention: UST Notification
Solid and Hazardous Waste/Ground
  Water Branch
West Virginia Department of Natural
  Resources
1201 Greenbriar Street
Charleston, West Virginia 25311

Wisconsin (State Form)
Bureau of Petroleum Inspection
P.O. Box 7969
Madison, Wisconsin 53707
608/266-7605

Wyoming (EPA Form)
Water Quality Division
Department of Environmental Quality
Herschler Building, 4th Floor West
122 West 25th Street
Cheyenne, Wyoming 82002
307/777-7781

INDEX

Note: Page numbers in italics refer to illustrations.

Abandoned Mines Land Fund, 300
Acetone, 17
Activated carbon treatment, 186
Acute exposure, 18–19
Administrative orders of federal agencies, 217–18, 258, 268, 273–74
Administrative process, action in the, 87–99
  general advice, 87–88
  making the agency enforce the law, 95–98
  participation in rulemaking process, 88–90
  permit process, 90–95
  review of agency action by the courts, 98–99
Agency for Toxic Substances and Disease Registry (ATSDR), 109, 168
  health assessments, 179–80
  list of 100 most hazardous contaminants, 168, 169
Agricultural Chemicals in Ground Water Proposed Pesticide Strategy, 285
Agricultural drainage wells, 29
Agricultural practices causing contamination, 32–34, 278–89, 364–68
Air stripping, 184
Alabama, 328
Aldicarb, 286–88, 366
Alternate concentration limits (ACLs), 221
Ambient monitoring, 51, 55–56, 251
American Association of Laboratory Accreditation, 64
American Council of Independent Laboratories, 64
American Petroleum Institute, 360
Amicus curiae, participation as, 97, 98, 284
Anaconda Copper Corporation, 370
Anaerobic lagoons, 186

Animal feedlots, 289, 364, 367
Annotated Bibliography on Wellhead Protection Areas, 156
Appalachian Research Defense Fund, 303
Appalachian Student Health Coalition, 173
Appealing EPA decisions, 94, 189, 260–62
Appealing permit decisions, 94, 302, 303, 338, 371–73
Aquifer protection programs, 152–61
Aquifers, 3–6
  bedrock or consolidated, 5–6
  mapping and contamination, 71–79, 315
  preventing exemptions of, under Safe Water Drinking Act, 262
  unconsolidated, 5, 6
  Wellhead Protection Program, 124, 152–59
Area of influence, 10
Arizona, 345, 372
Arizona Daily Star, 13
Army, U.S., Toxic and Hazardous Materials Agency, 188
Asbestos, 245
Ashland Oil, 265n.
Atomic Energy Act, 58
Automotive service station waste wells, 29

Barnstable, Massachusetts, 345
Basic information about groundwater, 1–10
  aquifers, 3–6
  containment movement, 6–10
  hydrologic cycle, 3, 4
Belleville, Indiana, 264
Benzene, 17, 264
Berkeley, California, 368
Berms, 185
Biological contaminants, effects on health of, 15
Biological reactors, 186
Bioreclamation, 186

405

Birth defects, 15, 17
Blue-baby syndrome, 17, 44
Bunker Hill, Idaho, 370
Bureau of Land Management, 56, 292, 293
Butte, Montana, 370

California, 335, 351, 352, 353, 354, 360,
    365, 366
  Hazardous Substance Control Act, 332–
    33
  Proposition 65, 345–46
Cancer, 12–13, 15, 17, 19
Cape Cod Planning and Economic
    Development Commission, 345
Caps, 186
Carbon absorption, 186
Carcinogens, 19
Causation, proof of, 109
Center for Community Action, Lumberton,
    North Carolina, 128
Centers for Disease Control, 42
Central nervous system disorders, 17
CERCLA, see Comprehensive
    Environmental Response,
    Compensation, and Liability Act
CERCLIS (Comprehensive Response
    Liability Information Service), 163
Channels and waterways, 185
Chemical Manufacturers Association, 346
Chemigation, 34, 367
Chickasaw Community Affairs Group
    (CCAG), 382
Chicot Aquifer, 248
Chlorination, 17–18, 186
Cholera, 15
Chronic exposure, 18–19
Ciba-Geigy, 246
Circulatory system disorders, 17
Citizen action, 83–119
  see also specific forms of citizen action
Citizen lawsuits, 102–106, 164, 172, 190,
    191, 240, 247, 335, 380
  coal mining permits and, 298, 302
  to enforce NPDES permit, 243
  forms of relief resulting from, 104–105
  imminent hazard, 190, 194–95, 197–98,
    210, 269
  limitations on, 105–106
    under Superfund, 189–90
  for Safe Water Drinking Act violations,
    149
  statute provisions concerning, 103
  types of, 102–104
  against underground storage tank owner
    or operator, 269, 274
  against waste disposal well operator,
    252, 253, 259

Citizen's Clearinghouse for Hazardous
    Waste, 37, 65
Citizens for Control of Toxic Wastes
    (CCTW), 219
Class V injection wells, 250, 354
Classification systems for groundwater,
    state, 317–18
Class II injection wells, 249, 355
CLEAN (Calcasieu League of
    Environmental Action Now), 248
Clean Water Act (CWA), 58, 65, 103, 123,
    126, 201, 289, 383
  imminent hazard actions under, 194–95
  industrial and commercial wastes and,
    233–34, 237, 242–45
  mining pollutants and, 291–92
  1987 amendments to, 283
  overview of, 124–25
  state programs to control nonpoint
    pollution sources under, 320
Coal mining, 40–41, 291–303
  Clean Water Act provision and, 291–92
  on federal lands, 292–93
  initiating enforcements actions, 297–98,
    299
  locations of mines in U.S., 295
  making sure mines are reclaimed, 298–
    300, 302
  participation in permit process, 300–
    303
  petitions to have land designated as
    unsuitable for surface mining,
    303
  policing operations to prevent
    contamination, 296–97
  state regulation of, 370–73
  taking action against contaminating
    mines, 294–96
Coastal areas, 328
Coast Zone Management Act, 59
Code of Federal Regulations, 128
College laboratories, 57, 64
Colorado, 371, 382
Columbia, Mississippi, 162
Commercial septic systems, 378–80
Commercial sources of groundwater
    contamination, see Industrial and
    commercial sources of groundwater
    contamination
Common-law actions, 106–109, 197
  negligence, 107, 270
  nuisance, 106–107, 191, 270
  problems with, 108–109
  relief resulting from, 108
  strict liability, 107
  trespass, 107–108, 270
Community action, see Grass-roots action

Community Awareness and Emergency Response Program (CAER), 346
Comprehensive Environmental Response, Compensation, and Liability Act (CERCLA, or Superfund), 194–95, 282, 294, 330
    action under, 162–91
        challenging agency removal action, 170–73
        getting potential Superfund sites evaluated by EPA, 166–69
        identifying spills and contaminated sites, 162–64
        participation in cleanup actions, 165
        participation in cleanups involving the polluter, 188–89
        participation in remedial actions, 173–88
        participation in removal actions, 170–73
        taking your own action, 191
        ways to get action to clean up spills and other emergencies, 164–65
    imminent hazard action and, 194–95, 197, 281
    industrial and commercial wastes and, 233, 237
    overview of, 124, 162
    state programs, 330–33
    waste disposal wells and, 252
Comprehensive planning on local level, 322, 323
Concerned Citizens to Save Fayette County (CCSFC), 173
Conditional use permits, 326
Cone of depression, 9, 10
Congressional documents, how to order, 387
Connecticut, 325, 335, 338, 345, 358–59, 383
Conservation Law Foundation, 360, 362
Constitution, U.S., 341
Contaminant movement and groundwater, 6–10
    direction of flow, 8
    human influence on flow, 9–10
    recharge and discharge areas, 6–8
    speed of flow, 9
Contaminant regulation by EPA:
    contaminants regulated since June 1989, 141
    contaminants to be regulated by 1997, 142–43
    monitoring for unregulated contaminants, 142–43, 144–45
    for 30 health-threatening contaminants, 136–40

Contamination sources, see Sources of groundwater contamination; specific sources
Cooperative Extension Service offices, county, 278
Cooper Tool Group, 348
Council on Environmental Quality (CEQ), 201
Court action, see Citizen lawsuits; Common-law actions
Court review of agency action, see Judicial review
Critical areas protection, 327–28

Department of Agriculture, 56
    Agricultural Research Service of, 279
    Agricultural Stabilization and Conservation Service, 288, 366
    Cooperative Extension Service, 288
    maps of, 74, 75
    Soil Conservation Service, 279, 288, 367
Department of Defense, 37–38, 56, 188
Department of Energy, 37, 38–40, 218
Department of Environmental Quality, state, 202, 330
Department of Public Health, state, 375
Department of Public Works, state, 383
Department of Transportation (DOT), 46, 306–309, 383
    Bureau of Motor Carrier Safety, 306
    Federal Highway Administration, 308, 382
    Federal Railroad Administration, 306, 308
    Office of Hazardous Materials Transportation, 308
    Office of Pipeline Safety, 306, 308
    publications, 383
Dikes, 185
Discharge areas, 6–7
Discharge Monitoring Reports (DMRs), 242
DNA, 15, 19
Dow Chemical, 341
D.R.A.S.T.I.C. system of aquifer assessment, 78–79
Dredging, 184–85, 186
Dust suppression devices, 186

East Setauket, Long Island, 265–66
Education programs, 288, 338, 366
Emergency information telephone numbers, 397–98
Emergency Planning and Community Right-to-Know Act (EPCRA), 306
    industrial and commercial wastes and, 232–40, 344, 346, 347

Emergency Planning and Community
    Right-to-Know Information Hotline,
    234
Emergency preparedness and response
    programs, state, 346
Enforcement actions by agencies:
    initiating, 95–97
    participating in, 97–98
    see also individual laws, agency actions,
        and contamination sources
Environmental Action Foundation, 275
Environmental and Energy Study Institute,
    278
Environmental Assessment (EA), 199, 200
Environmental impact report (EIR), 334
Environmental Impact Statement (EIS),
    199, 200, 201, 202, 334–35
Environmental Protection Agency (EPA),
    45, 46, 56, 57, 65, 66, 85, 87, 386
    laboratories certified by, 57, 64
    National Primary Drinking Water
        regulations, 281
    National Priorities List, 163, 174, 175
    Office of Drinking Water, 286
    Office of Groundwater Protection, 129–
        32
    Office of Pesticides and Toxic
        Substances, 279, 383
    permits, see Permit process; Permits
    publications, 156, 157, 354
    regional offices, addresses and telephone
        numbers of, 385
    requirements, 33
    state enforcement under supervision of,
        96, 125–28, 135
    statistics and surveys, 17, 24, 27, 157,
        203, 281
    testing by, 33
    see also specific laws and programs
        enforced by the EPA
Environmental Task Force, 275
Excavation, 184–85, 186
Expert help in mapping aquifers and
    contamination, 76–79

Fayette County, West Virginia, 173
Feasibility studies, 181
Federal facilities, Superfund cleanup of,
    188
Federal Highway Administration, 308,
    382
Federal Insecticide, Fungicide, and
    Rodenticide Act (FIFRA), 59, 123,
    125, 284
Federal Land Policy and Management Act,
    59
Federal lands, mining on, 292–93

Federal laws and regulations:
    existing, see specific laws and
        regulations
    finding, 128–29
    new, 130–31, 132
Federal programs and groundwater, 123–
    309
    aquifer protection programs, 152–61
    imminent hazard action, see Imminent
        hazard action
    industrial and commercial sources, 230–
        47
    introduction to federal groundwater
        protection, 123–32
    mining, 291–303
    National Environmental Policy Act,
        199–203
    pesticide and other agricultural
        programs, 278–89
    Safe Water Drinking Act water quality
        programs, 134–50
    Superfund action, 162–91, 194–95, 282,
        284
    transportation of hazardous materials,
        305–309
    treatment, storage, or disposal facilities,
        204–27
    underground storage tanks, 264–76
    waste disposal wells, 248–62
Federal Railroad Administration, 306, 308
Federal Register, 128–29, 197, 386
Feedlots, 289, 364, 367
Fertilizers, 33, 289, 366
Filipone, Ella, 161
Filtration, 186
Fish and Game Department, state, 202
Fish and Wildlife Service, U.S., 202
Floating covers, 185
Florida, 318, 341, 354, 357–58, 359, 360,
    362, 376–77, 379
Florida Underground Petroleum
    Environmental Protection Act, 359
Forest Service, 292, 293
Freedom of Information Acts, 83–85

Gas collection systems, 186
Gasoline, 264
    see also Underground storage tanks
Gastrointestinal illnesses, 15, 17
General Accounting Office (GAO), 40, 337,
    386–87
General Motors, 347
General plan, local, 322, 323
Generic environmental impact reports
    (GEIRs), 368
Genetic mutations, 17
Genoa Township, Michigan, 345

Gettysburg, Pennsylvania, 198
Government sources for reference
    materials, 386–87
Grading, 185
Grand View, Idaho, 219
GRANIT, 77
Grass-roots action, 111–18
    see also specific actions
Groundwater classification systems, state,
    317–18
Groundwater protection standards (GPLs),
    220–21
Groundwater pumping, 184
Groundwater quality standards, state, 318–
    20
Grout walls, 184
GSX Chemical Services, Inc., 127
Guidance for Applicants for Wellhead
    Protection Program Assistance
    Funds, 157
Guidelines for Delineation of Wellhead
    Protection Areas, 156

Hawaii, 326
Hazardous Liquid Pipeline Safety Act, 59
Hazardous Materials Transportation Act,
    59, 307
Hazardous waste:
    definition of, under RCRA, 207
    federal programs for disposal of, see
        Treatment, storage, or disposal
        facilities
    household, see Household hazardous
        wastes and products
    restricting disposal by origin of waste,
        341
    state and local programs for, 338–41,
        345–46
    transportation of, see Transportation of
        hazardous wastes and toxic
        chemicals
Hazardous Waste Treatment Council, 127
Hazard Ranking System (HRS), 174–79
Health and groundwater contamination,
    12–24
    common contaminants, 15–18
    conclusion, 21–22
    factors complicating link between health
        effects and contaminants, 18–20
    legal limits on contaminants, 20–21
    potential health effects of Superfund
        contaminants, 16
    testing contaminant's health effects, 14–
        15
Health risk assessment of potential
    Superfund site, 170–71
Hepatitis, infectious, 15

Highway runoff, state regulation of, 383
Hotlines, 234, 397
Housatonic Valley Association,
    Groundwater Action Plan (GAP),
    325
Household hazardous wastes and products,
    35–36
    categories of, 35–36
    chart of disposal methods for, 116–17
    reducing contamination by, 36, 114
    state and local programs for disposal of,
        340–41
    ways groundwater is contaminated by, 36
Hughes Aircraft Company, 13
Hydrologic cycle, 3, 4
Hydrologic reclamation plan, 301–302

Idaho, 371
Idaho Conservation League, 219
Idaho Fair Share, 219
Imminent hazard action, 104, 150, 190,
    193–98, 210, 281, 294, 309
    bringing a citizen suit, 190, 194–95,
        197–98, 210, 269
    determining when it is appropriate,
        193–96
    how to produce action, 196–97
Incineration, 184
Indiana, 335
Industrial and commercial sources of
    groundwater contamination, 230–
    47, 306
    Clean Water Act and, 233–34, 237, 242–
        45
    emergency planning for your
        community, 238–40
    identifying possible sources, 231–36
    materials regulated by federal statutes,
        232–33
    Resources Conservation and Recovery
        Act and, 231–33, 237, 240–41
    responding to threatening spills or
        chemicals, 236–37
    state regulation of, 344–49
    Toxic Substance Control Act and, 233,
        237, 245–47
    see also Transportation of hazardous
        wastes and toxic chemicals;
        Treatment, storage, or disposal
        facilities; Underground storage
        tanks
Industrial Directory, 231
Industrial septic systems, 378–80
Inhalation of contaminants, 19
Injection wells, see Waste disposal wells
Inorganic contaminants, health effects of,
    15–17

Inventory of Class V Wells, 354
Iowa, 354
Irrigation, 34, 367

Jay Woods Oil, Inc., 260
Jefferson County, Wisconsin, 367
Judicial review, 90, 98–99, 226
Justice Department, 218

Kansas, 355, 367
Karst limestone, 6
Kidney damage, 17
King, Jonathan, 41
Kingston, New Hampshire, 196

Laboratory testing of tap water, 57,
    64–66
Lagoons, *see* Surface impoundments
Lake Charles, Louisiana, 248
Land application, 34–35, 377
Landfills, 24–25, *26*
Land use controls, 322–28, 344, 373
    local, 322–26
    sensitive areas protection, 327–28
    state development legislation, 326
    see also Wellhead Protection Program
Land-use inventory, preparing a, 75
Lauderdale Citizens for a Clean
        Environment (LCCE), 115–18
Lawyer, selection of a, 102
Leachate control, 184
League of Women Voters, 341
Left Beaver Coal Company, 303
Legal limits on contaminants, 20–21
Legislation, *see* Federal laws and
        regulations
Liners, 186
Liver damage, 17
Livestock, 33
Livingston County, Michigan, 345
Local programs, *see* State and local
        programs for groundwater
        protection
Louisiana, 328, 330–31
Lung damage, 17
LUST (Leakage Underground Storage Tank)
        program, 264–76

Maine, 326, 359, 373
Mapping aquifers and contamination, 71–
        79, 315
    compiling your own water supply maps,
        72–75
    getting started, 72
    seeking the help of experts, 76–79
Maryland, 328, 341, 377, 379
Massachusetts, 318, 325, 327, 331, 332,

339, 359, 360, 366, 368, 377, 379,
        383
Massachusetts Clean Water Act (CWA), 187
Master plan, local, 322, 323
Material Safety Data Sheets (MSDSs), 234–
        35, 239
Maximum contaminant level (MCL), 135,
        142–45, 285
    policing variances and exemptions to,
        146–47
    for 30 contaminants, 136–40
Maximum contaminant level goal (MCLG),
        135, 182
    for 30 contaminants, 136–40
Metal contaminants, health effects of, 15–
        17
Methemoglobinemia, 17, 44
Methylene chloride, 17
Michigan, 335
Michigan Department of Public Health, 64
Microbiological degradation, 184
Military compounds, RCRA and, 218
Military toxics, 36–40
Mining, 40–41, 291–303
    Clean Water Act provisions and, 291–92
    on federal lands, 292–93
    state regulation of, 370–73
    see also Coal mining
Mini-superfund-type cleanups, 191, 359
Minnesota, 333, 335, 347, 376, 378
Minnesota Health Department, 188
Minnesota Pollution Control Agency, 188
Minnesota Project, 367
Mississippi, 328, 338
Missouri, 331, 345
Monitoring groundwater quality, *see*
        Testing groundwater quality
Montana, 316, 318, 372
Multiple pollutants, 19–20
Mutagens, 19

Nassau County Department of Health, 35
National Academy of Sciences, 21
National Association of Attorneys General,
        341
National Contingency Plan, 124, 165n., 191
National Environmental Policy Act
        (NEPA), 57, 199–202, 292, 293, 334
National Pollutant Discharge Elimination
        System (NPDES), 125
    permit program, 125, 242–43, 344
    permits under, 233, 242–43, 244, 246,
        291–92, 344
National Priorities List, 163, 174, *175*
National Response Center (NRC), 163
    reporting spills to, 164, 237, 267, 282,
        306

Nebraska, 367
Negligence, common-law action based on, 107, 270
Nervous system impairments, 17
Nevada, 371, 372
New federal legislation, 130–31, 132
New Hampshire, 359, 360, 379
  aquifer-mapping efforts in, 76–77
New Jersey, 319, 333, 335, 337, 339, 354
New Jersey Department of Environmental Protection, 281, 347–49
New Jersey Environmental Cleanup Responsibility Act (ECRA), 347–49
New Jersey Pollution Discharge Elimination System permit program, 344
New Jersey Underground Storage Tank Laws, 362
New Mexico, 371
New Mexico Water Quality Control Commission, 264
New York, 328, 335, 339, 354
  State Pollutant Discharge Elimination System (SPDES) permit, 379
New York City, 382
Nitrates/nitrogen, 17, 33
Nonhazardous waste dumps (solid waste facilities):
  federal programs, 204, 226–27
  state programs, 337–38
Nonpoint pollution sources, state programs to control, 320
North Alabama Lauderdale Citizens for a Clean Environment (LCCE), 115–18
North Carolina, 127–28, 332, 347
North Dakota, 367
Northville Industries, 265
Nuclear Regulatory Commission, 56, 294
Nuisance theory, common-law action based on, 106–107, 191, 270

Occupational Safety and Health Administration (OSHA), 234–35
Ocean County Citizens for Clean Water, 179, 246
Office of Groundwater Protection, Environmental Protection Agency, 129–32
Office of Surface Mining (OSM), 56, 293, 298–99
Office of Technology Assessment (OTA), 15, 17
Ohio, 354, 359, 360, 361, 372
Ohio Environmental Protection Agency, 316
Oklahoma, 353
On-Scene Coordinators, 164, 167, 168

On-site heating of soil, 184
Organic compounds as contaminants, health effects of, 17
Organizations providing information and services, 388–96
Organizing for community action, 112–13
Ottawa County, Michigan, 376
Oxidation, 186

Passaic River Coalition (PRC), 161
Passaic River Watershed Association (PRWA), 362
PCBs (polychlorinated biphenyls), 125, 173, 245, 346
Pennsylvania, 345
People Against Hazardous Landfill Sites (PAHLS), 225
People United to Restore the Environment, 187
Percolation tests, soil, 74–75
Permeable soils, identifying, 74–75
Permeable treatment beds, 184, 186
Permit process, 90–95, 243, 292, 371, 372
  challenging a permit decision, 94, 302, 303, 338, 371–73
  for coal mines, 298, 300–303
  commenting on the draft permit, 91–93
  important early steps in, 90–91
  petitioning for modifications or termination of the permit, 95, 243, 298
  reviewing the agency decision, 93–94
  for treatment, storage, or disposal facilities, 220–23
  for waste disposal well, 259–62
Permits, 326
  conditional use, 325
  mining, 293, 294, 296, 298, 300–303, 372–73
  National Pollutant Discharge Elimination System (NPDES), 233, 242–43, 244, 246, 291–92, 344
  state, 318, 337, 344–45, 360, 361, 370–73, 376, 379–80, 382
  for waste disposal wells, 252–53, 258
  see also Permit process
Pesticides, 17, 32–33, 278–88
  changes in, or cancellations of regulation, 284–88, 366
  disposal, 282, 283, 288
  education of users of, 288
  identifying the problem, 278–81
  label violations, 282–84
  preventing future contamination, 282–88
  remedying current contamination, 281–82

Pesticides (*continued*)
  state regulation of, 364–66
    government-applied pesticides, 368
Pesticides in Groundwater: Background Document, 279
Physicians, ability to detect contaminant-caused illnesses, 20
Pipelines, 307, 308, 381
  see also Transportation of hazardous wastes and toxic chemicals
Pipeline Safety Office, state, 306
Pipe vents, 186
Pits, *see* Surface impoundments
Plume formation of groundwater contaminants, 8, 9
Point-of-contamination monitoring, 51, 56
  agencies responsible for, 56
Political actions, 113, 153
Ponds, *see* Surface impoundments
Portland, Oregon, 382
Precipitation, 186
Preliminary assessments of potential Superfund sites by EPA, 166–67
Premanufacture notices (PMNs), 245
Probable hydrologic consequences (PHC) determination, 301
Property transfer and cleanup statutes, state, 347–49
Provincetown, Massachusetts, 264
Public hearings:
  Department of Transportation, 308–309
  on permits, 93, 209, 300, 302, 371, 372
  state, 318, 338, 339
Publicly owned treatment works (POTWs), 243–44

Quality Criteria for Water 1986, 243
Quality standards, state groundwater, 318–20

Radionuclides, 18
Railroads, 307, 308, 381
  see also Transportation of hazardous wastes and toxic chemicals
Raw-sewage disposal wells, 29
RCRA Groundwater Monitoring Technical Enforcement Guidance Document, 211
Recharge areas, 6–8
  identifying, 75
  protection of, 326, 327
Reclamation Act, 60
Reclamation of mines, 298–300, 303
Recycling, 114–15, 337, 338, 347
  chart of household hazardous waste suitable for, 116–17

Regional Aquifer System Analysis (RASA), 55
Regulation of pesticides, 284–88
Regulations of administrative agencies:
  commenting on proposed, 88–89
  initiating, 89
  judicial review of, 90
  negotiated, 89
Remedial actions, Superfund, 173–88
  federal funds to help understand the problem, 179
  following design and implementation of, 187–88
  obtaining health effects information, 179–80
  participation in EPA hazard ranking process, 174–79
Remedial investigation, 180–87
  choosing among remedial measures, 183
  critical issues in feasibility study analysis, 181–82
  influencing the feasibility study, 181
  other important considerations, 183
  participation in, 180–81
  remedial options, 183–86
  reviewing the final decision, 186–87
Removal actions, 170–73
Resource Conservation and Recovery Act (RCRA), 60–61, 103, 123, 126, 194–95, 288, 309, 337, 338, 340
  federal facilities and, 218
  imminent hazard actions and, 190, 194–95, 197–98, 269
  industrial and commercial wastes and, 231–33, 237, 240–41
  1984 amendment to, and land disposal, 217
  overview of, 123
  provision concerning citizen suits, 103, 104, 190, 269
  requirement for waste treatment, storage, or disposal, *see* Treatment, storage, or disposal facilities
Revegetation, 185
Rhode Island, 347
Right-to-know law, federal toxics, 163
Risk assessment and management, 21
Rockefeller, John D., IV, 173
Roos, Carol, 13
RPM (remedial project manager), 180
Rutgers University Law School, 246

Safe Drinking Water Act (SDWA), 56, 61–62, 65, 104, 123, 126, 127, 152, 285, 289

checking compliance with monitoring requirements and contamination limits, 135–46
imminent hazard situations, 150, 194–95, 281
participation in enforcement, 148–49
policing variances and exemptions, 146
waste disposal wells and, 249, 251, 252, 259–60
water quality programs, 134–50
San Antonio, Texas, 327
Save Our Ocean Committee, 246
Sedimentation basins and ditches, 186
Seepage basins and ditches, 185
Sensitive areas protection, state, 327
Septage disposal, 377
Septic disposal wells, 378
Septic systems:
contamination by commercial industrial systems, 45, 378–80
contamination by home systems, 44
leaching of, 15
as source of contamination, 41–45
state regulation of, 375–80
typical, 42–44
Septic system wells, 29
Site inspections of potential Superfund sites, 167–68
Site plan reviews, 325, 326
Skin, entry of contaminants through the, 19
Slurry walls, 183–84, 185
Soil Conservation Service, 56
Soil flushing, 184
Sole Source Aquifer Program, 158–61
Solid waste facilities, see Nonhazardous waste dumps
Solvents, 17
Sources of groundwater contamination, 24–46
agricultural practices, 32–34
household hazardous wastes and products, 35–36
land application, 34–35
landfills, 24–25
military toxics, 36–40
mining, 40–41
septic systems, 41–45
surface impoundments, 25–27
transportation, 45–46
underground storage tanks, 27–29
waste disposal wells, 29–32
South Carolina, 328
South Dakota, 335, 361, 371
Speed of flow of groundwater, 9
human influence on, 9–10

Spontaneous abortion, 17
Spraying of water on waste, 186
Staggers, Harley, Jr., 173
Standard Industrial Classification (SIC), 231
Standards:
groundwater protection (GPLs), 220–21
Safe Water Drinking Act, see Safe Drinking Water Act
state groundwater quality, 318–20
Stanley Furniture, 348
State agency enforcement of federal laws, 96, 259
agreement with EPA for, 96
citizen oversight of, 125–28
participating in state authorization process, 126–27
state program withdrawal, 127–28
see also individual laws, agencies, and problems
State and local programs for groundwater protection, 314–383
agricultural sources, 364–68
comprehensive, 315–20
general groundwater policies and strategy, 315–16
groundwater classification systems, 317–18
groundwater quality standards, 318–20
protecting groundwater through control of nonpoint pollution sources, 320
environmental policy and protection acts, 334–35, 368, 373
highway runoff, 383
industrial and commercial facilities, 344–49
land use controls, 322–38
LUST program, 270, 274–76
mining, 370–73
septic systems, 375–80
solid and hazardous waste disposal regulation, 337–41
Superfund programs, 330–33
transportation of hazardous wastes, 381–83
underground storage tanks, 357–62, 399–404
waste disposal wells, 351–55
State development laws, 326
State environmental policy acts (SEPAs), 334–35, 368, 373
State environmental protection acts, 335, 373
State Superfund programs, 330–33
going after responsible parties, 332

State Superfund programs (*continued*)
  identifying and ranking sites, 330–31
  participating in cleanup decisions, 331–
    32
  special liability rules, 332–33
Statute of limitations on common-law
    action, 108
Sterility, 17
Stinson Beach County Water District,
    California, 376
STOP (Stop Toxic Onsite Pollution), 162
Strict liability theory, common-law action
    based on, 107
Student Environmental Health Project
    (STEHP) service, Vanderbilt
    University, 64, 173
Subsurface drains, 184
Suffolk County, New York, 360, 362
Superfund:
  federal, *see* Comprehensive
    Environmental Response,
    Compensation, and Liability Act
  state, 330–33
Superfund Amendments and
    Reauthorization Act (SARA), 162n.
Superfund Comprehensive
    Accomplishments Planning (SCAP)
    reports, 163
Surface capping, 184
Surface impoundments, 25–27
Surface Mining Control and Reclamation
    Act (SMCRA), 63, 103, 123, 126,
    291, 293–303
  enforcement procedures, 297–98, 299
  overview of, 125
  see also Coal mining; Mining

Tap water monitoring, 56–57, 64–66
TCE, 13, 17, 188, 281
Telephone numbers, important, 385, 397–
    98
Terotogens, 19
Testing groundwater quality, 51–69, 251,
    294, 320
  monitoring underground sources of
    water, 51–56, 58–63
  monitoring water at the tap, 56–57, 64–
    66
  scrutinizing water test results, 66–69
Testing health effects of a contaminant,
    14–15
Texas, 352, 364
3M, 347, 348
Three Rivers, Michigan, 361–62
Threshold level of a chemical, 20–21
TOC (total organic carbon) tests, 67
Toluene, 17

Toms River, New Jersey, 246
Topographic maps, 73
TOX (total halogenated organics) test, 67
Toxic Substance Control Act (TSCA), 62,
    103, 123, 194–95, 346
  industrial and commercial wastes and,
    233, 237, 245–47
  overview of, 125
Transportation of hazardous wastes and
    toxic chemicals, 45–46, 305–309
  manifest requirements, 309
  obtaining information about, and
    potential problems, 305–306
  preventing contamination related to,
    307–309
  responding to spills and other releases
    resulting from, 306
  state regulation of, 381–83
Treatment, storage, or disposal facilities
    (TSDFs), 204–27
  definitions, 205
  enforcement of regulatory requirements,
    216–19
  gathering information on, 206–209
  hazardous waste, definition of, under
    RCRA, 207
  investigating violations of regulatory
    requirements, 210–16
  closure and post-closure care, 216
  design and operation requirements,
    211, 212–15
  financial responsibility, 216
  groundwater monitoring, 211
  limitation on land disposal of certain
    wastes, 211–16
  loss of interim status, 216
  nonhazardous waste dumps, 204, 226–
    27
  remedying releases into groundwater,
    209–10
  setting of regulatory requirements, 220–
    26
  closure and post-closure, 223–25
  exemptions, 225–26
  judicial review of permit decisions, 226
  operating permits, 220–23
Trenton, New Jersey, 281
Trespass, common-law action based on,
    107–108, 270
Trichloroethylene (TCE), 13, 17, 188, 281
Trihalomethanes, 17–18
Troubled Water (King), 41
Trucks, 307, 308, 381
  see also Transportation of hazardous
    wastes and toxic chemicals
Tuberculosis, 15
Typhoid, 15

Underground Injection Control, (UIC)
  program, 351, 352, 353, 355
Underground sources of drinking water
  (USDWs), 249, 351, 352, 354
Underground storage tanks, 17, 27–29,
  264–76, 282
  corrosion of, 27, 28
  discovering locations of, 265–67
  enforcing LUST regulations, 270–74
  with hazardous waste, see Treatment,
    storage, or disposal facilities
  leakage prevention, 28
  state regulation of, 357–62
  taking action against immediate threats,
    267–70
U.S. Code, 128
United States Geological Survey (USGS),
  55, 56, 72, 77, 289
  maps of, 73, 74, 75
  Water Resources Division, 76, 279
United States Public Interest Research
  Group (U.S. PIRG), 65
University laboratories, 57, 64
Uranium Mill Tailing Radiation Control
  Act, 62–63, 294

Vermont, 326
Virginia Citizens for Better Reclamation
  (VCBR), 300

Washington, 327, 335, 364–65
Waste disposal facilities, see Hazardous
  waste; Treatment, storage, or
  disposal facilities
Waste disposal wells, 29–32, 248–62
  citizen suits, 252, 259
  ensuring compliance with operating
    federal regulations, 252–57

EPA classification of, 249–50
how to find, 248–50
preventing aquifer exemptions, 262
preventing future problems by using the
  permit process, 259–62
questioning waste management, 261
state regulation of, 351–55
taking action against releases and
  contamination, 250–52
ways to use agency enforcement
  procedures, 257–59
Waste Management, Inc., 382
Waste reduction, 14
  see also Recycling
Wastewater treatment technology, 186
Water quality criteria, 182
Water Research and Development Act, 63
Water Resources Action Program (WRAP),
  77
Watershed boundaries, defining, 73, 74
Wellhead protection area, definition of the,
  153–56
Wellhead Protection Program, 124, 152–59
  initiating state action under, 152–53
  participating in implementation and
    enforcement of, 157–59
  participating in program development,
    153–57
Wells, influence on groundwater flow rate
  of, 9–10
Western Nebraska Resources Council, 127
West Oakland, Michigan, 358
Wetlands, 328
Wind fences, 186
Wisconsin, 319, 366, 377
Woods, Edmund, 260

Zoning ordinances, 322, 323–24, 341, 344

Pacific Northwest Office
216 First Avenue S., Suite 330
Seattle, Washington 98104
(206) 343-7340

Rocky Mountain Office
1600 Broadway, Suite 1600
Denver, Colorado 80202
(303) 863-9898

San Francisco Office
2044 Fillmore Street
San Francisco, California 94115
(415) 567-6100

Washington, D.C., Office
1531 P Street, N.W., Suite 200
Washington, D.C. 20005
(202) 667-4500

Also Available from Island Press

**Americans Outdoors: The Report of the President's Commission
   The Legacy, The Challenge**
Foreword by William K. Reilly
1987, 426 pp., appendixes, case studies, charts
Paper: $24.95 ISBN 0-933280-36-X

The Challenge of Global Warming
Edited by Dean Edwin Abrahamson
Foreword by Senator Timothy E. Wirth
1989, 350 pp., tables, graphs, index, bibliography
Cloth: $34.95 ISBN: 0-933280-87-4
Paper: $19.95 ISBN: 0-933280-86-6

A Complete Guide to Environmental Careers
By the CEIP Fund, Inc.
1989, 275 pp., photographs, index, references
Cloth: $24.95 ISBN: 0-933280-85-8
Paper: $14.95 ISBN: 0-933280-84-X

Crossroads: Environmental Priorities for the Future
Edited by Peter Borrelli
1988, 352 pp., index
Cloth: $29.95 ISBN: 0-933280-68-8
Paper: $17.95 ISBN: 0-933280-67-X

**Down by the River: The Impact of Federal Water Projects and
   Policies on Biodiversity**
By Constance E. Hunt with Verne Huser
In cooperation with The National Wildlife Federation
1988, 256 pp., illustrations, glossary, index, bibliography
Cloth: $34.95 ISBN: 0-933280-48-3
Paper: $22.95 ISBN: 0-933280-47-5

Hazardous Waste Management: Reducing the Risk
By Benjamin A. Goldman, James A. Hulme, and Cameron Johnson for
   Council on Economic Priorities
1985, 336 pp., tables, glossary, index
Cloth: $64.95 ISBN: 0-933280-30-0
Paper: $34.95 ISBN: 0-933280-31-9

Our Common Lands: Defending the National Parks
Edited by David J. Simon
Foreword by Joseph L. Sax
In cooperation with the NPCA
1988, 575 pp., index, bibliography, appendixes
Cloth: $45.00 ISBN: 0-933280-58-2
Paper: $24.95 ISBN: 0-933280-57-2

Reopening the Western Frontier
From *High Country News*
1989, 300 pp., illustrations, photographs, index
Cloth: $24.95 ISBN: 1-55963-011-6
Paper: $15.95 ISBN: 1-55963-010-8

Rush to Burn: Solving America's Garbage Crisis?
From *Newsday*
1989, 225 pp., photographs, graphs, tables, index
Cloth: $22.95 ISBN: 1-55963-001-9
Paper: $14.95 ISBN: 1-55963-000-0

Saving the Tropical Forests
By Judith Gradwohl and Russell Greenberg
Preface by Michael H. Robinson
Smithsonian Institution
1988, 207 pp., index, tables, illustrations, notes, bibliography
Cloth: $24.95 ISBN: 0-933280-81-5

Sierra Nevada: A Mountain Journey
By Tim Palmer
1988, 352 pp., illustrations, appendixes, index
Cloth: $31.95 ISBN: 0-933280-54-8
Paper: $14.95 ISBN: 0-933280-53-X

Western Water Made Simple
By the editors of *High Country News*
1987, 256 pp., illustrations, maps
Paper: $15.95 ISBN: 0-933280-39-4

Wildlife of the Florida Keys: A Natural History
By James D. Lazell, Jr.
1989, 225 pp., photographs, illustrations, index
Cloth: $31.95 ISBN: 0-933280-98-X
Paper: $19.95 ISBN: 0-933280-97-1

For additional information about Island Press publishing services and a catalog of current and forthcoming titles, contact Island Press, P.O. Box 7, Covelo, California 95428

Island Press